PROCLUS

A COMMENTARY ON THE
FIRST BOOK OF
EUCLID'S ELEMENTS

PROCLUS

A COMMENTARY ON THE

FIRST BOOK OF

EUCLID'S ELEMENTS

Translated

with Introduction and Notes by

GLENN R. MORROW

Adam Seybert Professor Emeritus
of Moral and Intellectual Philosophy
University of Pennsylvania

PRINCETON UNIVERSITY PRESS 1970
PRINCETON, NEW JERSEY

Published by Princeton University Press, 41 William Street,
Princeton, New Jersey 08540
In the United Kingdom, by Princeton University Press, Oxford

Library of Congress Card No. 73-90955

Library of Congress Cataloging-in-Publication Data
Proclus, ca. 410–485.
[Eis prōton Eukleidou stoicheiōn biblon. English]
A commentary on the first book of Euclid's Elements / Proclus;
translated with introduction and notes by Glenn R. Morrow.
p.   cm.
Includes bibliographical references and index.
ISBN 0–691–02090–6
1. Mathematics, Greek. 2. Geometry—Early works to 1800.
I. Morrow, Glenn R. (Glenn Raymond), 1895–1973. II. Title.
QA31.P7413   1992
516.2—dc20   92–11366

This book has been composed in Linotype Times Roman

First Princeton Paperback printing,
with a foreword by Ian Mueller, 1992

Princeton University Press books are printed on acid-free paper,
and meet the guidelines for permanence and durability of the
Committee on Production Guidelines for Book Longevity of the
Council on Library Resources

10   9   8   7   6   5   4   3   2

Printed in the United States of America

TO MY WIFE

τῇ ψυχῇ μου

Table of Contents

Foreword to the 1992 Edition
Ian Mueller

PROCLUS' COMMENTARY on book I of Euclid's *Elements* is almost certainly a written version of lectures which he presented to students and associates in Athens in the mid-fifth century.[1] The work was presumably circulated among philosophers in the Roman (Byzantine) empire and used as the basis for other people's lectures, just as Proclus made use of various written sources in the composition of his own commentary. Readers of the commentary should always bear in mind that, although it is the work of Proclus, it is also a record of an educational and intellectual tradition. Modern scholars refer to that tradition as Athenian Neoplatonism, but for Proclus and all members of the tradition it was simply Platonism, the philosophy of Plato.

Friedlein's standard edition of the Greek text of our commentary describes it as the work of Proclus Diadochus, or Proclus the successor. The exact meaning of "successor" is elusive,[2] but in the case of philosophy the basic idea of calling someone a successor is to identify that person as a member of a sequence of "head" teachers of a philosophical position; so Proclus was one of a sequence of head teachers of Platonism. We can trace a relatively clear line of succession in that sequence, starting with Proclus' own first teacher in Athens:

Plutarch[3] of Athens (ca. 350–ca. 432)
Syrianus (head from death of Plutarch until his own death)
Proclus (head from death of Syrianus until his own death in 485)
Marinus (head from death of Proclus until his own death)
Isidore (head from death of Marinus until his own death sometime before 526)
Damascius (head from death of Isidore until ?)[4]

---

[1] This introduction is intended to supplement Glenn Morrow's original introduction and notes, which it presupposes. There is, however, some unavoidable overlap. I cite Proclus' commentary according to the pagination of Friedlein, which is printed in the margin of Morrow's translation.

[2] On this topic see John Glucker, *Antiochus and the Late Academy* (*Hypomnemata* 56) (Göttingen, 1978), esp. pp. 146–58. This work is fundamental for understanding the history of late Platonism.

[3] Not to be confused with the more famous Plutarch of Chaironeia, who lived approximately three centuries earlier.

[4] Three other people are sometimes inserted in this chain: Domninus (an approximate contemporary of Proclus), Hegias, and Zenodotus (approximate contemporaries of Isi-

In 529 the Emperor Justinian issued laws barring pagans and other heretics from military service, public office, and teaching, with the penalty for offense being confiscation of property and exile. We do not know anything about the enforcement of this law.[5] However, according to the historian Agathias, Damascius and six other philosophers, the best known of whom is Simplicius, went to Persia, but soon were permitted by an agreement between Persia and Constantinople—reliably dated to late 532—to return home and live in peace. Scholars have disagreed about where home might be, and it has been suggested that at least Simplicius returned to Athens, where he composed his major commentaries on works of Aristotle. However, we have no evidence that pagan philosophy was taught in Athens after 529.

In his *Life of Proclus* (Par. 29)[6] Marinus says that his master inhabited the house in which Plutarch and Syrianus had lived before him. Marinus also tells us that the house was located near the temple of Asclepius and the theater of Dionysos, and visible from the Acropolis. Archaeologists have now identified what they call the "house of Proclus."[7] Whether or not this identification is correct, it is likely that Proclus lived in a house which was handed down through the succession and also served as the school where Proclus and his associates talked and held classes for an audience including both "mere listeners" and others eager to become bona fide Platonists.[8] This school had no physical connection with the

dore). For an attempt to fit them into the succession as "seconds-in-command" see Glucker, *Antiochus*, p. 155 n. 122.

[5] It is, then, somewhat misleading to speak simply of Justinian closing the schools and confiscating their properties, as is frequently done. On the subject of Justinian's laws and their effect on Athens see Alan Cameron, "The Last Days of the Academy at Athens," *Proceedings of the Cambridge Philological Society* 195 (n.s. 15) (1969): 7–29.

[6] The Greek text is now available in *Marino di Neapoli, Vita di Proclo*, critical text, introduction, translation, and commentary by Rita Masullo (Naples, 1985). Although Marinus' *Life* does provide some biographical and historical information, it is an example of pagan hagiography, and should be read as such. For discussion of it see H. J. Blumenthal, "Marinus' Life of Proclus: Neoplatonist Biography," *Byzantion* 54 (1985): 469–94. For general background see Garth Fowden, "The Pagan Holy Man in Late Antique Society," *Journal of Hellenic Studies* 102 (1982): 33–59.

[7] See pp. 42–44 of Alison Frantz, *Late Antiquity: A.D. 267–700 (The Athenian Agora* 24) (Princeton, 1988), an excellent source for the history of Athens as a "provincial university town" from the so-called Herulian invasion to its dark ages.

[8] Marinus, Par. 38. In Par. 22 Marinus gives a brief description of Proclus' working day, on which see Otmar Schissel, "Der Stundenplan des Neuplatoniker Proklos," *Byzantinische Zeitschrift* 26 (1926): 265–72. It is clear that Proclus had access to a considerable number of written works, but private libraries attached to private schools are not

so-called Academy founded by Plato in the fourth century B.C.E., an institution which almost certainly ceased to exist about three centuries after its founding.[9] The Neoplatonic "school" of Athens was a privately funded, self-perpetuating group of pagans who strove to keep alive the truth of the Hellenes by recruiting and teaching students. The teachers and pupils were in many cases wealthy aristocrats. Marinus (Par. 4) says that Proclus' parents were very rich, and that Proclus' inheritance made him indifferent to money. Proclus gave many gifts to his friends and to Athens, which, along with his native Xanthus (Kinik, Turkey) inherited his fortune (Par. 14). Of greater significance is the fact that the school had a very substantial endowment with which to support its members.[10] The income from the endowment would have been supplemented by student payments.[11] Proclus, who was active in civic affairs (Par. 15), asked the rulers to provide support for students (Par. 16); we are not told whether his request was granted.

Proclus, then, was a teacher, the head of an endowed "private school," which supported other teachers as well, and received income from student payments.[12] The sources of information about "higher education" in antiquity are scattered, but they enable us to put together a fairly clear picture of its broad outlines.[13] The core of the system was

unknown; Philostratus, *Lives of the Sophists*, II.21 (604) provides an example from ca. 200 C.E.

[9] See especially Glucker, *Antiochus*.

[10] The annual income from the endowment is given as at least one thousand gold *nomismata* or *solidi* (over fourteen Roman pounds of gold). On the value of the *solidus* see A.H.M. Jones, *The Later Roman Empire* (Oxford, 1964), pp. 445–48. Jones describes as liberal an allocation of six *solidi* a year for the monks of a monastery in the Jordan valley. The Justinian Code gives seventy *solidi* a year as the salary of a teacher of grammar or rhetoric in Carthage.

[11] On student payments see Alan Cameron, "Roman School Fees," *Classical Review* n.s. 15 (1965): 257–58.

[12] Because of the great difficulty of dating Proclus' works, which he revised over time, we cannot be sure that he did not write the Euclid commentary before he became head of the school. The evidence mentioned by Morrow on this question (p. lvi) is inconclusive, since Proclus can refer to Syrianus as "our head" while speaking of Syrianus' past achievements; see, for example, Proclus' commentary on the *Republic*, I.133.5–7. On the difficulties of dating Proclus' works see *RE*, 45th half-volume (1957), cols. 190–91.

[13] The best work on this subject remains H. I. Marrou, *A History of Education in Antiquity* (London, 1956), a translation of the third edition of Marrou's *Histoire de l'education dans l'antiquité*, of which there is a sixth edition (Paris, 1965). Fritz Schimmel provides an organized collection of relevant passages on Athens in late antiquity in "Die Hochschule von Athen im iv. und v. Jahrhundert p. Ch. n.," *Neue Jahrbücher für das klass-*

the individual teacher who made a living by teaching pupils who were young and mainly, but not exclusively, male. A successful teacher would have an inner circle of cohorts who also taught, and his students would come from all over the Roman world. Teaching could be done in private, but it was also done in public buildings. Some teachers held imperial or municipal appointments, but they did not normally have a monopoly on instruction in a locality. Sometimes these teachers supplemented their salaries with student payments, but the only income of others seems to have been their salaries, which could be quite substantial.

Given the individualistic nature of higher education in antiquity, it is not surprising that the line between a teacher and, say, a hierophant might be hard to draw. The main officially recognized subjects of higher education were letters (*grammatikê*), rhetoric, philosophy, medicine, and law. For the purposes of this introduction it is simplest to treat the last two of these as technical disciplines taught by practitioners. The other three formed the more general part of higher education—what we might call liberal education. Of these, letters, the study of literary classics, was the most fundamental and the least esteemed. The struggle between rhetoric and philosophy for the hearts and minds of young men goes back at least to the rivalry between Plato and Isocrates in the fourth century B.C.E. In general the rhetorician insisted on his ability to act effectively in the public arena, whereas the philosopher insisted on his deeper knowledge and greater purity. The competition between the rhetoricians and philosophers for students and for esteem led inevitably to a blurring of the distinction between their two fields. Aristotle had written a treatise on rhetoric, and the Stoics divided their logic into dialectic and rhetoric (Diogenes Laertius, VII.41). According to Cicero (*Tusculan Disputations* II.9), Philo, the last known head of Plato's Academy, divided his teaching equally between rhetoric and philosophy. There survive two commentaries by Proclus' master Syrianus on rhetorical treatises by Hermogenes;[14] Damascius studied rhetoric for three years, and taught it for nine (Photius, *Library* 181 [127a]).

---

ische Altertum 22 (1908): 494–513. See also Alan Cameron, "The End of the Ancient Universities," *Cahiers d'histoire mondiale* 10 (1966–1967): 653–73, and Garth Fowden, "The Platonist Philosopher and His Circle in Late Antiquity," *Philosophia* 7 (1977): 359–83.

[14] *Rhetores graeci*, vol. 16 (Leipzig, 1892). At the beginning of his commentary on Hermogenes' *On Ideas* Syrianus remarks that many sophists and Platonic philosophers had commented on it. Proclus remarks on the way mathematics provides a model for rhetoric at 24.21–25.5 and 247.6–248.15.

Athens appears to have been the preeminent philosophical center of the Greek world from at least the fifth until the first century B.C.E. Although there certainly were philosophical teachers and schools elsewhere, the major schools—Platonic, Epicurean, Stoic, and, for at least the beginning of the period, Aristotelian—were taught most authoritatively in Athens. Wars, and particularly the sacking of Athens by Sulla in 86 B.C.E., appear to have changed this situation. For the first four centuries of the common era we know more about the philosophers living in other cities than we do about ones living in Athens, and more about rhetors and sophists at Athens than about philosophers there.[15] However, it appears that at least by the end of the second century, when Roman interest in reviving Athens produced a very ambitious building program, the city reemerged as an educational center. In 176 Marcus Aurelius established four high-paying chairs of philosophy in Athens, one for each of the four major sects, Platonists, Peripatetics, Stoics, and Epicureans, as well as at least one chair in rhetoric. These chairs were official appointments, sustained by public funding; they were not part of a self-perpetuating school like Plato's Academy or the succession of Athenian Neoplatonists. There were also imperial appointments in letters and rhetoric in other cities, and localities made their own official appointments as well. The evidence suggests that, while public support for teachers of rhetoric and grammar continued into late antiquity, support for the teaching of philosophy dwindled.[16] The factors involved in this decline are undoubtedly complex, but two of them would seem to be the "impracticality" of philosophy and the sense that philosophy was less easily assimilated to Christianity than rhetoric. It is well-known that ancient education remained essentially pagan under the Christians for a long time, but it is one thing to take Thucydides or Demosthenes as a model of literary style, another to take Plato or Aristotle as a model of cosmic or theological understanding.

In his life of Plotinus (Par. 15) Porphyry refers to a correspondence between Plotinus and a "Platonic successor in Athens" named Eubulus,

---

[15] See especially the *Lives of the Sophists* by Philostratus (for the second century) and by Eunapius (for the fourth). For discussion of the second century see James H. Oliver, "The *diadochê* at Athens under the Humanist Emperors," *American Journal of Philology* 98 (1977): 160–78, and John Patrick Lynch, *Aristotle's School* (Berkeley and Los Angeles, 1972), pp. 168–207.

[16] The major exception is Alexandria. See H. D. Saffrey, "Le chrétien Jean Philopon et la survivance de l'école d'Alexandrie au VIᵉ siècle," *Revue des études grecques* 66 (1953): 396–410.

who would have been alive around the year 267, when Athens suffered a devastating attack from a people known as Herulians.[17] Unfortunately, we do not know exactly what the term "successor" means in this context. For us the Platonic succession begins again with Plutarch at a time when, after a long period in which most of the Agora lay in ruins, Athens had begun to rebuild. At the beginning of *The Platonic Theology* Proclus gives—in somewhat high-flown language—his own account of a "spiritual" Platonic succession. According to this account, the philosophy of Plato—that is, the truth—was understood in an imperfect way by unnamed early philosophers,[18] and expressed in a perfect way by Plato; it then became for the most part invisible to persons who called themselves philosophers, until there emerged a new set of true Platonic "exegetes": Plotinus, his pupils Amelius and Porphyry, their pupils Iamblichus and Theodorus of Asine, and finally Proclus' own teacher Plutarch, whom Proclus eulogizes without naming.

Plotinus, who died in Rome in 270, is generally taken to be the founder of Neoplatonism. But the history of Neoplatonism and its relation to earlier forms of Platonism and the philosophy of Plato himself are matters of considerable controversy. Amelius and Theodorus are relatively obscure figures who need not concern us.[19] Porphyry was Plotinus' successor and editor. He is mentioned several times by Proclus in the Euclid commentary.[20] Although Proclus is much more scrupulous than most or all of his contemporaries in the matter of naming sources, it is overwhelmingly likely that there is more Porphyry in the commentary—particularly in the commentary on the propositions—than the six occurrences of his name would suggest. Porphyry died in Rome at the very beginning of the fourth century. His most famous pupil was Iamblichus, who returned from Rome to his native Syria, where he established a very successful school and died around 330. Although details

---

[17] Later (Par. 20) Porphyry quotes Longinus, who refers to Eubulus and Theodotus as successors. We know nothing more about Theodotus. According to Longinus, Eubulus wrote on Plato's *Philebus* and *Gorgias* and on Aristotle's objections to the *Republic*; according to Porphyry (*On Abstinence* 4.16.8–9; cp. *On the Cave of the Nymphs* 6.10), he wrote on the god Mithras.

[18] The most important of these would probably be Pythagoras, Parmenides, and Empedocles.

[19] For discussion see: Luc Brisson, "Amélius, sa vie, son oeuvre, sa doctrine, son style," in *Aufstieg und Niedergang der römischen Welt*, vol. 2, part 36,2, pp. 793–860; and Werner Deuse, *Theodoros von Asine: Sammlung der Testimonien und Kommentar* (*Palingenesia* 6) (Wiesbaden, 1973). Theodorus was also a student of Iamblichus.

[20] To the references in the index to the present volume add "and to XXVI, 272–75."

of transmission are unclear,[21] it is universally accepted that Iamblichus is the major intellectual progenitor of the flourishing of Neoplatonism in fifth-century Athens. His Neoplatonism is marked by a multiplication of speculative ontological postulates, Pythagorean number mysticism, and the glorification of Greek and oriental polytheism and various magical practices known as theurgy. Proclus never mentions Iamblichus in his Euclid commentary, but the many parallels between the first part of the prologue of the commentary and Iamblichus' work *On mathematics in general* (*De communi mathematica scientia*) make it impossible to deny that Iamblichus was also one of Proclus' sources.[22]

Porphyry's philosophical views strike the modern reader as generally more rationalistic and less wildly speculative than those of Iamblichus. But the two men were united with each other and with Proclus in their vigorous espousal of paganism and opposition to Christianity, which became the religion of the emperors in the early fourth century and eventually the official religion of the empire. Porphyry's work *Against the Christians* was, as far as the Christians were concerned, the major polemic to be refuted. The writings of the "divine Iamblichus" were undoubtedly an important source of inspiration for Julian the Apostate,[23] who studied philosophy with a student of a student of Iamblichus, Maximus of Ephesus, and attempted to reverse the religious direction of the empire in the mid-fourth century. Although paganism survived throughout the empire long after Julian's death, by the mid-fourth century observance of pagan rites was undoubtedly a risky business.[24] It is not surprising that we do not find many references to Christians in Proclus. In those we do find, the Christians are named obliquely as, e.g., "strangers to our world," "the ignorant," "the godless."[25] Any at-

---

[21] For an attempt to explain the transmission see Alan D. Cameron, "Iamblichus at Athens," *Athenaeum* n.s. 45 (1967): 143–53.

[22] On this topic see Ian Mueller, "Iamblichus and Proclus' Euclid commentary," *Hermes* 115 (1987): 334–48.

[23] See especially Julian's letter 12 (Bidez).

[24] See Walter E. Kaegi, "The Fifth-Century Twilight of Byzantine Paganism," *Classica et Mediaevalia* 27 (1966): 243–75. Jean Gaudemet, *L'église dans l'empire romain (ive and ve siècles)*, 2d ed. (Paris, 1989), pp. 646–52 surveys the antipagan legislation of the fourth and fifth centuries.

[25] See Henri-Dominique Saffrey, "Allusions antichrétiennes chez Proclus: le diadoque platonicien," *Revue des sciences philosophiques et théologiques* 59 (1975): 553–63. Proclus did attack the doctrine of the temporal creation of the world in a work which Philoponus refers to as "eighteen arguments for the eternity of the world against the Christians."

tempt to understand the commentary as an intellectual production should take into account Proclus' view of himself as the defender and preserver of a great cultural tradition under attack by a godless and ignorant group of people who were taking over the world.

We have seen that for Proclus the tradition which we call Neoplatonism was simply a reemergence of the true understanding of Plato. Philosophy for Proclus is Platonic exegesis because Plato knew all the philosophy there is to know. Scholars disagree about the accuracy of Neoplatonic conceptions of Platonic philosophy, but there is fairly general agreement that Neoplatonic exegesis depends heavily on texts and traditions which postdate or are independent of Plato. These texts and traditions include: the works of Aristotle, which the Neoplatonists used to fill perceived gaps in Plato's discussions of particular topics; the scientific tradition and its exegesis and development in all philosophical schools including the Stoic school;[26] a curious body of ''Neopythagorean'' literature in which Platonic ideas are blended with number mysticism;[27] and syncretic spiritual/magical ideas, which for Proclus are most fully embodied in the *Chaldean Oracles*.[28] Broadly speaking one may say that in expounding Platonism Neoplatonists were willing to use anything in the Mediterranean and Middle Eastern tradition which they took to be true. Neoplatonists were guided not only by the principle that what Plato said was—in one way or another—true, but also by the principle that if something was true, Plato—in one way or another—expressed it, referred to it, or took it for granted.

It is not the purpose of this foreword to discuss the ways in which Proclus' Platonism coincided with or diverged from Plato's. But since Proclus' Platonism differs in significant ways from the views ascribed to Plato in standard scholarly works, an outline of some of its basic features which bear on the Euclid commentary may be useful.[29] Partic-

---

[26] As an example of the use of Stoic logic I mention Proclus' reference to ''the second type of hypothetical argument'' at 256.1–8.

[27] On Neopythagoreanism see especially Walter Burkert, *Lore and Science in Ancient Pythagoreanism*, trans. Edwin L. Minar, Jr. (Cambridge, Mass., 1972), a work which provides the foundations for a proper perspective on Proclus' many references to Pythagoras and Pythagoreans. For more detailed discussion of Pythagoreanism in Athenian Neoplatonism see Dominic J. O'Meara, *Pythagoras Revived* (Oxford, 1989).

[28] The versified record of divine revelations, probably written in the 2d century C.E. by ''Julian the Chaldean.'' See *The Chaldean Oracles*, Ruth Majercik, ed. and trans. (Leiden and New York, 1989).

[29] In what follows it is important to see that I am only occasionally concerned to distinguish between what Plato (or Aristotle) actually says and how Neoplatonists like Proclus

ularly important in this regard is the metaphysical and educational picture presented in parts of books V, VI, and VII of Plato's *Republic*. The metaphysical picture starts from a dichotomy between the intelligible world of being, the Forms, apprehended by intellect or mind (*nous*) independently of the senses, and the world of becoming, apprehended through the senses.

In his description of the so-called Divided Line at the end of book VI of the *Republic* Plato complicates the bifurcation between sensible and intelligible by subdividing the two realms and correlating certain "conditions of soul" with them (figure 1).

FIGURE 1

Realm	Level	Object	"Epistemic correlate"
Being			Knowledge (*epistêmê*)
	Upper	Forms	*nous, noêsis* ("intellection")
	Lower	?	*dianoia* ("understanding")
Becoming			Opinion
	Upper	sensibles	belief, sensation, perception
	Lower	images of sensibles	*eikasia* ("conjecture")

I have left the nature of the objects of *dianoia* unspecified because Plato is not entirely explicit about their nature. In his summary of this schema at 10.15–11.9 Proclus refers to these objects with the uninformative term *dianoêta*, but he subsequently calls them *logoi* (translated "ideas" by Morrow), the term I shall use. The lowest level in this division will not concern us further, but it is important to bear in mind that the relationship between the lower and upper levels in becoming is in a general sense the model for the relationship of lower to higher levels of reality: just as a sensible object is the cause of its reflected image, for Proclus a higher level produces the next lower level and the lower level is a copy of the higher "in another medium." This relationship does not just apply to becoming and being and to the lower and upper levels in the realm of

understand him. It is also important to realize that the metaphysical schemata applied by Proclus in different works are not easily reconcilable in all their details and that they are a good deal richer than my outlines of them. A similar comment applies to the views of various Neoplatonists whom I cite as evidence for one or another feature of a general doctrine; a great deal of work still needs to be done in sorting out the relationships among their positions. Two useful works for getting into these subjects are A. H. Armstrong, ed., *The Cambridge History of Later Greek and Early Medieval Philosophy* (Cambridge, 1967), and R. T. Wallis, *Neoplatonism* (London, 1972).

being. It also applies to being and a realm above it, a realm which Proclus and other Neoplatonists identify with the Good beyond being of *Republic* 509b, with the featureless One of the first hypothesis of Plato's *Parmenides*, and with the ultimate God.[30] Adding this realm and making some terminological adjustments to bring vocabulary somewhat closer to Proclus', we obtain the schema of figure 2.

FIGURE 2

Domain	Object	"Epistemic correlate"
the One	the One	union, assimilation, identification
Being	Forms	*nous, noêsis*
Mathematics	*logoi*	*dianoia*
Becoming	sensibles	opinion, sensation

At this point I want to introduce another, standard Neoplatonist hierarchy, which does not involve mathematics in an explicit way. In it the nonsensible world is divided into three realms, the One, Being or *Nous*, and Soul, where Soul serves in a quite complex and not easily fathomed way as the intermediary between the nonsensible and sensible realms. A version of this hierarchy—which Proclus invokes briefly at 115.12–16—is represented in figure 3.

FIGURE 3

One

Nous/Being

Soul

World Soul	individual souls
nature, the world or cosmos	individual living things

Scholars often call the upper three levels of this schema, hypostases, nonsensible realms of reality. The fourth level is derived from Plato's *Timaeus* (34cff., 41dff.), where the soul is said to be fashioned out of a third kind of being intermediate between the intelligible and the sensi-

---

[30] The One does not play a prominent role in the Euclid commentary because of the place of mathematics in Neoplatonist ontological and epistemological hierarchies. Proclus refers to the "indescribable and utterly incomprehensible" causal efficacy of the One at 5.19–20.

ble.[31] The fifth level is just the sensible world and its most important component, living things, conceived Platonistically as unions of soul and body. Neoplatonist discussions do not distinguish clearly and uniformly among the embodied soul existing at the lowest level, the soul conceived as separate or separable, and the hypostasis Soul, which sometimes even seems to be a form of *Nous*. For Proclus and other Neoplatonists the crucial point is that the human being has a soul which is derived from the hypostasis Soul and ultimately from the One. The goal of a human being is to rise above the conditions of ordinary existence and to rise as far as possible in the hierarchy just described. Although magic and superstition are an important part of the Neoplatonic tradition, Proclus clearly believes that education—and, in particular, education in mathematics of the kind represented by the commentary—is a component of this ascent.

We must suppose that schemas 2 and 3 somehow fit together in Proclus' mind, but it would, I think, be a mistake to try to combine them into a single schema. Proclus' general conceptions tend to be rather fluid compositions of a variety of components. Rather than trying to freeze those compositions, I want to add two other sets of components derived from Aristotle. The first is a hierarchy derived from Aristotle's *De anima*. Aristotle's divisions of the soul are themselves fluid, but the Neoplatonists focused on Aristotle's basic division of the soul's faculties into nutritive, sensitive or perceptive, and intellective. Aristotle's obscure description of *nous* at the end of III.5 is a major source for the Neoplatonic understanding of *Nous*:

> *Nous* in this sense of it is separable, impassible, unmixed, since it is in its essential nature activity. . . . Actual knowledge is identical with its object. . . . It does not sometimes think and sometimes not think. When separated it is alone just what it is, and this alone is immortal and eternal (we do not remember because, while this is impassible, passive *nous* is perishable); and without this nothing thinks. (Revised Oxford translation)

---

[31] Plato also uses mathematical ratios to describe the construction of the soul. Aristotle already knows of Platonists who call the soul a self-moving number. And there are indications of a tradition stretching down to Proclus (see 16.16–18.4) in which there is, in some sense, an identification of the domain of Soul and the domain of mathematics. See Philip Merlan, *From Platonism to Neoplatonism*, 3d ed. (The Hague, 1968), pp. 11–33, 222–26. The evidence, which is very obscure, seems to me to justify only the claim that some Neoplatonists analogized Soul and mathematical objects, not that they identified them.

For the Neoplatonists *Nous* is not, as we have seen, a part of Soul, but something above Soul. Sensation can be associated with the embodied soul, but not with the hypostasis Soul. If the embodied soul is going to be led away from the material world to the Forms, there is presumably going to have to be some transitional psychic activity corresponding to Soul in figure 3 and mathematics in figure 2. Plato provides one label for this activity with his term *dianoia*, which Proclus (10.15–11.25) declares to be the "criterion" of mathematics, related to the objects of mathematics as sensation and opinion are to sense objects.[32] The primary Neoplatonic contrast between *dianoia* and *nous* is the contrast between discursiveness, which can be thought of as the feature of ordinary reflection and thought, and the nondiscursive all-at-once grasping of a totality which is definitory of noetic apprehension.[33]

Aristotle provided the Neoplatonists with another psychic activity or faculty to associate with the transition from sensation to *nous*, imagination.[34] Aristotle's discussion of imagination in *De anima* III.3 is very cryptic. Perhaps its most crucial aspect for the Neoplatonists was the positioning of this discussion between the treatments of sensation and intellect. In Neoplatonist philosophy imagination itself occupies the analogous intermediate position. It serves as a kind of depository for sensations and thus provides the basis for an account of empirical knowledge. But more importantly, particularly in Proclus' Euclid commentary, it serves as a kind of movie screen on which *dianoia* projects images for mathematical reflection.[35] These images are ultimately derived from Forms, but since Forms are the objects of *Nous*, Proclus uses the term *logoi* to refer to what might be called dianoetic expressions of Forms. In Proclus' view *dianoia* studies these *logoi* by projecting im-

[32] For a brief discussion of this sense of criterion with references see Gisela Striker, Κριτήριον τῆς ἀληθείας, *Nachrichten der Akademie der Wissenschaften in Göttingen Philologisch-Historische Klasse* (1974), vol. 2, pp. 55–57.

[33] See, for example, 4.11–14 and 44.19–23 of the commentary.

[34] The topic of imagination in Neoplatonism is much more complicated than my discussion suggests. See the following papers by Henry J. Blumenthal: "Plutarch's Exposition of the *De anima* and the Psychology of Proclus," in Olivier Reverdin, ed., *De Jamblique à Proclus* (*Entretiens sur l'antiquité classique* 21) (Geneva, 1974), pp. 123–47; "Neoplatonic Interpretations of Aristotle on *Phantasia*," *Review of Metaphysics* 31 (1977–1978): 242–57; and "*Nous Pathêtikos* in Later Greek Philosophy," in Henry Blumenthal and Howard Robinson, eds., *Aristotle and the Later Tradition* (*Oxford Studies in Ancient Philosophy*. Supp. Vol.) (Oxford, 1991), pp. 191–206.

[35] Proclus uses the image of a mirror at 121.1–7.

ages of them onto the imagination (which he also calls passive *nous*).[36] In this sense we can say Proclus associates *dianoia* and imagination more or less inseparably—at least when he is thinking about mathematics.

Aristotle (*Metaphysics* E.1.1026a18–19; cp. K.7.1064b1–3) gave the Neoplatonists one other important classification, a division of theoretical philosophy into physics, mathematics, and theology.[37] Since Aristotle identified theology with first philosophy or the study of being qua being, the Neoplatonists had no difficulty in assimilating theology to the apprehension of Forms by *Nous*.[38] Mathematics clearly fits into schema 2, and so does physics, once we take it to be the study of sensibles. When physics is fitted into this scheme, the idea that it is mere opinion or sensation is no longer appropriate. In his commentary on the *Timaeus* (1.223.16–30) Proclus explains what Timaeus means when he characterizes the sensible world he is going to discuss as an object of opinion (27dff.). After dividing the rational soul into *nous*, *dianoia*, and opinion, Proclus explains that *nous* has converse with divinities, *dianoia* puts forward sciences, but opinion "brings forward things into other things." He goes on to explain that "opinion receives a scientific method of making distinctions from *dianoia* and applies it to other things. Opinion [in this sense] is not uncertain, it is not divided up by the variety of sensibles, and its knowledge (*eidêsis*) is not limited to mere suppositions; rather it receives its content from *nous* and *dianoia*,

---

[36] For the term "passive *nous*" see 52.3–12, 55.23–56.22, and 185.25–186.7. Proclus' account of mathematical reasoning in chapter VI of the first part of the prologue and particularly in chapter I of the second are philosophical classics which have often been discussed. I refer the reader to my own discussion in "Mathematics and Philosophy in Proclus' Euclid Commentary," in Jean Pépin and H. D. Saffrey, eds., *Proclus, lecteur et interprète des anciens* (Paris, 1987), pp. 305–18. This book contains the proceedings of a general conference on Proclus held to commemorate the fifteen-hundredth anniversary of his death; another such work is G. Boss and G. Seel, eds., *Proclus et son influence* (Zürich, 1987).

[37] On the Neoplatonic use of this trichotomy see Dominic J. O'Meara, "Le problème de la métaphysique dans l'antiquité tardive," *Freiburger Zeitschrift für Philosophie und Theologie* 33 (1986): 3–22. Among important Neoplatonic passages on this topic I signal Ammonius, *Commentary on Porphyry's* Eisagôgê 11.22–12.11, and Simplicius, *Commentary on Aristotle's* Physics 1.14–2.7. My discussion leaves out of account the Aristotelian distinction among theoretical, practical, and productive sciences. I briefly discuss the role of the theoretical/practical distinction in Neoplatonic pedagogy in n. 42 below.

[38] For the terminology see 9.14–25 and 20.3–14. Cp. Syrianus, *Commentary on Aristotle's* Metaphysics, 61.20–21.

contemplates the plan of the Creator, and judges the nature (*physis*) of things.''

I am now in a position to give the more precise account of the Proclean hierarchy toward which I have been working. At the highest level there is the One, which is apprehended only by a merging of the self which transcends knowledge. Below this are two kinds of knowledge. The higher kind is called theology and apprehends the Platonic divinities, the Forms, in a nondiscursive way by means of a "faculty" called *nous*. The lower kind of knowledge is mathematics, which deals in a discursive way with *logoi*, using imaginative representations of them. Beneath these is physics, which apprehends the sensible world as a whole and in its parts, using ideas derived from the two higher forms of knowledge; Proclus associates physics with a faculty he calls opinion.[39]

It remains to fit this cognitive hierarchy into the Neoplatonist educational program. For although the basic idea of this program is to lead the student up to the One through physics, mathematics, and theology by means of commentary on major texts, the progression does not appear to have been conceived in a linear fashion. Nor, it seems, could it be, since physics as conceived by Proclus presupposes some mathematics. Moreover, complications are introduced because students also need training in morals and logic. Our texts on the subject of education are somewhat diverse and discrepant. I shall base my account on Marinus' life of Proclus, supplemented with other materials.

According to Marinus (Par. 8ff.), Proclus originally intended to follow his father into the legal profession and so studied rhetoric in Constantinople (Istanbul). He accompanied his rhetoric instructor to Alexandria, where his patron goddess Athena exhorted him to study philosophy. In Alexandria Proclus studied mathematics and Aristotelianism—Marinus mentions especially logic[40]—the former subject with Heron, a very pious person who strengthened Proclus' piety, the latter with Olympiodorus.[41] Dissatisfied with his instruction in Alexandria,

---

[39] It should be clear that Proclus has substituted Aristotelian theology for what Plato in the *Republic* calls dialectic. I discuss Proclus' notion of dialectic in the Euclid commentary below.

[40] For at least the later Neoplatonists, Aristotle's logic included the *Rhetoric* and *Poetics* as well as the works contained in what we call the *Organon*. For them the core of logic is the *Posterior Analytics*, the theory of scientific proof. See, for example, Olympiodorus, *Prolegomena* 8.4–10.

[41] Marinus, our only source of information on these two teachers of Proclus, says next

Proclus left for Athens, where he read Aristotle's *De anima* and Plato's *Phaedo* with the aged Plutarch. After Plutarch's death he spent two years studying all of Aristotle—logic, ethics, politics, physics, and theology (metaphysics)—with Syrianus.[42] These, Marinus says (Par. 13), constituted a kind of preliminary initiation into the lesser mysteries, after which Syrianus led Proclus into the real mysteries: the doctrines of Plato.

If we take Proclus' educational journey to indicate a general educational plan, it seems that mathematics and logic were treated as preliminaries to higher philosophical study. The reading of *De anima* and the *Phaedo* was almost certainly intended to improve Proclus' understanding of the nature of the soul; in the case of the *Phaedo* I suspect that the arguments for immortality were particularly important. Since Marinus (Par. 9) says that in Alexandria Proclus "had absolutely no difficulty in understanding Aristotle's logical works in just one reading," it seems unlikely that much time was spent on them with Syrianus. Physics and theology were presumably studied a second time in connection with Plato. But the approach to the two authors would be quite different. Simplicius indicates the probable difference of approach to the two:

> There are two kinds of enlightenment which produce conviction; one proceeds from *nous*, one from perception. Aristotle prefers the latter since he is speaking to those who live by the senses. In his case compulsion lies in proofs (just as we force a person to be silent when he is not persuaded because of certain unfortunate preconceptions). Aristotle never wants to withdraw from nature; rather he investigates even what transcends nature in terms of its relation to nature. Conversely, Plato, following the Pythag-

---

to nothing about them. The Heron mentioned here is probably not the Heron to whom Proclus refers in the Euclid commentary, on whom see Morrow's note to 41.10 and A. G. Drachmann, *Ktesbios, Philon, and Heron* (Copenhagen, 1948), pp. 74–77. Similarly the Olympiodorus who taught Proclus is not the Olympiodorus of the previous note; he lived in the sixth century.

[42] The educational position of ethics and politics was a matter of dispute among the Neoplatonists. They seem to have resolved the dispute in a way which has its roots in Aristotle: virtue is first inculcated in a student through maxims and examples; then, after the student has learned logic, he is taught ethics in a theoretical way through Aristotle and Plato. See, for example, Simplicius, *Commentary on Aristotle's* Categories 6.1–5. Simplicius' commentary on the *Enchiridion* of Epictetus and Hierocles' commentary on the so-called *Golden Verses* of Pythagoras are extant examples of the way elementary ethics was taught. Perhaps Proclus received his rudimentary ethical training when he learned rhetoric, or perhaps we should associate this training with Heron's instilling piety in him.

orean manner, investigates natural things insofar as they participate in what transcends nature. Aristotle did not use myths or symbolic enigmas in the way some of his predecessors did, but he preferred obscurity of formulation to every other form of concealment. (Simplicius, *Commentary on Aristotle's* Categories 6.22–33; on Plato compare 22.9–16 of the Euclid commentary.)

Thus the course of Proclus' education was an initiation which started from more mundane perceptual matters, but led to the higher mysteries wrapped in Platonic enigmas.

There is an obvious tension between the division of theoretical philosophy into physics, mathematics, and theology, and the curriculum, which adds logic to these subjects. The Neoplatonists handled this difficulty by treating logic in a standard Peripatetic way as a tool (*organon*) for doing philosophy, something whose use had to be learned before one could reason at all.[43] However, there remains the problem that in Proclus' education training in mathematics coincided with training in logic, so that it, too, would seem to be a preliminary to, rather than a part of, philosophy. We know that some Neoplatonists felt this way. They cited in their favor the alleged inscription over the door of Plato's Academy: "Let no one who doesn't know geometry enter."[44] But the standard Neoplatonic view seems to have been that mathematics was a bridge or ladder between the sensible world of physics and the intelligible world of theology.[45] It is clear that Proclus shares this view,[46] so that the audience for the commentary should be thought of as students who have at least read all of Aristotle and, I would imagine, Plato's *Timaeus*. Of course, they well may have read more. The important point is that Proclus sees the study of mathematics as preparing the soul for an ascent to Platonic theology. One shouldn't think of that theology as just a matter of penetrating Platonic symbolic enigmas. It is certainly that, but Proclus is perhaps best known as the person who axiomatized theology in his *Elements of Theology*. Training in mathematics prepares the soul for theology both by leading the soul from material to spiritual things and by teaching it to reason about spiritual things.

[43] See, for example, Philoponus, *Commentary on Aristotle's* Categories 4.23–35.

[44] See, for example, Olympiodorus, *Prolegomena* 8.39–9.1. On this inscription see H. D. Saffrey, "ΑΓΕΩΜΕΤΡΗΤΟΣ ΜΗΔΕΙΣ ΕΙΣΙΤΩ: une inscription légendaire," *Revue des études grecques* 81 (1968): 67–87.

[45] See, for example, Ammonius, *Commentary on Porphyry's* Eisagôgê 11.23–13.7.

[46] See, for example, chapter VIII of part one of the prologue.

Having situated Proclus' commentary in the context of Athenian Neoplatonism, I want to say a few words about its content, focusing on some of the important chapters of the prologue.[47] Proclus begins the prologue with one of his expositions of the Divided Line passage of Plato's *Republic* discussed above. He emphasizes that mathematics deals with a realm intermediate between being and sensibles. Proclus discusses the intermediate character of geometry in chapter III of the second part of the prologue.[48] He indicates that although the principal concern of mathematics is "dianoetic forms," it also impinges on physics at its lower level and at its higher "it looks around upon the region of genuine being, teaching us through images the special properties of the divine orders." This three-level conception of geometry is fully reflected in the commentary on the propositions of book I, much of which focuses on Euclid's "dianoetic" reasoning. But Proclus thinks of the whole *Elements* as directed toward the construction of the regular solids used by Plato in the *Timaeus*, and he frequently refers to physical applications of geometrical results.[49] Moreover, the commentary, particularly the part devoted to Euclid's definitions, is full of indications of the metaphysical and theological truths imaged in geometrical concepts and propositions.[50]

In chapter II of part one of the prologue Proclus describes the Limit and the Unlimited as the common principles of mathematics because they are fundamental principles of all beings, a doctrine adapted by the Athenian Neoplatonists (and earlier Neopythagoreans) from Plato's *Philebus*. The material should be interpreted as an attempt to find mathematical facts to which a given metaphysical scheme can be applied. For example, that the sequence 2/1, 3/2, 4/3, . . . , $n + 1/n$, . . . exhibits ever-changing ratios is for Proclus an indication of the role of the Unlim-

[47] There are many parallels between the two parts of the prologue, since the first concerns mathematics in general, the second geometry.

[48] There is a briefer discussion of the intermediate character of mathematics in general at 19.6–20.7.

[49] See, for example, the very obscure claim about the astronomical use of proposition VIII at 268.15–269.6, and the remarks that follow. Sometimes the applications of mathematics to the sensible world are less scientific: see, for example, 149.8–150.12.

[50] See, for example, the discussion of the angle at 128.26–131.2, of the circle at 146.24–151.12 and 153.12–156.5, and of the square at 173.2–174.21, noting especially its concluding remark. 187.4–18 offers a spiritual interpretation of the first three postulates. Examples of such interpretations in the discussion of propositions are 290.15–291.19 and 293.15–294.14.

ited in mathematics, but the constancy of 2/1, 4/2, 6/3, . . . , 2n/n, . . .
indicates the role of the Limit. These mathematical facts do not strike us
as profound, but for Proclus they are a way of introducing the student to
deep metaphysical truths.[51]

Having discussed the common principles of mathematics, Proclus
turns in chapters III and IV to what he calls the common theorems of
mathematics, truths such as that things equal to the same thing are equal
to each other or that if a:b :: c:d then a:c :: b:d. These truths apply not
just to a single scientific domain (e.g., just to numbers or just to geometric magnitudes), but to all scientific domains in common. The basic
idea of common theorems in mathematics can be traced back to Aristotle. In *Metaphysics* M.1–3 Aristotle develops his own account of
mathematical ontology, which the Neoplatonists understood as "ab-
stractionism"—the view that mathematical objects are mental conceptions derived from sensibles.[52] In M.3 he defends this view by saying
that we no more need to suppose that there are mind-independent numbers or geometric magnitudes than we need to assume that "the universal parts of mathematics" deal with special objects other than numbers,
magnitudes, etc. For Proclus, Aristotle is totally wrong on this point:
what is more universal is ontologically and apodeictically prior to what
is less universal, just as geometric and arithmetic objects are prior to the
sensibles from which, according to Aristotle, they are abstracted. Proclus develops his "projectionist" account of mathematical reasoning, to
which I have already referred, by opposition to Aristotelian abstraction-
ism.[53]

Chapters IX and X of part one of the prologue throw considerable
light on the generally unscientific—or even antiscientific—intellectual
climate in which Proclus taught. Proclus has to argue against people who

---

[51] See also 131.9–132.17 for the Pythagorean "justification" of the division of plane,
rectilinear angles into acute, obtuse, and right. Proclus' reference to the "geometrical
number" of *Republic* 545eff. (8.14–20; cp. 23.12–24.3) is a good index of the way passages of Plato which are now usually treated as either mumbo jumbo or heavy-handed
humor were taken to be deep "symbolic enigmas" by the Neoplatonists.

[52] See my "Aristotle's Doctrine of Abstraction in the Commentators," in Richard So-
rabji, ed., *Aristotle Transformed* (London and Ithaca, 1990), 463–80.

[53] At 56.24 Proclus indicates that his projectionism does not agree with the views of
Porphyry and "most of the Platonists," presumably indicating the prevalence of abstrac-
tionism among his contemporaries. Iamblichus and Syrianus certainly held views like Proclus'; see Iamblichus, *On Mathematics in General* 34.9–12 or Syrianus, *Commentary on
Aristotle's* Metaphysics 91.11–92.10.

disparage mathematics because it doesn't teach anything of moral significance (*to kallos*) or of practical value in the "real" world. We might well accept the first charge and reject the second. Proclus argues in the reverse way: mathematics familiarizes us with order, symmetry, and definiteness, three preeminent characteristics of *to kallos*; and mathematics ought to be studied for its own sake, or, if an external motivation is needed, in order to purify the soul for higher apprehension. Chapter XII shows that even some Neoplatonists doubted the value of studying mathematics, and cited Plato in their defense. Proclus' own counterexegesis of Plato is a model of good sense. However, the point I would like to stress is that for the most part Proclus is probably dragging students through mathematics in the way that some modern students are dragged through the science requirements of a liberal arts curriculum. Proclus is not teaching the philosophical implications of a science that his students understand and appreciate. Rather he is trying to expose his students to the rudiments of Greek science and to get them to see that the study of mathematics contributes to reaching the Platonist goal of human perfection.[54] The first of Proclus' purposes—along with the character of his audience—explains the tedious detail with which he goes through the propositions of *Elements* I in the last half of the commentary.[55] The second explains his emphasis on the uplifting effect of mathematical study.

In chapters XII and XIII Proclus turns to the division of mathematics into different branches. Interest in classification is part of the scholasticism which colors late Platonism. However, classification of the mathematical sciences is also supposed to have its ground in the nature of reality: each science is identified by a relation to some feature or part of reality which it apprehends. Proclus offers two classifications. He calls the first Pythagorean, where we would call it Neopythagorean. The division is a rationalization of the five-part mathematical curriculum of the *Republic* with geometry and stereometry combined to produce the so-called quadrivium of arithmetic, geometry, harmonics, and astronomy (for which Proclus uses the less empirical–sounding name of "spherics"). The basic idea of this classification is that things are either dis-

[54] At 113.3–8 Proclus encourages "ambitious" and "able" students to pursue some less elementary topics which he cannot discuss further. Cp. 272.12–14.

[55] For a brief discussion of this material see pp. 290–93 ("Sur l'usage 'scholaire' des *Eléments*") of volume 1 of the French translation of the *Elements* by Bernard Vitrac (Paris, 1990).

crete or continuous, a multiplicity (*plêthos*) or a magnitude (*mege-thos*).[56] Both of these have an aspect of unlimitedness: there is no largest multiplicity, and no smallest magnitude. Science studies only limited multiplicity and magnitude, that is, the "how many" (*poson*) and the "how much" (*pêlikon*). The former can be divided into the in-itself and the relative, the latter into the stationary and moving, producing the qua-drivium of arithmetic, music (i.e., mathematical music theory), geom-etry, and astronomy (i.e., spherics, the study of rotating spheres). Pro-clus goes on to associate this classification with the creation of the World Soul in Plato's *Timaeus* and with the role of the Limit and the Unlimited as ultimate principles.

Given Proclus' adulation of Pythagoras and Plato, one might have expected him to be satisfied with the Pythagorean division alone. But he gives another, which he ascribes to Geminus.[57] Geminus' classification is, one might say, more realistic than the Pythagorean one. It makes a distinction between pure and applied mathematics, it includes more sci-ences, and it gives a more detailed account of them. For these reasons Proclus' inclusion of Geminus' classification is a reflection of his own reasonableness. And it is surely a part of Proclus' motivation to preserve the memory of the multiple achievements of Greek science mentioned by Geminus. But I think his most important motivation is the simple existence of Geminus' classification as part of the extant body of philo-sophical exegeses of the mathematical sciences. As Proclus says at the end of his presentation of the classification, "Such are the traditions we have received from the writings of the ancients regarding the divisions of mathematical science." (42.7–8; cp. 64.3–7) We must be grateful to Proclus for recording these and other traditions since he is, in many cases, our only source of information about them. On the other hand, because of Proclus' eclecticism and interest in preserving the knowledge

---

[56] The earliest example of the classification is in Nicomachus (*Introduction to Arith-metic* I.2 and 3), who sets it out more clearly than Proclus does. It has seemed to me best to diverge from Morrow's translation here because it misses some of the nuances of what Proclus says.

[57] On Geminus see, in addition to Morrow's note on 38.4, the introduction to Géminos, *Introduction aux phénomènes*, Germaine Aujac, ed. and trans. (Paris, 1975), and Otto Neugebauer, *A History of Ancient Mathematical Astronomy*, 3 vols. (Berlin and New York, 1975), pp. 579–80. Some scholars have claimed that Geminus was a major source for Proclus' commentary, but the issue remains moot.

of Greek achievements, one must not expect that everything in the commentary will cohere as a philosophical whole.

Having described the division of mathematics into branches, Proclus turns in chapter XIV to discuss what holds the branches together. He does this in terms of the reference to dialectic as like the capstone of the mathematical sciences in Plato's *Republic* and the reference to the unifying bond of the mathematical sciences in the pseudo-Platonic *Epinomis*.[58] Proclus denies Eratosthenes' likely interpretation that proportion is the unifying bond, and substitutes a hierarchy of unifying bonds: universal mathematics; dialectic; and *nous*, "the completion of the upward journey and of gnostic activity." Although in some places (see, for example, the commentary on the *Parmenides* 648.1–656.14) Proclus seems to accept the identity of dialectic and the apprehension of Forms, in chapter XIV and elsewhere in the Euclid commentary he clearly sees dialectic as preliminary to the noetic apprehension of Forms. For Proclus in the commentary dialectic is basically the understanding and use of the methods of mathematical reasoning, methods which he identifies with analysis, demonstration or synthesis, division, and definition—by contrast, notably, with causal and symptomatic argument (69.9–19).[59]

Probably the most frequently cited passage in the Euclid commentary is chapter IV of the second part of the prologue, where Proclus gives an outline of the history of geometry down to Euclid—a history focused on the role of Plato and the Academy. It is often said that the ultimate source of this passage is the work of Aristotle's pupil Eudemus, perhaps with Geminus as the major intermediary.[60] Recent work has placed more emphasis on post-Eudemian and even Neoplatonic aspects of the passage.[61] Some scholars have questioned the historical reliability of even Eudemus, but—speaking generally—one can say that the rule of thumb

---

[58] Proclus denied (probably correctly) the authenticity of the *Epinomis*; see *Anonymous Prolegomena to Platonic Philosophy*, L. G. Westerink, ed. and trans. (Amsterdam, 1962), X.25.4–10. His doing so is quite compatible with his assigning the dialogue a certain authority and thus wanting to determine its doctrines accurately.

[59] For an attempt to give a unified account of Proclus' notion of dialectic see Alain Lernould, "La dialectique comme science première chez Proclus," *Revue des sciences philosophiques et théologiques* 71 (1987): 509–35.

[60] Neither Eudemus nor Geminus is mentioned in chapter IV. In the index to the translation one should add three further references to Eudemus (233, 260, 275) and one to Geminus (190).

[61] See Conrado Eggers Lan, "Eudemo y el 'catálogo de geómetras' de Proclo," *Emerita* 53 (1985): 127–57.

is that the more likely Eudemus is to be the source of a historical remark by Proclus about early Greek mathematics, the more likely the remark is to be true. Unfortunately, unless Proclus cites him explicitly, we are on very shaky ground in invoking Eudemus as an authority. The moral for the reader of the commentary is always to be wary of taking what Proclus reports as history; every claim has to be weighed against other available evidence—if there is any.[62] I do not, of course, mean to imply that Proclus is worthless as a historical source or that he made up "facts." It is clear that one of Proclus' purposes in teaching geometry was to convey what information he had about the history of elementary geometry down to Euclid's time. If he hadn't done this, we would know virtually nothing about the subject.

At the end of his history of geometry Proclus introduces Euclid, a person about whom he clearly has no direct information other than the names of books assigned to him, a (questionable) reference to him by Archimedes, and an anecdote connecting him with Ptolemy the First. Proclus says that Euclid was a professing Platonist and that he organized the *Elements* to culminate in the treatment of the regular solids because of Plato's use of these solids in the physics of the *Timaeus*. He may be right, but there is no reason to think he had evidence for these claims which we lack, and good reason to think he is accommodating Euclid to his own philosophical program. There is no philosophy expressed in any work assigned to Euclid; the *Elements* looks like a work of pure mathematics of the kind we are all familiar with. Proclus wants the work to be Platonist because he wants to use it for Platonist purposes. There is nothing objectionable about his doing so, but it would be wrong to infer from his doing so that the *Elements* or its author was Platonist in any interesting sense.

To understand a philosophical or scientific text is to make sense of it, and what makes sense is relative to an outlook. Proclus' own outlook and the understanding of Plato on which it is based are not ours. So naturally his interpretation of Euclid is not always ours. But his attempt to read Euclid in the light of his own philosophical outlook is not importantly different from a modern philosopher/teacher reading an ancient text in terms of his or her own philosophical perspective. Nor are

---

[62] The most disputed of Proclus' historical references concern members of the pre-Euclidean tradition, in particular Thales and Pythagoras or other Pythagoreans. In general Proclus' accounts of post-Euclidean material are taken to be reliable.

Proclus' methods of teaching the text of Euclid fundamentally different from the methods we use: he pursues a general line of interpretation, a reading, while presenting a great deal of material about the history of his subject and of interpretations of his text and related matters. As an extension of this comparison between Proclus and the contemporary teacher I would like to recall the position of the "Academy" in the fifth century. Proclus taught as a preserver of a noble intellectual heritage in a society increasingly indifferent and even hostile to that heritage. Many members of today's academy see themselves in a similar position. It is unlikely that this similarity of structure has no reflection in content. About eight hundred years separate Proclus from Socrates, Plato, and Aristotle; only about two hundred years separate our "postmodern" world from the Enlightenment. Proclus is not a postmodernist, but reflection on his ways of thinking and their relation to his time may shed light on the intellectual turmoil of our own.[63]

[63] I would like to thank the American Council of Learned Societies and the John Simon Guggenheim Memorial Foundation for the financial support which enabled me to write this introduction. I would also like to thank Chris Bobonich, Richard Kraut, and Janel Mueller for valuable comments on an earlier version of this foreword.

Preface

THE JUSTIFICATION for offering to modern students a translation of a fifth-century commentary on a mathematical textbook already more than seven hundred years old when the commentary was composed is given in the chapters of the Introduction. Here it is enough only to mention the twin themes of these chapters: the greatness of the work on which the commentary was written, and the talent, learning, and sympathetic understanding which the commentator brought to the exposition of it. Among the many creations of the Greek mind, Greek geometry is one of the most splendid, and Euclid has been its honored exponent and spokesman for more than two thousand years. Yet the respect universally accorded him and the science he stands for is seldom accompanied by an understanding of the intellectual principles on which this science was built and the procedures by which its details were developed. Such lack of understanding does not prevent our being impressed by the final product. But just as our appreciation of the work of Ictinus, for example, in designing the Parthenon can be immensely enriched by a knowledge of the difficulties the Greek architect had to overcome, the materials with which he worked, the instruments that were available for handling them, and the traditions of his craft that served him both as restraints and as inspiration, so also a knowledge of the empirical and conceptual materials at the disposal of the Greek mathematician, the conventions and norms that implicitly guided him from the days of Thales and Pythagoras and finally came to clear consciousness in the work of Euclid, and the difficulties with which the advance of geometrical science was confronted and the ingenious devices by which these difficulties were in the main surmounted can contribute enormously to our understanding and enjoyment of this monument of the Greek mind.

Such an understanding Proclus is peculiarly fitted to give us. Though not himself a creative mathematician, he was an acute and competent critic, and immensely learned, his mind stored with the achievements of the preceding thousand years of mathematical inquiry. Living at the end of the Hellenistic Age, he could survey the accomplishments and errors of his predecessors with something like an Olympian eye. And since he possessed a due share of

Hellenic intelligence and precision, together with a Platonic devotion to mathematics as the liberating science, he is uniquely qualified to transmit to us something of this Hellenic and Platonic enthusiasm.

To this I must add my personal conviction that the student of Plato's later thought, if he hopes ever to comprehend the part that mathematics played in the development of Plato's thinking, and particularly the puzzling statements of Aristotle regarding the latest form of the theory of Ideas, must enter with sympathy and understanding into this climate of thought created by the mathematicians of Plato's and the following centuries. As a devout Platonist, Proclus is naturally a resource of the first order for enabling us to do this.

The purpose of this book, then, is to make available to English readers a treatise of unique value for the history of mathematics and of philosophy. Like the mathematics with which it deals, it has universally been treated with respect, and it has often been cited as evidence for this or that fact in the history of mathematics; but it has seldom been looked at as a whole, chiefly because of the obstacles presented by a Greek text of such length. These obstacles, I venture to say, are appreciated even by those who can read Proclus' language with facility; and for those who cannot they are insuperable barriers. I hope this translation will help to overcome these hindrances to acquaintance with a treatise of outstanding quality, almost the only one from antiquity that deals with what we today would call the philosophy of mathematics.

My notes do not presume to be a full commentary on the text. I have not essayed to deal with all the mathematical questions presented, still less with the issues involved in Proclus' excursions into Neoplatonic ontology and cosmology. What I have tried to do in them is to provide the explanations essential to understanding my translation (and the text of Friedlein on which it is based), to identify the passages in previous writers to which Proclus refers, to characterize briefly the personages that figure in his historical comments, and to furnish some preliminary aids to the further exploration of the philosophical and mathematical contents so richly deployed before us. I have labored to make my translation as clear a presentation of Proclus' thought as my powers and insight permitted; yet those who make use of it will certainly find that there is still much to explore.

In addition to my debt to previous translators and scholars, which is evident in my footnotes, I acknowledge with grateful appreciation my indebtedness to my colleague at the University of Pennsylvania, Charles H. Kahn, for constant advice and encouragement; to Robert S. Brumbaugh, of Yale University, who read my manuscript for the Princeton University Press and made many helpful suggestions, in particular calling attention to the significance of Iamblichus' *De Communi Mathematica Scientia*; and to Ian Mueller, of the University of Chicago, who read the manuscript in its entirety, saved me from many errors, and made numerous pertinent comments on the text, some of which I have presented in my notes, with his initials "I.M." to identify them.

To Mr. Sanford G. Thatcher, Social Science Editor of the Princeton University Press, I am greatly indebted for the assiduous attention he has given to this manuscript, and for the advice and assistance he has so generously rendered in editing it and in overseeing its progress through the Press. To him and to his associates I hereby express my warm and sincere thanks. Mrs. Georgia Minyard, of the University of Pennsylvania, has also given me much competent help in editing and proofreading, for which I am most grateful.

Glenn R. Morrow

Swarthmore, Pennsylvania
May 30, 1969

Abbreviations

THE FOLLOWING abbreviations have been used in the footnotes for authors and works frequently cited.

Barocius = *Procli Diadochi Lycii in Primum Euclidis Elementorum Commentariorum Libri IV a Francisco Barocio Patritio Veneto Editi*, Padua, 1560.

CAG = *Commentaria in Aristotelem Graeca*, 23 vols., Berlin, 1882-1909.

Diels[6] = Hermann Diels and Walther Kranz, *Fragmente der Vorsokratiker*, 6th edn., Berlin, 1951-1952.

Dodds = E. R. Dodds, *Proclus: The Elements of Theology*, 2nd edn., Oxford, 1963.

Friedlein = *Procli Diadochi in Primum Euclidis Elementorum Librum Commentarii ex Recognitione Godofredi Friedlein*, Leipzig, 1873.

Gow = James Gow, *History of Greek Mathematics*, Cambridge, 1884; reprinted, New York, 1923.

Grynaeus = *Commentariorum Procli editio prima quae Simonis Grynaei opera addita est Euclidis elementis graece editis*, Basel, 1533.

Heath = Thomas Heath, *Greek Mathematics*, 2 vols., Oxford, 1921.

Heath, *Euclid* = Thomas Heath, *The Thirteen Books of Euclid's Elements*, Translated from the text of Heiberg with Introduction and Commentary, 2nd edn., 3 vols., Cambridge, 1926.

Heiberg = *Euclidis Elementa edidit et Latine interpretatus est I. L. Heiberg*, 5 vols., Leipzig, 1883-1888. Supp. Vol. VI (1896), ed. H. Menge.

Kroll = Wilhelm Kroll, *De Oraculis Chaldaicis*, Breslau, 1894.

RE = Pauly-Wissowa-Kroll, *Realencyclopädie der Classischen Altertumswissenschaft*, Stuttgart, 1894-1952.

Rosán = Lawrence J. Rosán, *The Philosophy of Proclus*, New York, 1949.

Schönberger = Leander Schönberger, *Proklus Diadochus: Kommentar zum Ersten Buch von Euklids Elementen*, a translation into German, with Introduction, Commentary, and Notes by Max Steck, Halle, 1945.

Tannery = Paul Tannery, *Mémoires Scientifiques*, 17 vols., Paris, 1912-1950.

Taylor = Thomas Taylor, *The Philosophical and Mathematical Commentaries of Proclus on the First Book of Euclid's Elements*, 2 vols., London, 1788, 1789.

Van der Waerden = B. L. Van der Waerden, *Science Awakening*, New York, 1961.

Ver Eecke = Paul ver Eecke, *Proclus de Lycie: Les Commentaires sur le Premier Livre des Eléments d'Euclide*, a translation into French, with an Introduction and Notes, Bruges, 1949.

Von Arnim = Ioannes von Arnim, ed., *Stoicorum Veterum Fragmenta*, 4 vols., Leipzig, 1903-1924.

Introduction

PROCLUS: HIS LIFE

AND WRITINGS

PROCLUS was born at Byzantium in the early years of the fifth century, just as the Hellenistic Age was drawing to a close. His parents were natives of Lycia in southern Asia Minor, hence he is known in the ancient catalogues as Proclus Lycius. His early education was acquired at Xanthus, a little city on the southern coast. From there he proceeded to Alexandria with the intention of following the profession of his father, who had earned a considerable reputation as a pleader in the courts of the imperial city. But on a trip to Byzantium that he made during these student days he experienced a divine call, so his biographer Marinus[1] tells us, to devote himself to philosophy; and on his return to Alexandria he began to attend the lectures of Olympiodorus[2] on Aristotle and of Heron[3] on mathematics, distinguishing himself among his fellow students and teachers by his capacity for rapid assimilation and by his extraordinary memory. But the teachers at Alexandria failed to satisfy him, and while still under twenty years of age he went to Athens, where the School of Plato had recently experienced a notable revival under Plutarch.[4] And it is with Athens and the School of Plato that he was associated during his entire later career.

The first philosopher he met at Athens was Syrianus,[5] who later became head of the School after the death of Plutarch. Syrianus

[1] Marinus was a pupil of Proclus and author of a commentary on the *Data* of Euclid which is extant (see Heiberg VI). His *Vita Procli* is the chief, and almost the only, source of information about his master's life. For a translation of this *Life* see Rosán, 13-35; and for the Greek text see the supplement in the Didot volume of Diogenes Laertius (Paris, Firmin-Didot, 1829, pp. 151-170). The "divine call" is mentioned in Pars. 6, 9, and 10.

[2] This is not the Aristotelian commentator of the same name, who lived in the following century.

[3] This is not the famous Heron of Alexandria, who lived two centuries earlier.

[4] Plutarch the Athenian, whose commentary on Aristotle's *De Anima*, no longer extant, was highly esteemed in antiquity.

[5] Syrianus' commentary on Books ΒΓΜΝ of Aristotle's *Metaphysics* is still extant. See *CAG* VI, Kroll.

introduced him to the great Plutarch, who was so taken with the zeal and aptitude of the young man that he took him under his personal charge, although his advanced age prevented him from taking many students. He read with Proclus the *De Anima* of Aristotle and the *Phaedo* of Plato, encouraging the young Proclus to keep a written record of what was said, with his own comments upon it. Thus Proclus began early the practice of writing commentaries, a habit to which is due so much of his prodigious literary production. Plutarch died not long afterwards, but under Syrianus during the next two years, so we are told, Proclus read all the writings of Aristotle in logic, ethics, politics, and physics; and then after going through these "lesser mysteries," as Syrianus called them, he was initiated into the greater mysteries of Plato's philosophy. Syrianus also introduced him to the Orphic writings and the Chaldaean Oracles. These were the beginnings of a career of study, teaching, and writing that lasted almost without interruption until the death of Proclus at the age of seventy-five.[6] In his later years he was himself head of the School, whence comes the title Diadochus (Successor) which has become attached to his name.

Marinus tells us little about the events of Proclus' mature life but makes obscure references to a "storm" that gathered around him at one time and forced him to retire from Athens for a while.[7] It is natural to interpret this episode as connected in some way with Proclus' devotion to the ancient religion. Pagan beliefs were still tolerated in the schools, although (since Christianity was now the official religion of the Empire) the practice of the cults was forbidden. Proclus, like most of the philosophers at Athens, was still attached to the ancient faiths. Such was his piety that he personally observed all the sacred days of the Egyptians as well as the Greeks and devoted himself assiduously to the study of mythology and religious observances. He declared that it becomes the philosopher not to observe the rites of one city or of a few cities only, but to be the common hierophant of the whole world. Perhaps it was the

---

[6] So Marinus tells us (Pars. 3 and 26). His death occurred, as we can see from other facts recorded by Marinus (Pars. 36, 37), in the year 485. If he was seventy-five years old when he died, he must have been born in 410. But Marinus also describes a horoscope taken at his birth which by astronomical calculation indicates February 18, 412, as the date of his birth.

[7] Marinus, *Vita Procli*, Par. 15.

private practice of unauthorized rites that stirred up the storm against him. Whatever the explanation, we are told that he spent the year of his exile in Lydia investigating the ancient religious customs of that region and interpreting to their worshippers the deeper meaning of their practices. When the year was over, he returned to Athens and resumed his study and teaching. The School was financially independent, for although the munificent salaries provided by Marcus Aurelius for professors at Athens had long been discontinued, the Academy was supported—and amply, we may assume—from the endowments that had accumulated during the eight centuries since Plato's death.[8]

Proclus was handsome in appearance, even more so, says Marinus, than his numerous portraits show him to have been (evidently Proclus did not share Plotinus' scruples against having his portrait painted), of vigorous health, with abounding yet disciplined energy. He had a pleasing voice and a command of language and thought that captivated his hearers, so that when he was speaking a radiance seemed to flow from him. And according to his eulogist he exhibited all the virtues—physical, moral, political, ascetic, intellectual, and "theurgic"[9]—in a preeminent degree. But of all the traits mentioned in this eulogy the most important for the student of the history of thought was Proclus' "unbounded love of work."[10] Besides conducting five or six classes, holding other conferences, delivering lectures, and discharging his religious obligations, he customarily wrote seven hundred or more lines a day, Marinus tells us. This assertion, incredible as it may seem, is borne out by the volume of the writings attributed to him. And their scope, as well as their volume, is enormous, extending from philosophy and theology through mathematics, astronomy, and physics to literary criticism and poetry. They constitute a priceless resource for the student of this latest period of ancient philosophy. We must be content to mention only a part of the items in this treasury.[11]

[8] Damascius records that the annual revenue from its property amounted to a thousand (gold?) pieces at this time, as compared with three in Plato's day. See the *Vita Isidori*, Par. 158, in the Didot volume cited in note 1 above.
[9] The term comes from Iamblichus, says Marinus (*Vita Procli*, Par. 26).
[10] *Ibid.*, Par. 22: ἄμετρος φιλοπονία.
[11] See Rosán for a full list, with a bibliography of the various editions and translations of individual works during the modern period to 1940.

Proclus' most systematic philosophical works are his *Elements of Theology* and a later treatise on *Platonic Theology*. The texts of both have survived.[12] The former, like the much later *Ethics* of Spinoza, presents the "elements" in geometrical fashion, in a series of propositions each supported by its proof. The *Platonic Theology* likewise, despite its title, is an exposition of Proclus' own system, which of course he regarded as Platonic. Much better known to the student of Plato is his impressive series of commentaries on the dialogues: a commentary on the *Parmenides*, one on the *Timaeus* (which, Marinus reports, was Proclus' favorite among his commentaries), another on the *First Alcibiades*, and another on the *Republic*. The texts of all these have been preserved.[13] His commentary on the *Cratylus* survives only in excerpts, and the commentaries that he wrote on the *Philebus*, the *Theaetetus*, the *Sophist*, and the *Phaedo* have all been lost. We possess only a fragment of his commentary on the *Enneads* of Plotinus, and we know of his *Eighteen Arguments for the Eternity of the World* (a tract against the Christians) because it is extensively quoted in Philoponus' book written to refute it.[14] His treatises on *Providence*, on *Fate*, and on the *Subsistence of Evil* were long known to us only through the Latin translation of William of Moerbeke; but large portions of the Greek text have recently been recovered and edited by Helmut Boese.[15]

His scientific works include the commentary on Euclid here translated, a short elementary treatise on astronomy entitled *Sphaera*, his *Hypotyposes* (or Outlines) *of the Hypotheses of the Astronomers*, a paraphrase of and perhaps also a commentary on Ptolemy's *Tetrabiblus*, and a book on *Eclipses* surviving only in a Latin translation. We also know of an essay on the parallel postulate in

[12] The *Elements of Theology* has been elegantly edited and translated into English by E. R. Dodds (2nd edn., Oxford, 1963).

[13] The commentary on the *First Alcibiades* has recently been translated into English by William O'Neill (The Hague, 1965). A French translation by Festugière of the commentary on the *Timaeus* (Paris, 1968) has just been completed.

[14] John Philoponus, *De Aeternitate Mundi contra Proclum* (Leipzig, 1899), written probably before the middle of the sixth century.

[15] *Procli Diadochi Tria Opuscula*, Berlin, 1960.

geometry[16] which has not survived, but one can get an idea of its contents from the relevant parts of the commentary on Euclid (191.23-193.7; 365.7-373.2).[17] We also possess his *Elements of Physics*, mainly a summary of Books VI and VII of Aristotle's *Physics* and the first book of the *De Caelo*, presented in geometrical form with propositions and proofs. We have only fragments of his book on the *Objections of Aristotle to Plato's Timaeus*, to which he himself refers in his commentary on the *Timaeus*;[18] these fragments show that he was principally concerned in this work to answer Aristotle's objections to Plato's physical theories.

Proclus also wrote numerous works on religion, but none of them survives except in fragments. Like Aristotle,[19] he believed that ancient traditions often contain truth expressed in mythical form. Orphic and Chaldaean theology engaged his attention from his earliest days in Athens. He studied Syrianus' commentary on the Orphic writings along with the works of Porphyry and Iamblichus on the Chaldaean Oracles; and in his later life he undertook a commentary on this collection which, Marinus tells us, cost him five years of labor. Marinus' account describes this labor as both critical and constructive. Whatever in these traditions Proclus found to be empty or contrary to well-established principles he would reject as trivial or fraudulent, supporting his judgment by arguments and proofs; on the other hand, whenever he came upon anything he thought "fruitful," he would interpret it and bring it into harmony with other elements of doctrine, presenting the result, both in his lectures and in his writings, with great clarity and "enthusiasm."[20] We may well suppose that his criticism was as precise and exacting, according to the conventions governing theology, as we find it to be in his examination of Euclid and his critics in the commentary before us. In this commentary, as the reader will see, mathematical reasoning and exposition is often interrupted by digressions point-

---

[16] Referred to by Philoponus in his commentary on the *Posterior Analytics* (*CAG* XIII.2, 129.16, Wallies).

[17] These and later references to Proclus' *Commentary* are to the pages and lines of Friedlein's edition. See the Translator's Note.

[18] Ed. Diehl, I, 404.20; II, 279.2-14.

[19] *Met.* 1074b1ff.

[20] Marinus, *Vita Procli*, Par. 22.

ing out the moral or metaphysical significance of a figure or a theorem under discussion, and its value in directing the mind upward to the region beyond mathematical existence. So in his study of mythology we must presume that one of his main concerns was to determine the adequacy of a myth for imaging the higher grades of being and for assisting the soul in worship to reunite itself with the great source of all being. This religious motive is an inseparable part of his, as of all Neoplatonic metaphysics.

Proclus composed innumerable hymns to the gods, of which seven survive, written in Homeric language and exhibiting a considerable measure of literary quality as well as deep religious feeling. It is not surprising that his biographer should tell us that he lived in constant communication with the gods, addressing his adoration and aspirations in prayers and observances and receiving messages from them in dreams. One incident of the many reported by Marinus is particularly striking. The great gold and ivory statue of Athena that had stood in the Parthenon since the days of Phidias and Pericles was removed from the Acropolis during Proclus' lifetime. Evidently deeply shocked by this action of the Christians, Proclus dreamed that a beautiful woman appeared to him and commanded him to get his house in order quickly, "because the lady of Athens wishes to dwell with you."[21]

Even if we count only those works that can indisputably be attributed to Proclus, the range and volume of his production is still enormous, constituting, in Cousin's opinion, a veritable fifth-century *Encyclopédie*.[22] And in clarity, in precision of expression, and in the logical articulation of the various parts of the individual works, these writings manifest qualities exceedingly rare in this twilight period of Hellenic culture. If his religious works, to judge from the fragments that survive, often reveal a tendency to superstition and

[21] *Ibid.*, Par. 30. Proclus' house, which had been the residence of Syrianus and of Plutarch before him, stood on the south slope of the Acropolis ("easily seen from it," says Marinus), adjacent to the Theater of Dionysus and the Temple of Asclepius. Recent excavations in this area have brought to light the foundations of a house with an exedra, which has been labelled the House of Proclus, although there are no inscriptions or artifacts to support this identification. An unidentified bust has been found here also, and it may be a bust of Proclus.

[22] In the Introduction to his edition of Proclus (xviii), 1864, quoted by Rosán, 57n.

magic (or theurgy, as these later Platonists called it), we must re-
member that this characteristic was almost inseparable from the
thought of the Neoplatonists of this time. Dodds has aptly remarked: "Proclus' qualities were all but unique in an age when his
defects were all but universal."[23] We are told that, when Proclus
first arrived as a youth at Athens, he reached the Acropolis late in
the day, in fact just as the keeper was preparing to close the gates.
"Really if you had not come," he said, "I would have locked up."[24]
This story picturesquely illustrates, as we see the situation today,
Proclus' part in delaying for a time the closing of the classical
age. He was indeed the last creative mind in Greek philosophy. It
was less than half a century after his death that the voice of pagan
philosophy was stilled at Athens with the edict of Justinian in 529
closing the schools and confiscating their properties.

Proclus left instructions that he should be buried beside his
master in a double tomb constructed by Syrianus on the slope of
Lycabettus. This was the epitaph that he composed for himself:[25]

> Proclus was I, of Lycian race, whom Syrianus
> Beside me here nurtured as a successor in his doctrine.
> This single tomb has accepted the bodies of us both;
> May a single place receive our two souls.

EUCLID AND THE
ELEMENTS

THERE ARE few books that have played a larger part in the thought
and education of the Western world than Euclid's *Elements*. For
more than twenty centuries it has been used as an introduction to
geometry, and only within the last hundred years has it begun to
be supplemented, or supplanted, by more modern textbooks. "This
wonderful book," writes Sir Thomas Heath, "with all its imperfec-
tions, which indeed are slight enough when account is taken of the
date at which it appeared, is and will doubtless remain the greatest
mathematical textbook of all times. Scarcely any other book except
the Bible can have been circulated more widely the world over, or

[23] In the Preface, xxvi.
[24] Marinus, *Vita Procli*, Par. 10.
[25] *Ibid.*, Par. 36.

been more edited and studied."[26] Euclid's was not the first attempt
by Greek mathematicians to formulate and organize the materials
of geometry into an elementary text. But so perfect was his mastery
of the tradition and so skillful his arrangement of his materials that
his work very soon replaced all its predecessors; indeed, Euclid's
success has almost blotted out our view of the men who preceded
him, whose contributions we now know of mainly from the references in Proclus and in other ancient mathematicians. The *Elements* was without a rival in the ancient world. The greatest mathematicians of the following centuries occupied themselves with it.
Heron, Porphyry, Pappus, Proclus, and Simplicius wrote commentaries. Theon of Alexandria reedited it, though not altogether to its
advantage. Apollonius of Perga attempted a restatement of its
principles but did not succeed in replacing it. So great was the
prestige of the *Elements* in comparison not only with the work of
Euclid's predecessors but also with the other writings of Euclid that
in later times the author was seldom referred to by name but
instead usually by the title ὁ στοιχειωτής, "the elementator."[27]

Yet about the author of this remarkable work we have very little
exact information. Even Proclus appears to be unsure of his birthplace and of the dates of his birth and death. He tells us (68.10)
only that he belonged to the philosophy, or school, of Plato and
that he lived after Plato's immediate disciples but earlier than the
time of Eratosthenes and Archimedes. From other sources we learn
that he had a school at Alexandria;[28] and Proclus indicates that his
floruit must be placed under Ptolemy I, who reigned from 306 to
283 B.C. He composed numerous other mathematical treatises,
most of which are referred to by Proclus. The Greek text of his *Data*
is extant, together with a commentary on it by Proclus' pupil Marinus. So also are the greater part of his *Phaenomena* and the *Divisions* in an Arabic version. But the *Pseudaria*, on fallacies, has been
lost and so have the three books of *Porisms* and his treatises on
Surface Loci and on *Conics*. He also wrote an *Elements of Music*,
fragments of which are extant.

[26] Heath I, 357ff.

[27] I have avoided the awkward English term by uniformly translating
ὁ στοιχειωτής as "the author of the *Elements*." For the meaning of στοιχείωσις
and στοιχεῖον see Proclus' explanation below (71.24ff.).

[28] Pappus VII, 678, 10-12, Hultsch. On Pappus see note at 189.12.

The *Elements of Geometry* now exists in fifteen books, of which the last two are clearly not by Euclid. No suspicion, however, attaches to the other thirteen, although they have been subject to alteration, addition, and rearrangement by later editors and scribes. It is now possible for us, on manuscript authority, to distinguish an older and more authentically Euclidean text from a later one made by Theon of Alexandria sometime during the fourth century, from which most of the Latin and English translations until very recent times have been derived. But it is likely that both versions circulated in ancient times, since Proclus, though later than Theon, frequently follows the older version. There is reason to believe that there was a Latin translation of Euclid, in part at least, in late antiquity;[29] but if such a translation existed, it played no part in the transmission of Euclid to modern Europe. The *Elements* was translated into Arabic in the eighth century, one of the first Greek books to be translated into that language; and this translation was followed by at least two other Arabic versions from the Greek in the following centuries.[30] It was from the Arabic that the first medieval Latin translations were made, those of Athelhard of Bath in the twelfth century and of Johannes Campanus in the thirteenth. The first Latin translation from the whole of the Greek text was that of Bartolomeo Zamberti, which appeared in a printed book at Venice in 1505. The *editio princeps* of the Greek text was edited by Simon Grynaeus the elder and published at Basel in 1533; this edition, though based on only two manuscripts, and those very inferior ones, itself served as the basis for all later texts and translations until the nineteenth century. With the discovery and utilization of better Greek manuscripts than those used by Grynaeus, and particularly with the identification of Vatican 190 (Heiberg's P) as representing the text of Euclid prior to its recension by Theon, it was possible for Heiberg to replace the imperfect text of Grynaeus with what appears now to be a definitive text. Since the appearance of Heiberg's edition in 1888 it has been translated into English with a voluminous introduction and rich commentaries by Sir Thomas Heath. This edition, together with Heath's earlier *Greek*

[29] Heath, *Euclid* I, 91-93.
[30] For more information about the Arabic versions see the full account in Heath, *Euclid* I, 75-90.

Mathematics, has been invaluable to me at almost all stages of the present undertaking, as my footnotes will show.

PROCLUS AS A COMMENTATOR
ON EUCLID

MATHEMATICS was naturally one of the fundamental subjects of instruction in the Platonic school at Athens in Proclus' time. In the *Republic* Plato had made himself the champion of mathematical study as the main avenue of approach to the world of Forms, the region of intelligible entities that can alone provide objects of knowledge, as contrasted with fragile and uncertain opinions about the objects in the sense world. Such sure and certain knowledge is indispensable, Plato thought, both for the statesman who hopes to give competent guidance to his city and for the philosopher who aspires to apprehend genuine, that is, stable and eternal, being. The new Platonism that arose and flourished in late antiquity had in the main given up the hope of regenerating society, but it held fast to this belief in a higher realm of supersensible realities and to this insistence on mathematics as the primary way of access to that higher world.

Proclus must therefore have taken part in the mathematical instruction of the members of the school, and it is evident that this commentary is, in part at least, the result of his lectures to them. There are references to "my hearers" (210.19, 375.9) and to the needs of beginners which require the lecturer to pass over "for the present" certain matters that are too difficult, further explanation being reserved for a later occasion (113.6, 272.12). Proclus sometimes explains quite elementary matters, such as the meaning of "subtending" (238.13) or the distinction between adjacent and vertical angles (298.14), as if to persons unfamiliar with geometrical terminology. At the beginning of his comments on the propositions, he sets forth systematically the six parts into which every fully expressed theorem or problem can be analyzed (203.1-15), applies these distinctions to the first proposition (208.1-210.16), and advises the student for the sake of practice to do the same thing for each of the remaining ones (210.17-28). Then follows an explanation of various technical terms—lemma, case, porism, objection, and reduction. At the proper time the meaning of geometrical conversion is made clear and the various kinds of conversion distinguished (252-254, 409.1-6); and likewise the procedure of reduction to impossibility (254.21-256.9). Locus-theorems and their kinds come up for explanation later (394.11-396.9). And when the student is introduced to the method of application of areas, that "invention of the Pythagorean Muse," only its simplest use is expounded and illustrated, but the student is apprised that from this procedure is derived the later use of "parabola," "hyperbola," and "ellipse" to denote the curves produced by conic section (419.15-420.6), which are only mentioned but not elucidated here.

Yet it is also evident that much, perhaps the greater part, of the *Commentary* as we have it is addressed to a more advanced audience. Besides frequent references to constructions and theorems, surfaces and curves, that are beyond the scope of Euclid's first book but are presumed to be familiar to the reader, there are everywhere refinements of analysis and comment that can hardly be intended for beginning students. Such are the learned discussion of the meaning of "element" (71.25-73.14) and the appraisal of the superior merits of Euclid's *Elements* (73.14-75.26). Proclus emphasizes repeatedly the basic character of geometry as a hypothetical science, which starts from first principles (ἀρχαί) which it does not demonstrate—namely, axioms, definitions, and postulates. He expounds Aristotle's explanation of the difference between these three types of ἀρχαί (76.6-77.6) and later gives a résumé of the controversy over the distinction between postulates and axioms (178.1-184.10). Euclid's famous parallel postulate comes in for especially prolonged examination. Proclus believes with Ptolemy that it can be demonstrated, but he finds Ptolemy's proof fallacious (365-368) and attempts one of his own (371.10-373.3). The distinction between problem and theorem receives repeated attention (77-81, 201.3-15, 210.5-10, 220.6-222.19, 241.19-244.9), the discussion culminating in the doctrine that problems are concerned with establishing existence, theorems with demonstrating properties. Similarly we are shown the difference between strict demonstration, which reveals the cause of a connection of attributes, and demonstration by "signs," which merely establishes the fact of the connection

(206.14-207.3). Although all demonstrated conclusions appear to be universal, we are warned that some are truly universal because they reveal intrinsic connections between subject and predicate, while others are partial only, in that they do not comprehend the full range of the predicate character (245.24-246.5, 251.11-19, 253.20-254.3, 390.12-392.8). The role of the infinite and of infinite divisibility in mathematical discourse is taken up in two lengthy passages (277.25-279.11, 284.4-286.11). And we are constantly being introduced to differences of opinion among recognized authorities, to proofs alternative to those that Euclid employs, and to criticisms of his selection or ordering of propositions.

Even more indicative of Proclus' addressing a wider audience than elementary students are his excursions into the region that is of interest especially to the philosopher of mathematics. These portions of Proclus' doctrine will be discussed more fully in the following section. Here we will only remark that for Proclus mathematics is a part of a larger system of thought and for this reason carries the student beyond its ostensible boundaries into the paths of cosmological and metaphysical speculation. And Proclus is always eager to lead the way into these paths.

The selection of Euclid's *Elements* as the text for exposition is easily understood when we recall the prestige that this introductory treatise enjoyed throughout antiquity. But for Proclus, the Platonist, the choice of Euclid has another and deeper justification. He regards Euclid as having been a member of the ancient Platonic school, as in fact the mature product of that effort to organize and systematize geometrical science which, according to his account, characterized the mathematical research of the fourth century and which had its chief center and inspiration in the Academy of Plato's time (66.8-68.10). He even takes the ultimate purpose of the *Elements* to be the construction of the five "cosmic elements," the five regular solids that figure so prominently in Plato's *Timaeus* and with which in fact the *Elements* concludes (68.21-23, 70.19-71.5, 82.25-83.2). This interpretation of the *Elements* is most certainly mistaken. There are large parts of the treatise that have no direct connection with the cosmic figures and some that have no relation at all.[31]

---

[31] See Heath, *Euclid* I, 2.

Proclus' judgment here is warped by his enthusiasm for Plato and his particular admiration for the *Timaeus* among the Platonic dialogues. Whether Proclus was right in regarding Euclid as having been a member of the Academy we cannot say; but it is fairly certain that he got his first instruction from mathematicians who had been trained in the tradition of the Academy. The evidence Proclus gives of the role played by the Academy in the development and refinement of mathematical procedure in the fourth century is sufficient foundation for his association of Euclid with Plato. For Proclus, therefore, Euclid is not merely the recognized master of the discipline that he expounds, but also an authentic adherent of the Platonic theory of knowledge.

But Proclus was a Pythagorean as well as a Platonist, if indeed those two designations can be taken to mean different things at this period of ancient thought. For the Pythagoreans, numbers and figures were images, or symbols, of nonmathematical kinds of being— of the moral virtues, for example, or political relationships, or theological and supercosmic realities. One of the striking features of the *Commentary* is the way in which its discussion of Euclidean theorems and constructions is interrupted by expositions of Pythagorean lore. These Pythagorean passages do not affect the quality of what we would regard as the strictly mathematical material, and they are usually designated as digressions by Proclus himself, sometimes with apologies for the length to which these speculations have carried him (e.g. 91.11, 142.8, 151.13). Proclus was a competent mathematician, but he could not regard mathematics as a self-enclosed field, without implications for the cosmic philosophy that he espoused (see especially 174.17, 214.14). As Heath puts it, Proclus regarded mathematics as a handmaid to philosophy.

Many commentaries on Euclid had been composed prior to Proclus' time. He not only acknowledges their existence but declares that he is going to draw upon them freely, albeit critically. "We shall select the more elegant of the comments made on them by the ancient writers, though we shall cut short their endless loquacity, and present only what is most competent and relevant to scientific procedures, giving greater attention to the working out of fundamentals than to the variety of cases and lemmas which, we observe, usually attract the attention of the younger students of the subject"

(200.11-18). To judge from the very last words in the *Commentary* (432.15-19) and from occasional comments elsewhere (e.g. 84.10, 289.12, 328.16), he has a low opinion of some of the commentaries in circulation. There were, however, some distinguished commentators among his predecessors, such as Heron, Porphyry, and Pappus, all of whom Proclus cites frequently and with respect. He also draws heavily upon Aristotle, Archimedes, Apollonius, Posidonius, Geminus, and above all, for the early period, upon Eudemus, Aristotle's pupil who wrote a history of geometry that has since been lost. He refers to opinions of his two revered teachers, Plutarch (125.16) and Syrianus (123.19). He cites at length a controversial passage from Carpus of Antioch (241.19-243.11). He gives us verbatim Ptolemy's proof of the parallel postulate and points out the fallacy it contains (365.7-368.1; cf. 362.14-363.18). Eratosthenes, Speusippus, Zenodotus, Zenodorus, Menaechmus, Menelaus of Alexandria, Philolaus, Philon, Theodorus of Asine, Nicomedes, Oenopides of Chios, Perseus, and numerous others from all periods, some of whom are now only names to us, come in for mention as authors of mathematical discoveries or sponsors of opinions that Proclus either commends or disapproves.[32] The sceptical contentions of Zeno of Sidon are subjected to prolonged criticism and refutation (214.18-218.11). The Epicurean contempt for geometry, as laboring to prove what is obvious even to an ass (that is, that one side of a triangle is shorter than the other two), is philosophically rebuked (322.4-323.3), and the Aristotelian contention, later espoused by the Stoics, that points, lines, and mathematical objects generally are merely abstractions from sense objects refuted (12.2-18.4; cf. also 89.15-92.1). These numerous and sometimes very extended references to opinions and accomplishments of his predecessors, taken together with the material rescued from Eudemus' early history of geometry, make Proclus' *Commentary* a priceless source of information regarding the geometry of the previous nine or ten centuries.

It is evident that Proclus' students are expected to have the text of Euclid before them, or readily available. But only for the first proposition does Proclus follow Euclid's proof step by step. He does

---

[32] For page references to persons without supporting references in the text consult the Index.

so in this instance to provide an example of what the student should do himself for each of the following propositions. Proclus' later procedure is first to expound the exact meaning of the theorem or problem with which he is concerned, frequently calling attention to the precision of its language, and then to justify the place in the order of the exposition to which Euclid assigns it. Instead of a detailed examination of Euclid's proof there is usually a reminder, sometimes quite extended, of the axioms, postulates, and previous theorems by means of which the theorem in question is established (for a good example, see 240-241). He turns then to objections that have been brought against Euclid's construction or his proof, usually identifying the author of an objection. Again he normally meets the objection not from his own resources, but by citing previous mathematicians who have dealt with it. His modesty and his fairness in giving credit to others are notable—and are even noted by an unknown scribe in the margin of one of our manuscripts (at 352.14). When in the proof of a theorem Euclid uses only one of two or more possible cases, as is his custom, Proclus will often prove one or more of the omitted cases; sometimes he simply calls attention to them and recommends that his readers, "for the sake of practice," prove them for themselves. Sometimes he gives an alternative proof of a theorem devised by one of his predecessors for the obvious purpose of showing the superior elegance or appropriateness of Euclid's demonstration (e.g. 280.9, 282.20, 335.15).

Proclus does not often find occasion himself to criticize Euclid. For him "our geometer," or "the author of the *Elements*," is usually above reproach, when properly understood. He seems occasionally to sense a lapse from rigor in an enunciation or a proof and endeavors to smooth over the defect, sometimes by specious reasoning, as in his answer to the objection raised at 223.16ff., and in one instance, it seems, by the insertion of an extra and unneeded term in Euclid's text, namely ἑξῆς in the enunciation of XIV[33] (294.16). In 361.17-362.11 he voices a criticism (misguided, it seems) of Euclid's division of his theorems about parallel lines; and in 393.5-394.7 he questions whether Euclid's restriction of "parallelogram"

---

[33] Propositions in Euclid's first book will normally be referred to hereafter by Roman numerals only.

to four-sided figures is justified. Despite his twice-announced intention (84.12, 200.16), he deals with a great variety of "cases" in his comments, and some of them are evidently inserted as necessary supplements to Euclid's demonstrations. This is especially true of his comment on XLIII, where he says (417.2f.) we must consider the case in which the internal parallelograms do not meet at a point (thus making the complements nonparallelogrammic areas) in order "to see that the enunciated conclusion holds." But the most striking addition to Euclid is his alternative solution of the problem of XXIII—"On a given straight line and at a point on it to construct a rectilinear angle equal to a given rectilinear angle"—where he sees that the use of XXII for the construction in XXIII requires a certain modification that Euclid does not indicate; Proclus introduces the modification and thus constructs the triangle in what he says is a "more instructive fashion" (334.7).[34]

Coming to more substantive matters, we find him criticizing Euclid's definition of plane angle. To call it "the inclination to one another of two lines" is to put angle under the category of relation, an opinion which is open to numerous objections (122.21-123.13, 128.3-22). Again Euclid's fourth postulate—"that all right angles are equal to one another"—states an intrinsic property of right angles, hence is either an axiom, according to Geminus, or a theorem, according to Aristotle (182.21-183.13). Likewise the famous fifth postulate—"If a straight line falls upon two straight lines and makes the interior angles in the same direction less than two right angles, the straight lines, if produced indefinitely, will meet in the direction in which are the angles less than two right angles"—ought, he says, to be stricken altogether from the list of postulates (191.21); for it is a theorem capable of proof, which Ptolemy even claims to have demonstrated in one of his books. Moreover, it is a proposition whose converse he maintains Euclid himself proves later;[35] and surely a proposition whose converse is demonstrable can itself be demonstrated. Later he examines Ptolemy's demonstration, finds it defective, and attempts a different one of his own (371.10-373.2), which unfortunately, like Ptolemy's, succeeds only in begging the question.

---

[34] This irregularity in Euclid is also noted by Heath, *Euclid* I, 295.
[35] In XVII. See 364.21-25 and note.

This attempt to prove Euclid's fifth postulate is Proclus' most ambitious contribution to the elements of geometry. He apparently thought highly enough of it to write a book on the topic, if this is the correct interpretation of a reference in Philoponus.[36] But there are also other evidences of his mathematical independence. He points out (405.6-406.9), contending that this has previously escaped notice, that XXXV-XXXVIII are special cases of the first proposition in Book VI, which proves not only these theorems but also a much wider range of quantitative relationships between triangles than the equality that these four propositions establish.[37] With regard to the contested fourth postulate—that all right angles are equal—he argues for its validity on the ground that the right angle is the standard by which all other angles are measured and as such is always exactly a right angle (188.11-20). This is an idea that suggests a generally more satisfactory conception of postulate than those over which his contemporaries disputed. Again in commenting on Euclid's proof of IV—in which Euclid invokes the principle that two straight lines cannot enclose an area, although this is not listed among his axioms—Proclus asserts that Euclid was not required to prove this, since it is implied by his definition of straight line and by his first postulate (239.15ff.); and he ends by bringing forward a consideration similar to that which he had used to defend the fourth postulate, namely, that a straight line is the measure of all other lines, as the right angle is the measure of all other angles, and by this fact presumably is entitled to a place among the undemonstrated principles (240.3-10). Finally, he claims as his own "contribution to the construction given by the author of the *Elements*" (335.15) his method of solving the problem of XXIII which was described above. These examples show that Proclus was not an uncritical expositor, nor merely a repository for the comments and criticisms of his predecessors, but an independent thinker and mathematician in his own right.

There are hints in the text that Proclus thought of extending his commentary to the books after the first (272.14, 279.12, 398.18, 423.6, 427.10, 432.9). But there is no mention of such a continuation in the tradition, nor have any indisputably genuine fragments

[36] *CAG* xiii.2, 129.16, Wallies.
[37] Heath discounts Proclus' claim somewhat; see note at 406.9.

of commentaries on later books been uncovered in the manuscripts of Proclus.[38] Indeed at the very end of this commentary he expresses doubt whether he will be able to go through the remaining books in the same fashion (432.9) and contents himself with urging any who plan to write commentaries on later books to follow the example he has set in the work he has just completed. Still we need not assume that Proclus did not lecture on later books, for references to them abound in the commentary that we have; we can conclude only that he lacked the time to put his comments into a form suitable for a larger circle of philosophical readers.

Concerning the period of Proclus' life in which the present commentary was written we have one slight indication in the text. In the discussion of the definition of angle, after exploring the difficulty in classifying it as relation or as quantity or as quality, Proclus remarks, "But let us follow our 'head' ($\tau\hat{\omega}$ $\dot{\eta}\mu\epsilon\tau\dot{\epsilon}\rho\omega$ $\kappa\alpha\theta\eta\gamma\epsilon$-$\mu\acute{o}\nu\iota$) and say that the angle as such is none of the things mentioned but exists as a combination of all these categories" (123.19). The reference is clearly to Syrianus and points to a time when he was still head of the School. This does not help us much, since the date of Syrianus' death or retirement is unknown; yet it does confirm an a priori presumption that this work is a product of the earlier, rather than the later period of Proclus' career.

PROCLUS' PHILOSOPHY
OF MATHEMATICS

THE Prologue to this commentary is not only one of the most finished of Proclus' philosophical essays, but also one of the most valuable documents in ancient philosophy. We have already mentioned the priceless information it gives us regarding geometers and the history of geometry prior to Proclus' time. Yet the value of the matter it contains regarding the foundations of mathematics and of geometry in particular is even greater, though less widely recognized. As one would expect in an introduction to Euclid's *Elements*,

[38] C. Wachsmuth, *Rheinisches Museum* XVIII, 1863, 132-135, thought he had discovered some fragments of a comprehensive commentary, but see Heath, *Euclid* I, 32n.

it explains the meaning of "element" in geometry, states the theoretical and pedagogical purposes of an elementary treatise, and offers a striking evaluation of the excellence of Euclid's own work (69.4-75.4). Moreover, it contains a defense of pure mathematics and of geometry in particular against its critics, including a careful and illuminating interpretation of the attitude of Plato, who was sometimes invoked by the critics of mathematics in support of their contentions (25.15-32.20). More fundamental for the understanding both of Plato and of the science that Euclid expounds are the questions Proclus raises concerning the nature of the objects of mathematical inquiry and the character and validity of the procedures used by mathematicians in handling these objects. These are basic problems in what we would now call the philosophy of mathematics. Proclus' essay is the only systematic treatise that has come down to us from antiquity dealing with these questions; and for those not already acquainted with Proclus' commentary, it will perhaps be useful to preface the text itself with a brief statement of Proclus' answers to them.

The objects of mathematical inquiry—numbers, points, lines, planes, and all their derivatives—are neither empirical things nor pure forms. "Mathematical being," as Proclus puts it in his opening sentence, occupies an intermediate position between the simple immaterial realities of the highest realm and the extended and confusedly complex objects of the sense world. The superiority of mathematical objects to sensible things is evidenced by the exactness and stability of their natures and by the irrefutable character of the propositions establishing their attributes and relations to one another. In contrast with sense objects, "mathematicals" are devoid of matter and not subject to the changes that affect physical things; thus they are intelligible entities, capable of providing a foundation for the demonstrations that mathematics undertakes. On the other hand, they do possess a kind of extendedness. The series of numbers consists of discrete members, and geometrical figures are divisible into parts. And no mathematical object is unambiguously unique, since mathematical reasoning often involves a comparison of two or more lines, or of two or more circles or other figures. Their plurality and their extendedness thus show that they do have a kind of matter, mathematical matter, underlying them; and therefore they fall short of the immaterial and unextended beings that are the objects of pure intelligence.

Whence come these mathematical objects? It is obvious that they are not presented in sense-perception: our senses never show us points without parts, lines without breadth or depth, right-angled triangles all possessing a fixed ratio between their sides, nor any other of the perfectly precise objects that figure in mathematical discourse. Rather they issue from the mind itself, being "projected" by it from the inner store of forms that constitute its essence. The mind is "reminded" of these ideas by sense-perception, and its thinking is an "unfolding" of their content under the guidance of a higher and fully unified intelligence (12.1-13.26; cf. 49.4-56.22). Following Plato, Proclus calls the level of intelligence that is operative in mathematics the διάνοια, generally translated here as "understanding." This is the level of discursive thought, proceeding step by step from one factor to another and to their integration into a whole. Like the objects that it examines, this level of thought occupies a middle position between sense-perception and the highest intelligence, which Proclus, again following Plato, calls νόησις, or Nous. Sense-perception has but a dim apprehension of being; its knowledge is fragmentary, imprecise, and unstable, and amounts only to opinion (δόξα) about its objects. The understanding introduces clarity and precision, but its cautious discursive procedure, its treatment of its objects as extended, and its dependence upon higher knowledge for its principles reveals its inferiority to the immediate and total and sure comprehension of Nous (3.14-4.14). Since the understanding occupies this middle station, its function has a twofold reference. On the one hand, it develops the content of the pure forms that it receives from Nous, imitating their unity and simplicity as best it can by ordering and integrating the complex and diverse factors that it distinguishes; on the other hand, it furnishes the paradigms to which the changes and variety in the sense world conform. It is thus a genuine intermediary between the lowest and the highest levels of being, imitating the highest and furnishing intelligible patterns for understanding the world of physical processes below it (16.4-16).

In this activity of unfolding the content of pure ideas and tracing the details of their implications and mutual relations, the understanding relies upon a special capacity, the imagination or image-making faculty, for exhibiting the variety and complexity present in the mathematical forms that it explores (51.9-56.22). The ideas with which the understanding is equipped are, like the partless ideas in Nous, unextended, indivisible, uncompounded, containing the variety of their content in an undivided unity. But the imagination expresses these ideas as formed, extended, and divisible, providing them with an intelligible matter—space—in which they can be deployed. It thus presents not the pure idea itself, but a picture of it, or rather a series of pictures, all possessing the common character that constitutes the essence of the idea. The circle in the understanding, before it is projected on the screen of the imagination, is one only, without extension, having neither center nor circumference; but as depicted in imagination it is extended and may appear in any one of a variety of sizes and positions. Unless it were so presented, it would be impossible for the understanding to explore by reasoning its constituent elements and their relations to one another. It is these pictures in the imagination with which mathematical reasoning deals, using them as aids for discerning the universal character present in each and all of the imaginary circles and for demonstrating its properties and relations to other universals found in other similarly presented objects.

As the Platonist will readily see, this is an adaptation and elaboration of the doctrine illustrated by the divided line in the *Republic*. The part played by the imagination is Proclus' main addition to the Platonic theory, an addition which anticipates, it need hardly be pointed out, Kant's doctrine of the schematism of the understanding. Again the relation between the understanding and the higher intelligence, as Proclus conceives it, goes far beyond the tentative formulations in the *Republic*. Nevertheless in characteristic fashion Proclus considers this derivative theory as remaining faithful to the source from which it comes; and the use of the metaphors of imitation and paradigm and of the concept of recollection (ἀνάμνησις), as well as the explicit references to the "leadership" of Plato, show that he regards it as genuinely Platonic. For him, as for Plato, the understanding and Nous alike are engaged in apprehending the intelligible world. And it is because mathematics opens the way to that world and thus emancipates the mind from the bondage to the senses and equips it to rise to superior levels of understanding that Proclus proclaims its importance for education and for all the higher life.

Besides having a world of objects that are the creations or projections of the mind's essential nature, mathematics follows a method of its own devising in examining these objects and arriving at certain conclusions about their properties and relations. This is the procedure of positing plausible starting-points (ἀρχαί) and establishing conclusions by deduction from these initially adopted first principles (75.6-26 and *passim*). The discovery and elaboration of this method of scientific inquiry is the great and distinctive contribution of the Greek mathematicians.[39] It was worked out, as we can see from the historical evidence that Proclus gives us, in part at least under the influence of Plato and certainly in close collaboration with the fourth-century Academy (66.8-68.4); hence it constitutes another, though less obvious because so all-pervading, Platonic note in Proclus' commentary.[40] These starting-points upon which all later reasoning depends appear in geometry as the familiar axioms, postulates, and definitions of the Euclidean method. The axioms underlying mathematical reasoning are posited because they are seen to be true in and of themselves—or at least "more evident than their consequences," Proclus adds in one passage (75.17)—having no need of confirmation by demonstration or construction, such as the axiom that things equal to the same thing

[39] Though the Greeks borrowed much in mathematics from the Babylonians and the Egyptians, this method of mathematical reasoning and demonstration is their own; nothing like it is found in the cultures from which they borrowed. See Kurt von Fritz in *Archiv für Begriffsgeschichte* I, 1955, 13f.

[40] Plato's contribution to the development of this mathematical method must have been considerable. It was formerly maintained that he was responsible for its chief features—the axiomatic structure of proof, the use of the method of analysis, and the restriction of mathematical existence to what can be constructed. More recent discoveries regarding the methods used by mathematicians before Plato compel us to reduce somewhat this estimate of his contributions. See Árpád Szabó, "Anfänge des Euklidischen Axiomensystems," in O. Becker (ed.), *Zur Geschichte der Griechischen Mathematik*, Darmstadt, 1965, 450ff.

are equal to each other. Postulates are demands that certain procedures be permitted, such as to draw a line from any point to any other, or to describe a circle about any point with any radius. These procedures are not as self-evident as the axioms, but it requires little effort of thought to see that they are obviously valid. Thus if one pictures a point "flowing" without deviation towards another point, the drawing of a straight line in imagination inevitably results; and if one thinks of a line revolving about a stationary end, the other end-point will easily be seen to describe a figure with a circumference everywhere equidistant from the stationary point. In addition to axioms and postulates, the mathematician needs clear definitions of the objects which he is considering. The circle just mentioned, for example, the mathematician defines precisely as a figure bounded by a line every point on which lies at an equal distance from a point within it, its center. These starting-points—definitions, axioms, and postulates—are collectively the hypotheses of the reasoning that follows; and the theorems that result are established by strict deduction, either immediately from the first principles or from them in conjunction with other theorems previously established.

To the Greek mathematicians must be credited also a host of supplementary devices and refinements of procedure for expounding and elaborating their increasingly complex material. There are important distinctions between the methods of analysis and synthesis, between types of definitions, between kinds of proof, between problems and theorems, between a theorem and its converse—entire or partial—and a variety of technical terms whose use must be mastered in the handling of these procedures. With each of these Proclus is familiar, and they all appear at appropriate points in his commentary. But the method mentioned in the preceding paragraph is basic to them all; it represents the distinctive character of mathematical inquiry, the positing of principles and the deduction of their consequences. This distinction between hypotheses and conclusions at each stage of the reasoning, and the orderly progression from propositions established to further conclusions based on them as premises, produce unity, coherence, and solidity in the system of mathematical propositions, which despite their variety and complexity are firmly rooted in and unified by the principles from which they are derived.

This cosmos[41] of mathematical propositions exhibits a double process: one is a movement of "progression" (πρόοδος), or going forth from a source; the other is a process of "reversion" (ἄνοδος) back to the origin of this going forth (18.17-28; 19.5-20). Thus Proclus remarks that some mathematical procedures, such as division, demonstration, and synthesis, are concerned with explicating or "unfolding" the simple into its inherent complexities, whereas others, like analysis and definition, aim at coordinating and unifying these diverse factors into a new integration, by which they rejoin their original starting-point, carrying with them added content gained from their excursions into plurality and multiplicity (57.18-26). For Proclus the cosmos of mathematics is thus a replica of the complex structure of the whole of being, which is a progression from a unitary, pure source into a manifold of differentiated parts and levels, and at the same time a constant reversion of the multiple derivatives back to their starting-points. Like the cosmos of being, the cosmos of mathematics is both a fundamental One and an indefinite Many, and it has the one character only because it has the other.

The body of mathematical knowledge is, then, the product of the activity of the mind, working on objects that it has drawn from within itself and handling them in accordance with principles of its own devising. But this picture of mathematics, unless we introduce certain qualifications, could give a distorted idea of Proclus' conception, suggesting that the mathematical world is an arbitrary creation, the product of imagination in the most fanciful sense of the term.

First we must recall that the understanding, and the imagination which aids it, occupy a middle station between sense-perception and pure reason. As an intermediary, the understanding has contacts and resemblances with the levels of experience on both sides of itself. From the one side it is stimulated by sense objects to bring forth ideas appropriate to understanding them, and it continues to be guided by sensory experience in producing further refinements of

[41] Proclus often uses the more pretentious term διάκοσμος to denote such an ordered system (e.g. 13.14, 17.1, 18.24, 61.27, and *passim*).

its initial recollections. It is methodologically significant, as well as historically true, that the first conceptions of numbers by the Pythagoreans pictured them as identical with empirically given points or constellations of points;[42] and likewise that the first conceptions of lines and planes arose from observation of visible objects, such as the boundaries between plots of arable land in the Nile valley. Mathematical reasoning in antiquity never completely lost touch with the empirical material that had stimulated its beginning, although the connection became more and more tenuous and indirect with the progress of generalization and abstraction. Nor was the complementary process of applying mathematical results to sensible experience by any means absent in antiquity. Proclus' frequent references to the many achievements of the mathematical sciences in serving the everyday needs of men (e.g. 19.21ff., 41.3-42.6)—in surveying the surface of the earth, in applying force to move objects, in building fortifications and instruments of war—show that he and the Greeks were well aware of the utility of mathematics, and these applications to empirical problems demonstrably influenced the direction of mathematical imagination. From the other side mathematical understanding is guided by the superior level of intelligence from which its ideas are derived, so that it is constantly required by the logic of the larger system of which it is a part to rise to ever higher levels of comprehension, that is, to more and more refined analyses and more and more general principles.

Furthermore, these ideas themselves are not passive contents but active agencies, as Proclus conceives them, developing of their own volition, so to speak, into the complex structures of the mathematical world, "unfolding"—to use once more Proclus' expressive term—as they are examined, and revealing ever new aspects of their character and relationships. The understanding does not manipulate them; rather they carry the understanding with them as they unfold (17.1-18.1). An example from antiquity would be the way in which the ideas of number, equality, ratio, and proportion became increasingly complex as they were employed successively by Pythagoras, Eudoxus, and Euclid and his successors. At

[42] On early Pythagorean conceptions of number see John Burnet, *Early Greek Philosophy*, 3rd edn., London, 1920, 99ff.

the same time these ideas maintain their unity and integrity, and their unfolding is a process whereby the richness of their content is explicitly displayed within their essential unity. Being thus under the guidance of Nous, the understanding in its dialectical activity is not arbitrarily creative but is dependent at every stage on the higher intelligence that presides over it and on the living ideas that it has derived from that higher intelligence (42.13-43.21).

Following Aristotle's lead, Proclus at times conceives of the mathematical sciences, arithmetic and geometry in particular, as distinct, each positing a certain genus of subject-matter that it proposes to examine and specific principles that it adopts for handling this subject-matter. Arithmetic has the priority over geometry, since it starts from the positionless unit, which is simpler than the point, a unit having position; and below them is ranged a series of other sciences dependent on them—astronomy, optics, and mechanics, to mention only the most important—each taking a more complex species of object and specific rules as principles for studying it. Each of them is in this sense bound by the presuppositions from which it starts. It is the mark of a tyro in geometry, for example, to employ principles and modes of reasoning that do not conform to the conventions of his science (58.3-61.24). "No science demonstrates its own first principles," says Proclus (75.14).

But Proclus is not content to leave the matter thus, as Aristotle appears to have been. These starting-points are not free postulates; just as astronomy and the other sciences start from principles established in arithmetic and geometry, so the starting-points of these two sciences come from a still higher science, from what Proclus calls "general mathematics" (ὅλη μαθηματική; 7.17-10.14, 18.10-20.7). The reality of this more general science is shown by certain concepts and principles that arithmetic and geometry have in common, such as the axioms regarding equals and the rules of proportion. The objects taken to be equal will differ for the arithmetician and the geometer; the one will demonstrate the theorem of alternate proportion in terms of numbers, the other in terms of continuous magnitudes, and each will regard as inadmissible in his science the mode of proof used by the other. But these particular uses indicate that above them stands the principle of alternation applicable alike to magnitudes and numbers; and it is inconceivable that we should have a knowledge of the particular applications yet no science at all of the universal principle itself. "Knowledge of those objects is by far the prior science, and from it the several sciences get their common propositions" (9.14-16).

The mathematical sciences therefore are not independent and self-contained within their starting-points. Just as arithmetic and geometry are dependent on general mathematics, so the principles of general mathematics in turn are dependent on a higher science than that which explicitly posits them. Of more universal import than the starting-points of mathematics are the principles of the Limit (πέρας) and the Unlimited (ἄπειρον; 5.15-7.12). The Limit expresses the character of boundedness or determinateness that belongs to every intelligible object; the Unlimited represents the fecundity of being, ever going beyond its own limits and developing into a world of derivatives which are other than itself but are everywhere constrained by the unitary source from which they have proceeded. The Limit and the Unlimited pervade all being, generating everything out of themselves (5.15-22). Closely connected with them are other pairs of principles, Likeness and Unlikeness, Equality and Inequality, Harmony and Disharmony. These principles themselves are derivatives of a still more eminent principle, "the indescribable and utterly incomprehensible causation of the One" (5.19). We are here in the region of the very highest science, the "science of being as being" (9.19), as Proclus calls it in Aristotelian terms. This science is truly universal and self-contained; it is the "unhypothetical science" (31.20), for hypotheses and consequences are one in the all-comprehending insight of Nous. But we cannot follow Proclus further into these metaphysical and epistemological reaches of his thought. Enough has been said to show that the mathematical sciences, with their specific presuppositions, are a part of and continuous with the larger enterprise of dialectic, not only intent on unfolding their ideas into a system of implications, but also insistent on giving an account of their presuppositions themselves, in the fashion of the dialectic portrayed in the Platonic dialogues.

Hence Proclus speaks of dialectic as the "unifying principle" of the mathematical sciences or—again borrowing a term from Plato—as their "capstone" (42.9-44.24). It is not only the central activity at work in mathematics itself, creating all the increasingly complex objects that mathematics examines and the variety of auxiliary methods that it employs, but also the crowning science, the realization of that immediate and complete understanding toward which all discursive thought moves. Proclus would hardly be surprised to hear of the astonishing expansion of mathematics that has taken place in modern times, an expansion that has carried it far beyond its ancient bounds. He was himself well aware of the remarkable development that had occurred during the golden age of Greek geometry, from the simple concepts and procedures of Pythagoras to the complexities of Archimedes and Apollonius, and had exerted an influence, through ever widening areas of mathematical analysis and increasing refinements of procedure, on Geminus, Ptolemy, Heron, Pappus, Diophantus, and many others down to Proclus' own time. He could only anticipate that this expansion would continue, assuming a continuance of the dialectical urge that had operated in the past. For the objects that mathematical understanding sets before itself are not given once for all, but are the product of the mind's own "inexhaustible resources" (18.3, 37.11), working under the stimulus of sense-perception and the guidance of the higher Nous.

Obviously for Proclus the mind is not a blank tablet, owing its content and capacities entirely to sense-experience; but neither is it, as some later idealists would have it, a faculty with a fixed store of a priori forms and concepts, and limited by this native equipment. "The soul therefore was never a writing-tablet bare of inscriptions; she is a tablet that has always been inscribed and is always writing itself and being written on by Nous" (16.8-10). This conception of the understanding as infinitely resourceful, capable through dialectic of transcending any given boundaries and exploring freely whatever new implications are opened up for it, is an answer to the criticism that Proclus' mathematical philosophy is archaic or provincial because of its reliance upon Plato, Aristotle, and Euclid. And in taking the Platonic dialectic as the heart of mathematical reasoning, Proclus has provided in advance the clue to understanding the rich development of mathematics that has occurred since Euclid's day, carrying it far beyond the explicit content of ancient mathematics to new worlds undreamed of by the Greeks. This dialectical capacity of the soul, which enables it to reach upwards as well as downwards, is the epistemological counterpart to the ontological station that Proclus assigns to it midway in the scale of being, participating in the highest and most absolute, yet descending into the lower grades of physical occurrence to guide and control. Because of its central status in the cosmos and the intellectual resources with which it is endowed, the soul is adequate to the explanation and understanding of all levels of being.

Since mathematics is the primary manifestation of these powers of the soul, the study of this science is the gateway to knowledge and blessedness. It is on this note that Proclus concludes the first part of his Prologue. Mathematics, he says (46.20-47.6), "arouses our innate knowledge, awakens our intellect, purges our understanding, brings to light the concepts that belong essentially to us, takes away the forgetfulness and ignorance that we have from birth, sets us free from the bonds of unreason; and all this by the favor of the god[43] who is truly the patron of this science, who brings our intellectual endowments to light, fills everything with divine reason, moves our souls towards Nous, awakens us as it were from our heavy slumber, through our searching turns us back on ourselves, through our birthpangs perfects us, and through the discovery of pure Nous leads us to the blessed life."

[43] For a suggestion regarding the identity of this god see note at 46.25.

Translator's Note

THE FIRST printed Greek text of Proclus' commentary was edited by Simon Grynaeus the elder and published at Basel in 1533 as an appendix to his edition of Euclid. Grynaeus used only a single manuscript, and that an inferior and mutilated one. The Latin translation of Francis Barocius (Barozzi), published at Padua in 1560, was based on a much better text established by the translator himself from five other manuscripts in addition to the one used by Grynaeus. Unfortunately Barocius did not publish his Greek text; nevertheless, the faithfulness with which he evidently followed his original makes it possible in most cases to infer the text from which he worked. The present translation is based on the text of Gottfried Friedlein (Leipzig, 1873), which replaced the inadequate text of Grynaeus. Friedlein's text, though not free of faults, is reasonably satisfactory, and is furnished with an excellent *Index verborum* that greatly facilitates the task of a translator.

The only previous English translation is that of Thomas Taylor (*The Philosophical and Mathematical Commentaries of Proclus on the First Book of Euclid's Elements*, 2 vols., London, 1788, 1789), which was based on Barocius and Grynaeus. The contributions of this indefatigable Platonist are so numerous and important that it is almost an act of impiety to presume to replace this early product of his industry with a new translation. But Taylor's work is long out of print and is difficult to obtain. What is more, the imperfect text from which he worked (he even failed to make full use of Barocius, though professing to regard him as an indispensable supplement to Grynaeus) and his numerous paraphrases and unaccountable omissions make his translation not truly representative of the original Proclus. It is good, however, to know that Taylor's merits and achievements have been recognized recently by the publication of a volume of selections from his other writings, with biographical and critical comments.[1]

Two other recent translations (the only ones of which I know) have been of great help to me. One is a translation into German by

---

[1] *Thomas Taylor the Platonist: Selected Writings*, edited, with Introductions, by Kathleen Raine and George Mills Harper, Princeton, 1969.

Leander Schönberger, published at Halle in 1945, to which I am much indebted for aid in interpreting difficult passages. A still more recent translation, into French, by Paul ver Eecke (Bruges, 1948) has been helpful in its notes and references to mathematical matters. But neither of these more modern translations, nor Taylor's earlier one, makes it easy to find one's way from the Greek text to the translation, or vice versa, a shortcoming which much reduces their value to the student. Since Friedlein's text is now readily available in a recent reprint, I have included his pagination in the margins of my translation; and for those who may wish to compare a particular rendering with the Greek text, I have included line numbers as well as page numbers in my footnotes. Our manuscripts also contain indications of the ancient division of the *Commentary* into four books. These also I have put in the margins, together with Barocius' helpful division of the Prologue into chapters and the titles he assigned to them. Finally, since Proclus' commentary presupposes that its readers will have ready access to the text of Euclid on which it comments, I have presumed to provide this help to the modern reader by giving in footnotes the Euclidean proofs of the various propositions as they appear in the commentary.

Wherever I have found it necessary or desirable to depart from the text of Friedlein, I have indicated in a footnote the reading that I have adopted. The occasional words and phrases enclosed in brackets are additions to Friedlein's text, drawn either from Barocius or from my own presumptions of what Proclus intended. I should like to think that my corrections and emendations, together with those in Schönberger's and ver Eecke's translations, will be of use to some future scholar who may undertake the task of bringing out an improved text.

PROCLUS
A COMMENTARY ON THE
FIRST BOOK OF
EUCLID'S ELEMENTS

PROLOGUE: PART ONE

Book I
Chap. I
The Inter-
mediate
Status of
Mathemat-
ical Being

MATHEMATICAL being necessarily belongs neither among the first nor among the last and least simple of the kinds of being, but occupies the middle ground between the partless realities[1]—simple, incomposite, and indivisible—and divisible things characterized by every variety of composition and differentiation. The unchangeable, stable, and incontrovertible character of the propositions about it shows that it is superior to the kinds of things that move about in matter. But the discursiveness of [mathematical] procedure, its dealing with its subjects as extended, and its setting up of different prior principles for different objects—these give to mathematical being a rank below that indivisible nature that is completely grounded in itself.

It is for this reason, I think, that Plato assigned different types of knowing to the highest, the intermediate, and the lowest grades of reality.[2] To indivisible realities he assigned intellect, which discerns what is intelligible with simplicity and immediacy, and by its freedom from matter, its purity, and its uniform mode of coming in contact with being is superior to all other forms of knowledge. To divisible things in the lowest level of nature, that is, to all objects of sense-perception, he assigned opinion, which lays hold of truth obscurely, whereas to intermediates, such as the forms studied by mathematics, which fall short of indivisible but are superior to divisible nature, he assigned understanding. Though second in rank to intellect and the highest knowledge, understanding is more perfect, more exact, and purer than opinion. For it traverses and unfolds the measureless content of Nous by making articulate its concentrated intellectual insight, and then gathers together again the things it has distinguished and refers them back to Nous.

[1] 3.5 ὑποστάσεις.
[2] 3.16 *Rep.* 511b-e and 533e-534c. On the renderings of the Platonic terms in this passage see note at 10.27 below.

As the forms of knowing differ from one another, so also are their objects different in nature. The objects of intellect surpass all others in the simplicity of their modes of existence, while the objects of sense-perception fall short of the primary realities in every respect. Mathematical objects, and in general all the objects of the understanding, have an intermediate position. They go beyond the objects of intellect in being divisible, but they surpass sensible things in being devoid of matter. They are inferior to the former in simplicity yet superior to the latter in precision, reflecting intelligible reality more clearly than do perceptible things. Nevertheless they are only images, imitating in their divided fashion the indivisible and in their multiform fashion the uniform patterns of being. In short, they stand in the vestibule of the primary forms, announcing their unitary and undivided and generative reality, but have not risen above the particularity and compositeness of ideas and the reality that belongs to likenesses; nor have they yet escaped from the soul's varied and discursive ways of thinking and attained conformity with the absolute and simple modes of knowing which are free from all traces of matter. Let this be our understanding, for the present, of the intermediate status of mathematical genera and species, as lying between absolutely indivisible realities and the divisible things that come to be in the world of matter.

5

Chap. II
The Common Principles of Mathematical Being. The Limit and the Unlimited.

To find the principles of mathematical being as a whole, we must ascend to those all-pervading principles that generate everything from themselves: namely, the Limit and the Unlimited.[3] For these, the two highest principles after the indescribable and utterly incomprehensible causation of the One, give rise to everything else, including mathematical beings. From these principles proceed all other things collectively and transcendentally, but as they come forth, they appear in appropriate divisions and take their place in an

---

[3] 5.18 On the Limit (πέρας) and the Unlimited (ἄπειρον, ἀπειρία) as cosmogonic principles see Proclus' *Elements of Theology*, Prop. 89, and Dodds' note (246-248). These speculations have their source in Plato's *Philebus* 16cff. and 23cff., which in turn professes its dependence on the inspired wisdom of the "ancients."

ordered procession,[4] some coming into being first, others in the middle, and others at tht end. The objects of Nous, by virtue of their inherent simplicity, are the first partakers of the Limit and the Unlimited. Their unity, their identity, and their stable and abiding existence they derive from the Limit; but for their variety, their generative fertility, and their divine otherness and progression they draw upon the Unlimited. Mathematicals are the offspring of the Limit and the Unlimited, but not of the primary principles alone, nor of the hidden intelligible causes, but also of secondary principles that proceed from them and, in cooperation with one another, suffice to generate the intermediate orders of things and the variety that they display. This is why in these orders of being there are ratios proceeding to infinity, but controlled by the principle of the Limit. For number, beginning with unity, is capable of indefinite increase, yet any number you choose is finite; magnitudes likewise are divisible without end, yet the magnitudes distinguished from one another are all bounded, and the actual parts of a whole are limited. If there were no infinity, all magnitudes would be commensurable and there would be nothing inexpressible or irrational,[5] features that are thought to distinguish geometry from arithmetic; nor could numbers exhibit the generative power of the monad, nor would they have in them all the ratios—such as multiple and superparticular—that are in things.[6] For every number that we examine has a different ratio to unity and to the number just before it. And if the Limit were absent, there would be no com-

[4] 5.24 We shall frequently encounter in the *Commentary* this doctrine of the procession (πρόοδος) of beings from their primary source. For the intricate details of the processional order, as Proclus conceives it, see his *Elements of Theology, passim.*

[5] 6.21 ἄρρητον, ἄλογον. This is a reflection of Euclid's distinction between two orders of irrationals (Bk. X, Deff. III and IV). Ἄρρητον denotes a line incommensurable in length with a given (rational) line, ἄλογον a line which is commensurable neither in length nor in square with the given line.

[6] 6.25 On the elaborate classification in Greek arithmetic of numerical ratios greater than unity see Nicomachus, *Introduction to Arithmetic*, Chaps. 17-23; and for a convenient English listing Heath I, 101ff.

mensurability or identity of ratios in mathematics, no similarity[7] and equality of figures, nor anything else that belongs in the column of the better.[8] There would not even be any sciences dealing with such matters, nor any fixed and precise concepts. Thus mathematics needs both these principles as do the other realms of being. As for the lowest realities, those that appear in matter and are moulded by nature, it is quite obvious at once that they partake of both principles, of the Unlimited as the ground that underlies their forms and of the Limit by virtue of their ratios, figures, and shapes. It is clear, then, that the principles primary in mathematics are those that preside over all things.

Just as we have noted these common principles and seen that they pervade all classes of mathematical objects, so let us enumerate the simple theorems that are common to them all, that is, the theorems generated by the single science that embraces alike all forms of mathematical knowledge; and let us see how they fit into all these sciences and can be observed alike in numbers, magnitudes, and motions. Such are the theorems governing proportion, namely, the rules of compounding, dividing, converting, and alternating; likewise the theorems concerning ratios of all kinds, multiple, superparticular, superpartient, and their counterparts;[9] and the theorems about equality and inequality in their most general and universal aspects, not equality or inequality of figures, numbers, or motions, but each of the two by itself as having a nature common to all its forms and capable of more simple apprehension. And certainly beauty and order are common

Chap. III
The
Common
Theorems
Governing
Mathematical Kinds

8

[7] 7.2 ταυτότης, translated here as similarity, i.e. identity of form, to differentiate it from ἰσότης, equality.

[8] 7.3 On the Pythagorean doctrine of the two columns (συστοιχίαι) see Arist. *Met.* 986a22-26. The ten members in each column, as listed by Aristotle, are

Limit	–	Unlimited	Rest	–	Motion
Odd	–	Even	Straight	–	Curved
One	–	Many	Light	–	Dark
Right	–	Left	Good	–	Evil
Male	–	Female	Square	–	Oblong

The column on the right (i.e. the reader's left) is the "better."

[9] 7.26 See note at 6.25.

to all branches of mathematics, as are the method of proceeding from things better known to things we seek to know[10] and the reverse path from the latter to the former, the methods called analysis and synthesis.[11] Likeness and unlikeness of ratios are not absent from any branch of mathematics, for we call some figures similar and others dissimilar, and in the same way some numbers like and others unlike. And matters pertaining to powers[12] obviously belong to general mathematics, whether they be roots or squares. All these Socrates in the *Republic* puts in the mouth of his loftily-speaking Muses, bringing together in determinate limits the elements common to all mathematical ratios and setting them up in specific numbers by which the periods of fruitful birth and of its opposite, unfruitfulness, can be discerned.

Chap. IV
In What
Way These
Common
Theorems
Subsist

9

Consequently we must not regard these common theorems as subsisting primarily in these many separate forms of being, nor as later born and deriving their origin from them, but as prior to their instances and superior in simplicity and exactness. For this reason, knowledge of them takes precedence over the particular sciences and furnishes to them their principles; that is, these several sciences are based upon this prior science and refer back to it. Let the geometer state that if four magnitudes are proportional they will also be proportional alternately[13] and prove it by his own principles, which the arithmetician would not use; and again let the arithmetician lay it down that if four numbers are proportional

[10] 8.6 I.e. from premises to conclusions, the characteristic feature of demonstration as formulated by Aristotle, *Post. Anal.* 71b9-72b4, whose language this passage echoes.

[11] 8.8 Proclus is fond of chiasmus and lists these two terms in the reverse order of their preceding description. Analysis proceeds from conclusions to the premises that will establish them, synthesis from premises to the conclusions that follow from them. See 43.18-21 and 69.16-19.

[12] 8.12 δυνάμεις. In the context, this term obviously means mathematical "powers," and since the following words δυναμένων and δυναστευομένων are echoes of *Rep.* 546b, to which Proclus refers in the next sentence, I have translated them in accordance with the meaning which Proclus himself gives them in his *Commentary on the Republic* II, 36.9-12, Kroll. On the passage in the *Republic* see note at 23.21.

[13] 9.4 The theorem referred to is Euclid V. 16.

they will also be proportional alternately and establish this from the starting-points of his own science. Then whose function is it to know the principle of alternation alike in magnitudes and in numbers and the principles governing the division of compound magnitudes or numbers and the compounding of separate ones? It cannot be that we have sciences of particular areas of being and knowledge of them but have no single science of the immaterial objects that stand much closer to intellectual inspection. Knowledge of those objects is by far the prior science, and from it the several sciences get their common propositions, our knowledge ascending from the more partial to the more general until at last we reach the science of being as being.[14] This science does not consider it its province to study the properties that belong intrinsically to numbers, nor those that are common to all quantities; rather it contemplates that single form of being or existence that belongs to all things, and for this reason it is the most inclusive of the sciences, all of which derive their principles from it.

Always it is the higher sciences that provide the first hypotheses for the demonstrations of the sciences below them, and the most perfect of the sciences out of its own store lends to all others their principles, more general principles to some and to others less general ones. This is why Socrates in the *Theaetetus*,[15] mingling play with seriousness, likens the forms of knowledge in us to doves. Some of them, he says, fly in groups and others separately from the rest. The more comprehensive and general sciences contain many special sciences in themselves, whereas those that handle generically distinct objects remain apart and have little to do with one another, because they start from different first principles. One science, however, must stand above the many sciences and branches of knowledge, that science which knows the common principles that pervade all kinds of being and furnishes

10

[14] 9.19 Aristotle's definition of the highest science; see *Met.* 1026a31, 1064a3, 28, and b7.
[15] 10.2 *Theaet.* 197d.

to all the mathematical sciences their starting-points. Here let us end our remarks about this highest science.

Chap. V
The Organ
of Judgment
in Mathe-
matics

Next we should see what faculty it is that pronounces judgment in mathematics.[16] On this doctrine let us again follow the guidance of Plato. In the *Republic* he sets on one side the objects of knowledge and over against them the forms of knowing, and pairs the forms of knowing with the types of knowable things. Some things he posits as intelligibles (νοητά), others as perceptibles (αἰσθητά); and then he makes a further distinction among intelligibles between intelligibles and understandables (διανοητά) and among perceptibles between perceptibles and likenesses (εἰκαστά). To the intelligibles, the highest of the four classes, he assigns intellection (νόησις) as its mode of knowing, to understandables understanding (διάνοια), to perceptibles belief (πίστις), and to likenesses conjecture (εἰκασία).[17] He shows that conjecture has the relation to per-

11

ception that understanding has to intellection; for conjecture apprehends the images of sense objects in water or other reflecting surfaces, which, as they are really only images of images, occupy almost the lowest rank in the scale of kinds, while understanding studies the likenesses of intelligibles that have descended from their primary simple and indivisible forms into plurality and division. For this reason the knowledge that understanding has is dependent on other and prior hypotheses, whereas intellection attains to the unhypothetical principle itself.

Since, therefore, mathematical objects have the status neither of what is partless and exempt from all division and diversity nor of what is apprehended by perception and is highly changeable and in every way divisible, it is obvious that they are essentially understandables and that understand-

[16] 10.17 κριτήριον, "court," "tribunal"; frequently used by Hellenistic writers to denote the mental faculty that has competence to judge a matter under consideration.

[17] 10.27 See note at 3.16. I have tried to preserve uniformity in all translations of these terms; i.e. νοῦς and νόησις are rendered as "nous," "intellect," or "intellection"; νοητά as "intelligibles"; διάνοια as "understanding" (for both the faculty and the activity); and διανοητά as "understandables."

ing is the faculty that is set over them, as perception is over sense objects and conjecture over likenesses. Hence Socrates describes the knowledge of understandables as being more obscure than the highest science but clearer than the judgments of opinion.[18] For the mathematical sciences are more explicative and discursive than intellectual insight but are superior to opinion in the stability and irrefutability of their ideas. And their proceeding from hypotheses makes them inferior to the highest knowledge, while their occupation with immaterial objects makes their knowledge more perfect than sense-perception. Such, then, is the criterion of judgment in all mathematics, as we have delineated it according to Plato's thought: the understanding, a faculty higher in rank than opinion but inferior to intellect.

12

Chap. VI
The Being of
Mathemati-
cal Genera
and Species;
the Mode of
Their Sub-
sistence

Next[19] we must ascertain what being can fittingly be ascribed to mathematical genera and species. Should we admit that they are derived from sense objects, either by abstraction, as is commonly said, or by collection from particulars to one common definition? Or should we rather assign to them an existence prior to sense objects, as Plato demands and as the processional order[20] of things indicates?

In the first place, if we say that mathematical forms are derived from sense objects—that the soul, from seeing material circles and triangles, afterwards shapes in herself the form of circle and the form of triangle—whence come the exactness and certainty that belong to our ideas?[21] Necessarily either from sense objects or from the soul. But they cannot come from sense objects, for then there would be far more

---

[18] 11.19   *Rep.* 533d.

[19] 12.2   Kepler esteemed this passage so highly (12.2-17.4) that he translated it in its entirety and incorporated it in his *Harmonice Mundi* (1619), Bk. IV, Chap. 1. See Johannes Kepler, *Gesammelte Werke*, ed. Max Caspar, VI, Munich, 1940, 218-221.

[20] 12.9   πρόοδος. See note at 5.24.

[21] 12.14   λόγοι. This term appears frequently in close proximity to εἴδη in the following pages, and usually it is hard to distinguish any difference in signification. Wherever they obviously refer to concepts, forms, or ideas, I have uniformly translated the former as "ideas," the latter as "forms," in order to preserve for the English reader whatever distinction Proclus intends between them.

precision in sense objects than there is. They come therefore from the soul, which adds perfection to the imperfect sensibles and accuracy to their impreciseness. For where among sensible things do we find anything that is without parts, or without breadth, or without depth? Where do we see the equality of the lines from center to circumference? Where the fixed ratios of the sides? Where the rightness of angles? Do we not see that all sensible things are confused with one another and that no quality in them is pure and free of its opposite, but that all are divisible and extended and changing? How, then, can we explain that very stability which unchangeable ideas have, if they are derived from things that are ever changing from one state to another? For it is admitted that anything which results from changing beings receives from them a changeable character. And how can we get the exactness of our precise and irrefutable concepts from things that are not precise? For whatever yields knowledge that is steadfast has that quality itself in greater degree. We must therefore posit the soul as the generatrix of mathematical forms and ideas. And if we say that the soul produces them by having their patterns in her own essence and that these offspring are the projections[22] of forms previously existing in her, we shall be in agreement with Plato and shall have found the truth with regard to mathematical being. If, on the other hand, she weaves this enormous immaterial fabric and gives birth to such an imposing science without knowing or having previously known these ideas, how can she judge whether the offspring she bears are fertile or wind eggs, whether they are not phantoms instead of truth? What canons could she use for measuring the truth in them? And how could she even produce such a varied mass of ideas without having their essence in herself? For thus we should be making their being come about by chance, without reference to any standard. If, therefore, mathematical forms are products of the soul and the ideas of the things that the soul produces are not

[22] 13.10 προβολαί. Proclus expounds at considerable length later (51-56) this doctrine of mathematicals as projections by the understanding upon the screen of mathematical space.

derived from sense objects, mathematicals are their projections, and the soul's travail and her offspring are manifestations of eternal forms abiding in her.

In the second place, if we collect our ideas in mathematics from the lower world of sense objects, must we not say that demonstrations using terms from the sense world are better than those based on more general and simpler forms? For we assert that the premises[23] must always be of the same family as the demonstrations we use in hunting a conclusion. If, then, particulars are the premises of universal conclusions, and sense objects the premises of conclusions about the objects of understanding, how can we take the universal as the standard of demonstration rather than particulars and pronounce the nature of understandables to be more akin to demonstration than sensibles?[24] If, we say, a man demonstrates that the isosceles triangle has the sum of its angles equal to two right angles and that the same is true of the equilateral and the scalene triangles, he does not properly understand these propositions; rather it is he who demonstrates about any triangle without qualification that knows in the strict sense of the term.[25] Again we say that a universal premise is better for demonstration than a particular, and next that demonstrations from universals are more truly demonstrative, and that the premises from which demonstrations proceed are prior and naturally superior to particulars as causes of what is demonstrated. Therefore it is far from true to say that the demonstrative sciences attend to the secondary and obscurer objects of perception, rather than being concerned with objects known by the understanding which are more perfect than those that are familiar to perception and opinion.

In the third place, we affirm that those who speak thus make the soul less honorable than matter. For in saying that

[23] 14.4 *αἰτία*; cf. Arist. *Post. Anal.* 71b20-32. This statement and the other reminiscences of Aristotelian doctrine in this paragraph show clearly against whom the present argument is directed.
[24] 14.11 *Ibid.* 85b23ff.
[25] 14.15 *Ibid.* 73b28ff., 85b5ff.

15

matter receives from nature the substantial things that are more truly existent and clearer, while soul out of them fabricates in herself secondary images and later-born likenesses— likenesses inferior in being to their originals, since the soul has abstracted from matter things that are by nature inseparable from it—do they not thereby declare the soul to be less important than matter and inferior to it?[26] For matter is the locus of embodied forms, soul the locus of ideas. Soul, then, is the locus of primary, matter of secondary realities; soul the locus of things preeminently real, matter of things derivative from them; soul the locus of essential beings, matter of things that come to be by afterthought. How, then, can we say that the soul, which is the primary partaker of Nous and intelligible being, imbibing her knowledge and the whole of her life from that source, is the receptacle for the murkier forms of what has the lowest seat in the scale of existence and is more imperfect in its being than all else? But it is superfluous to refute this doctrine, which has often before been brought to an accounting.

If, however, mathematical forms do not exist by abstraction from material things or by the assembling of the common characters in particulars, nor are in any way later-born and derivative from sense objects, of necessity the soul must obtain them either from herself or from Nous, or from both herself and that higher intelligence. Now if she gets them from herself alone, how can they be images of intelligible forms? And how can they fail to receive some increment of being from the higher realities, occupying as they do a middle position between indivisible and divisible nature? And how can the forms in Nous maintain their primacy as the first patterns of all things?[27] Yet if they come from Nous alone, how can the inherent activity and self-moving character of soul be preserved when she receives her ideas from elsewhere,

16

---

[26] 15.5 This third argument, like the two preceding ones, is aimed at Aristotle; ἔνυλοι λόγοι and τόπος τῶν εἰδῶν are Aristotelian phrases (see *De An.* 403a25, 429a27).

[27] 15.26 *Sc.* if mathematical concepts are not derivative from them.

like a thing moved by outside forces? And how will she differ from matter, which is only potentially all things and generates none of the embodied forms? There is left only the conclusion that soul draws her concepts both from herself and from Nous, that she is herself the company[28] of the forms, which receive their constitution from the intelligible patterns but enter spontaneously upon the stage of being. The soul therefore was never a writing-tablet bare of inscriptions; she is a tablet that has always been inscribed and is always writing itself and being written on by Nous. For soul is also Nous, unfolding herself by virtue of the Nous that presides over her, and having become its likeness and external replica. Consequently if Nous is everything after the fashion of intellect, so is soul everything after the fashion of soul; if Nous is exemplar, soul is copy; if Nous is everything in concentration, soul is everything discursively.

Realizing this, Plato constructs the soul out of all the mathematical forms, divides her according to numbers, binds her together with proportions and harmonious ratios, deposits in her the primal principles of figures, the straight line and the circle, and sets the circles in her moving in intelligent fashion.[29] All mathematicals are thus present in the soul from the first. Before the numbers the self-moving numbers, before the visible figures the living figures,[30] before the harmonized parts the ratios of harmony, before the bodies moving in a circle the invisible circles are already constructed, and the soul is the full company of them. This, then, is a second world-order which produces itself and is produced from its native prin-

17

[28] 16.6 πλήρωμα, i.e. "complement," "crew," "equipment." That this company of ideas constitutes the essence of the soul is reiterated frequently in the sequel; e.g. 16.27ff, 17.6, 36.15, 45.22, 55.18, 62.23.

[29] 16.22 *Tim.* 35a-36c. For Proclus' full explanation we must go to his *Commentary on the Timaeus*, esp. II, 237.11-246.11, Diehl, where we learn that the "straight lines" are the two longitudinal sections into which the Demiurge divided his "compound," and the ends of which he bent around to form the two circles of the Same and the Other.

[30] 16.24 The αὐτοκίνητοι ἀριθμοί and the ζωδιακὰ σχήματα are the paradigms in the soul and in Nous of the mathematicals in the understanding. On these "self-moving ideas" see below, 140.13ff.

ciple, which fills itself with life and is filled with life from the Demiurge, in a fashion without body or extendedness; and when it projects its ideas, it reveals all the sciences and the virtues. In these forms consists the essence of the soul. We must not suppose number in her to be a plurality of monads, nor understand the idea of interval as bodily extension,[31] but must conceive of all the forms as living and intelligible paradigms of visible numbers and figures and ratios and motions— as does the *Timaeus*, which in the construction and generation of the soul equips her with the mathematical forms and establishes in her the causes of all things. For the "seven terms"[32] contain in themselves the principles of all numbers, whether linear, plane, or solid; the "seven ratios"[33] exist in her as prior causes of all the ratios; the principles of the figures are fundamental in her composition; and her primary motion embraces all others and in its movement brings them into being, for of all moving things the circle and circular movement are the starting-point. The mathematical ideas that make up the complement of souls are therefore substantial and self-moving. By emitting and unfolding them the understanding brings into being all the variety of the mathematical sciences; and it will never cease generating, bringing more and more of them to light as it explicates the partless ideas within itself. For it contains in advance all mathematical concepts, since it is their originating principle, and by virtue of its boundless power projects from these previously known starting-points the varied body of mathematical theorems.

18

Chap. VII
The Function and Procedures of General Mathematics; Its Scope

But from the being of mathematical concepts let us go back to that unitary science which we showed to be prior to the several mathematical sciences. Let us consider its function, its powers, and the scope of its activities.

We must lay it down that the function of general mathematics is, as we said earlier, dianoetic thinking.[34] It is not the kind of thought that characterizes intellect, steadfastly based

---

[31] 17.8   See 53-54.   [32] 17.15   *Tim.* 35bc.
[33] 17.17   *Tim.* 36ab.
[34] 18.11   I.e. imaginative and discursive thinking, such as characterizes διάνοια.

on itself, perfect and self-sufficing, ever converging[35] upon itself. Nor is it such as goes with opinion and perception, for these forms of knowing fix their attention on external things and concern themselves with objects whose causes they do not possess. By contrast mathematics, though beginning with reminders from the outside world, ends with the ideas that it has within; it is awakened to activity by lower realities, but its destination is the higher being of forms. Its activity is not motionless, like that of the intellect, but because its motion is not change of place or quality, as is that of the senses, but a life-giving activity, it unfolds and traverses the immaterial cosmos of ideas, now moving from first principles to conclusions, now proceeding in the opposite direction, now advancing from what it already knows to what it seeks to know, and again referring its results back to the principles that are prior in knowledge. Moreover, it is not, like Nous, above inquiry because filled from itself, nor is it satisfied, like perception, with matters other than itself; rather it advances through inquiry to discovery and moves from imperfection to perfection.

And its powers are manifestly of two sorts. Some develop its principles to plurality and open up the multiform paths of speculation, while others assemble the results of these many excursions and refer them back to their native hypotheses. Because it is subordinate to the principles of the One and the Many, the Limit and the Unlimited, the objects under its apprehension occupy a middle station between the indivisible forms and the things that are through and through divisible. Consequently it is only natural, I think, that the cognitive powers operating in the general science that deals with these objects should appear as twofold, some aiming at the unification and collection of the manifold for us, others at dividing the simple into the diverse, the more general into the particular, and the primary ideas into secondary and remoter consequences of the principles. The range of this thinking

19

---

[35] 18.14 For συνεῦον in Friedlein read συννεῦον. This misspelling occurs quite frequently in Friedlein's text, but I shall ignore later instances.

extends from on high all the way down to conclusions in the sense world, where it touches on nature and cooperates with natural science in establishing many of its propositions, just as it rises up from below and nearly joins intellect in apprehending primary principles. In its lowest applications, therefore, it projects all of mechanics, as well as optics and catoptrics[36] and many other sciences bound up with sensible

20

things and operative in them, while as it moves upwards it attains unitary and immaterial insights that enable it to perfect its partial judgments and the knowledge gained through discursive thought, bringing its own genera and species into conformity with those higher realities and exhibiting in its own reasonings the truth about the gods and the science of being. So much for these matters.

Chap. VIII
The Utility
of Mathe-
matics

We can see at once that the applications of this science range from the most commanding knowledge to the most humble. The *Timaeus* calls mathematical knowledge the way of education,[37] since it has the same relation to knowledge of all things, or first philosophy, as education has to virtue. Education prepares the soul for a complete life through firmly grounded habits, and mathematics makes ready our understanding and our mental vision for turning towards that upper world. Thus Socrates in the *Republic*[38] rightly says that, when "the eye of the soul" is blinded and corrupted by other concerns, mathematics alone can revive and awaken the soul again to the vision of being, can turn her from images to realities and from darkness to the light of intellect, can (in short) release her from the cave, where she is held prisoner by matter and by the concerns incident to generation, so that she may aspire to bodiless and partless being. For the beauty and order of mathematical discourse, and the abiding and

21

steadfast character of this science, bring us into contact with the intelligible world itself and establish us firmly in the

[36] 19.27 From κάτοπτρον, "mirror." On this science, see 40.16ff.

[37] 20.10 This reference to the *Timaeus* appears to be a slip; Proclus probably intended to write *Republic*, to which he refers in the immediately following lines.

[38] 20.18 *Rep.* 527eff.

company of things that are always fixed, always resplendent with divine beauty, and ever in the same relationships to one another. In the *Phaedrus* Socrates presents us with three types of persons who are moving upwards, each of them in fulfillment of a primary vital impulse:[39] the philosopher, the lover, and the musician. The lover begins his upward journey from the appearance of beauty and uses the intermediate forms of lovely things as stepping-stones; the musician, who is third in rank, moves from harmonies that he hears to unheard harmonies and to the ratios that exist among them. Thus for one of these persons sight is the organ of recollection, and for the other hearing. But whence does the philosophic nature get its impulse toward intellectual understanding and its awakening to genuine being and truth? For it likewise needs help, since its native principle is imperfect; its natural virtue has its vision and its character undeveloped. Such a man does indeed excite himself and flutter about being, but he must be given mathematics, says Plotinus,[40] if he is to become familiar with immaterial nature; when he uses this as a model, he can be led to the practice of dialectic and to the contemplation of being in general.

From what we have said it is clear that mathematical science makes a contribution of the greatest importance to philosophy and to its particular branches, which we must also mention. For theology, first of all, mathematics prepares our intellectual apprehension. Those truths about the gods that are difficult for imperfect minds to discover and understand, these the science of mathematics, with the help of likenesses, shows to be trustworthy, evident, and irrefutable.

22

[39] 21.7   *Phaedr.* 249d ff. (?).

[40] 21.21   *Enneads* I. 3. 3. Plotinus, apparently Egyptian by birth, taught philosophy in Rome from 242 until his death in 270. He was the greatest of the Neoplatonists and perhaps the greatest philosophical mind between Aristotle and Aquinas. He regarded his philosophy as Platonism, but it was a Platonism considerably modified to meet the needs of the Hellenistic age. His discourses, each of them based on some theme or problem in Platonic philosophy, were edited after his death by his friend and disciple Porphyry, who grouped them into six books, each containing nine discourses, whence the title *Enneads*. See 56.24.

It proves that numbers reflect the properties of beings above being and in the objects studied by the understanding reveals the powers of the intellectual figures. Thus Plato teaches us many wonderful doctrines about the gods by means of mathematical forms,[41] and the philosophy of the Pythagoreans clothes its secret theological teaching in such draperies. The same trait is evident throughout the "sacred discourse,"[42] in the *Bacchae* of Philolaus, and in the whole of Pythagoras' treatise on the gods.

Mathematics also makes contributions of the very greatest value to physical science. It reveals the orderliness of the ratios according to which the universe is constructed and the proportion that binds things together in the cosmos, making, as the *Timaeus* somewhere says, divergent and warring factors into friends and sympathetic companions.[43] It exhibits the simple and primal causal elements as everywhere clinging fast to one another in symmetry and equality, the properties through which the whole heaven was per-

[41] 22.11 We have, I believe, no writing of Plato's in which such teachings can be found, and it is significant that Proclus does not name any. He may be referring to versions of Plato's "unwritten doctrines." Such accounts were easily subject to contamination in this era of revived Pythagoreanism.

[42] 22.14 This is certainly a reference to the ἱερὸς λόγος that is cited frequently by Iamblichus in his *Life of Pythagoras* (146-148 and *passim*), by Syrianus in his *Commentary on Aristotle's Metaphysics* (10.5, 123.2, 140.16, 175.6, Kroll), and by Hierocles, an Alexandrian contemporary of Proclus, in his *Commentary on the Carmen Aureum* (Mullach, *Fragmenta Philosophorum Graecorum* I, 464). See Holgen Thesleff, *The Pythagorean Texts of the Hellenistic Period*, Abo, 1965, 164-168. Iamblichus of Chalcis in Syria was a pupil of Porphyry and later the founder of a school of Neoplatonism in Syria. His writings had a great influence on the Athenian School which arose during the following century under Plutarch. Philolaus was a Pythagorean of Croton or Metapontum during the second half of the fifth century B.C. The authenticity of the fragments attributed to him is one of the most disputed questions in early Greek philosophy. For the fragments of the *Bacchai* see Diels[6] I, 415-416. The next line in Proclus' text I interpret as a general reference to the "Pythagorean" writings, rather than to a specific treatise distinct from the ἱερὸς λόγος. For a list of these writings, see Thesleff, *Introduction to the Pythagorean Writings of the Hellenistic Period*, Abo, 1961, 18-21.

[43] 22.22 Cf. *Tim.* 32c, 88e.

fected when it took upon itself the figures appropriate to its particular region; and it discovers, furthermore, the numbers applicable to all generated things and to their periods of activity and of return to their starting-points, by which it is possible to calculate the times of fruitfulness or the reverse for each of them. All these I believe the *Timaeus* sets forth, using mathematical language throughout in expounding its theory of the nature of the universe. It regulates by numbers and figures the generation of the elements, showing how their powers, characteristics, and activities are derived therefrom and tracing the causes of all change back to the acuteness or obtuseness of their angles, the uniformity or diversity of their sides, and the number or fewness of the elements involved.[44]

How, then, can we deny that mathematics brings many remarkable benefits to what is called political philosophy? By measuring the periods of activity and the varied revolutions of the All, it finds the numbers that are appropriate for generation, that is, those that cause homogeneity or diversity in progeny, those that are fruitful and perfecting and their opposites, those that bring a harmonious life in their train and those that bring discord, and in general those that are responsible for prosperity and those that occasion want. All this the speech of the Muses in the *Republic*[45] sets forth when it makes "the whole geometrical number" the factor that determines whether births will be better or worse, and thus whether the manners of a state will be preserved uncorrupted or a good polity degenerate into unreason and passion. Everyone can see that it belongs to mathematics as a whole—not to a part of it, such

[44] 23.11 *Tim.* 53ff.

[45] 23.21 *Rep.* 545e-547a. The description of the "geometrical number" in this passage early became proverbial for obscurity. For an account of its difficulties and a plausible resolution of them see James Adam (ed.), *The Republic of Plato*, Cambridge, 1929, II, 204-208, 286-312. Proclus himself deals with it at length in his *Commentary on the Republic* II, 36-46, Kroll, though this passage contains so many lacunae at critical points that Proclus' interpretation is itself obscure to us. See also "Le Nombre Géométrique de Platon" in J. Dupuis' edition of Theon of Smyrna, Paris, 1892, 365-400.

24 as arithmetic or geometry—to furnish the knowledge of this geometrical number that is spoken of here. For the ratios that govern fruitful and unfruitful generation pervade all mathematics.

Again it perfects us for moral philosophy by instilling order and harmonious living into our characters; it furnishes the gestures, songs, and dances appropriate to virtue by which, as we know, the Athenian Stranger wishes those who are to share in moral virtue to be perfected from their youth onwards;[46] it gives the proportions that characterize the virtues—now in numbers, now in movements, now in musical concords—and shows up by contrast the excesses and deficiencies of vice, thereby helping us to make our characters measured and ordered. For this reason Socrates in the *Gorgias*, when reproaching Callicles for his unordered and dissolute life, says "you are neglecting geometry and geometrical equality"; and in the *Republic* he finds the interval separating the pleasure of the tyrant from that of the king to be analogous to that between a plane and a solid number.[47]

Finally, how much benefit mathematics confers on the other sciences and arts we can learn when we reflect that to the theoretical arts, such as rhetoric and all those like it that function through discourse, it contributes completeness and orderliness, by providing for them a likeness of a whole made perfect through first, intermediate, and concluding parts; that to the poetical arts it stands as a para-

25 digm, furnishing in itself models for the speeches that the authors compose and the meters that they employ; and that for the practical arts it defines their motion and activity through its own fixed and unchangeable forms. In general, as Socrates says in the *Philebus*, all the arts require the aid of counting, measuring, and weighing, of one or all of them;[48] and these arts are all included in mathematical reasonings and are made definite by them, for it is mathe-

[46] 24.9 *Laws* 672-673.
[47] 24.20 *Gorg.* 508a, *Rep.* 587d.
[48] 25.7 *Phil.* 55e.

matics that knows the divisions of numbers, the variety of measures, and the differences of weights. These considerations will make evident to the student the utility of general mathematics both to philosophy itself and to the other sciences and arts.

Chap. IX
An Answer
to the De-
tractors of
Mathematics

26

There are nevertheless contentious persons who endeavor to detract from the worth of this science, some denying its beauty and excellence on the ground that its discourses say nothing about such matters, others declaring that the empirical sciences concerned with sense objects are more useful than the general theorems of mathematics. Mensuration, they say, is more useful than geometry, popular arithmetic than the theory of numbers, and navigation than general astronomy. For we do not become rich by knowing what wealth is but by using it, nor happy by knowing what happiness is but by living happily. Hence we shall agree, they say, that the empirical sciences, not the theories of the mathematicians, contribute most to human life and conduct. Those who are ignorant of principles but practised in dealing with particular problems are far and away superior in meeting human needs to those who have spent their time in the schools pursuing theory alone.

To those who say these things we can reply by exhibiting the beauty of mathematics on the principles by which Aristotle attempts to persuade us.[49] Three things, he says, are especially conducive to beauty of body or soul: order, symmetry, and definiteness. Ugliness in the body arises from the ascendancy of disorder and from a lack of shapeliness, symmetry, and outline in the material part of our composite nature; ugliness of mind comes from unreason, moving in an irregular and disorderly fashion,[50] out of harmony with reason and unwilling to accept the principles it imposes; beauty, therefore, will reside in the opposites of these, namely, order, symmetry, and definiteness. These characters we find preeminently in mathematical science.

[49] 26.13   See *Met.* 1078a33ff. But πείθειν suggests that Proclus is referring to one of the persuasive discourses, such as the *Protrepticus*.
[50] 26.19   An echo of *Tim.* 30a.

We see order in its procedure of explaining the derivative and more complex theorems from the primary and simpler ones; for in mathematics later propositions are always dependent on their predecessors, and some are counted as starting-points, others as deductions from the primary hypotheses. We see symmetry in the accord of the demonstrations with one another and in their common reference back to Nous; for the measure common to all parts of the science is Nous, from which it gets its principles and to which it directs the minds of its students. And we see definiteness in the fixity and certainty of its ideas; for the objects of mathematical knowledge do not appear now in one guise and now in another, like the objects of perception or opinion, but always present themselves as the same, made definite by intelligible forms. If, then, these are the factors especially productive of beauty, and mathematics is characterized by them, it is clear that there is beauty in it. How could it be otherwise when Nous illumines this science from above and its earnest endeavor is to spur us to move from the sense world into that intelligible region?

We do not think it proper, moreover, to measure its utility by looking to human needs and making necessity our chief concern. For so we should be admitting that theoretical virtue itself is useless, because it separates itself from the affairs of men and prefers not to know anything at all about the objects of their striving. Thus Socrates in the *Theaetetus*, in his truly inspired description of the "leaders of the philosophic chorus,"[51] withdraws them from connection with human life and lifts their thought to the mountaintop of being, emancipated from necessity and utility. We must therefore posit mathematical knowledge and the vision that results from it as being worthy of choice for their own sakes, and not because they satisfy human needs. And if we must relate their usefulness to something outside them, it is to intellectual insight that they must be said to be contributory. For to that they lead the way and prepare us by

27

28

[51] 27.23   *Theaet.* 173c-177a.

purifying the eye of the soul and removing the hindrances that the senses present to our knowing the whole of things. Just as we judge the usefulness or uselessness of the cathartic virtues in general by looking not to the needs of living, but rather to the life of contemplation, so we must refer the purpose of mathematics to intellectual insight and the consummation of wisdom. For this reason the cultivation of it is worthy of earnest endeavor both for its own sake and for the sake of the intellectual life. Evidence that it is intrinsically desirable to those who are engaged in it is, as Aristotle somewhere says, the great progress that mathematical science has made in a short time, although no reward is offered to those who pursue it, and the fact that even those who gain but slight benefit from it are fond of it and occupy themselves with it to the neglect of other concerns.[52] So those who despise mathematical knowledge are they that have no taste for the pleasures it affords.

Consequently instead of crying down mathematics for the reason that it contributes nothing to human needs—for in its lowest applications, where it works in company with material things, it does aim at serving such needs—we should, on the contrary, esteem it highly because it is above material needs and has its good in itself alone. In general it was when men had ceased to be anxious about the necessities of life that they turned to the study of mathematics. This is as we should expect; for men must first concern themselves seriously with the things that are kindred and of one blood with them in the world of generation, and afterwards with the things that release the soul from the world of generation and remind it of being. In this sense, then, necessities come before things intrinsically valuable; that is, we seek out the objects akin to perception before we pursue the ends apprehended by Nous. In fact the whole of generation and the soul-life that is implicated in it[53] are so constituted by

29

---

[52] 28.20 I have not been able to identify this passage in Aristotle. It is possible that Proclus is thinking of *Rep.* 528bc.

[53] 29.10 Reading (apparently with ver Eecke) ἐν αὐτῇ for ἐν αὐτῇ in Friedlein.

nature as to move from the imperfect towards the perfect. Let this be our answer to those who decry mathematical knowledge.

Chap. X
Answer to
the Charge
that Plato
Discredited
Mathematics

Some persons from our own household, citing Plato in support of their views, will perhaps try to induce disdain for mathematics among the more superficial students. For, they say, the philosopher himself in the *Republic* excludes mathematical knowledge from the company of the sciences and criticizes it for not knowing its starting-points. They cite the remark about the study "whose starting-point is unknown and whose middle premises and conclusions follow from what is unknown" and all the other accusations that Socrates throws out against mathematics in that book.[54] Now since we are arguing with friends, we shall remind them that Plato himself clearly affirms that mathematics purifies and elevates the soul, like Homer's Athena dispersing the mist from the intellectual light of the understanding, a light "more worthy of preservation than ten thousand bodily eyes," and thus dispenses Athena's gifts as well as those of Hermes.[55] We would remind them, furthermore, that Plato everywhere calls it science and declares it to be the source of the greatest happiness to those who pursue it.

30

But what does he mean by denying it the name of science in the *Republic* passage? I shall explain briefly, since my present discourse is addressed to scholars. Plato often gives the name "science" to any knowledge of universals, so to speak, whether the manner of this knowing be scientific or empirical, distinguishing it thus from perception, which apprehends particulars. It is in this sense, I think, that he uses the word "science" in the *Statesman* and the *Sophist*,[56] including among the sciences that "high-born sophistic" that Socrates in the *Gorgias* asserts to be only an empirical rou-

---

[54] 29.24 *Rep.* 533b-d; cf. 510c.
[55] 30.5 *Odyssey* XIII, 189-352, where Athena disperses the mist from the eyes of Odysseus so that he recognizes his native island. The words in quotation marks come from *Rep.* 527e.
[56] 30.16 *Soph.* 231b.

tine,[57] as well as the menial type of sophistry and many other procedures which are routines and not true sciences. This apprehension of universals itself he divides into a kind that knows and one that does not know causes; the former he thinks deserves to be called a science, the latter a routine. In this sense he sometimes gives the name "science" to the arts, but never to routines. For "how could what is without a logos be a science?" he asks in the *Symposium*.[58] Consequently every form of knowledge which apprehends the logos, or cause, of the things it knows is a science. Again this science that knows its objects through their causes he divides into two types, one proceeding by guesswork and aiming at a particular end, the other concerned with knowing abstract and unchanging being; and by this token he separates science from medicine and all studies of material things, whereas mathematics and in general all investigation of eternal realities he calls science. Once more he would divide this science, which we distinguish from the arts, into a part that does and a part that does not proceed from hypotheses. The unhypothetical science of the whole of things mounts upwards to the Good, to the cause high above all else, making the Good the goal of its ascent;[59] but that which shows what follows from previously determined starting-points moves not towards a principle, but to a conclusion. In this sense, then, he says, because mathematics uses hypotheses, it falls below the unhypothetical and perfect science. For genuine science is one, the science by which we are able to know all things, the science from which come the principles of all other sciences, some immediately and some at further remove.

Let us, then, not say that Plato excludes mathematics from the sciences, but only that he ranks it second to the one highest science; nor that he declares mathematics to be ignorant of its own principles, but says rather that it takes its principles from the highest science and, holding them

31

32

[57] 30.19 *Gorg.* 464cff.
[58] 31.1 *Symp.* 202a.
[59] 31.17 For the language and thought here see *Rep.* 511bc.

without demonstration, demonstrates their consequences. Similarly he sometimes allows that the soul, which is constituted of mathematical ratios, is a source of motion and sometimes that it receives its motion from intelligible things. These statements are in full accord with each other; for the soul is a cause of motion in things that receive their motion from outside themselves, but not therefore the cause of all motion. In the same way mathematics is second to the highest science and imperfect as compared with it but nevertheless is a science—not an unhypothetical science, but one which, being capable of knowing the specific ideas that are in the soul, exhibits the premises of its conclusions and thus has reasons for the matters known to it. So much regarding Plato's opinions about mathematics.

Chap. XI
The Qualities Required of the Mathematician

33

What may we require of the mathematician, and how can we properly judge him? These questions must be discussed next. The man who has had a general education, says Aristotle, can exercise critical judgment in any field,[60] but he who has been trained in mathematics is a proper critic of the correctness of mathematical reasonings. He must have acquired certain standards of judgment. In the first place, he should know when he can make his demonstrations general and when he must look to the properties of the species. Often things that differ in species have identical properties; for example, all triangles have the sum of their angles equal to two right angles. But there are many things that have the same name yet whose common character differs in different species; for example, similarity in figures and in numbers. We should not therefore demand of the mathematician a single demonstration in such cases, for the principles of figures and numbers are not the same but vary with the underlying genera. When, however, the essential property is one, the demonstration should also be one. The property of having angles equal to two right angles is common to all species of triangles, and that in which this property inheres is the same in them all, namely, the triangle and its definition.

[60] 32.24   Cf. *Nic. Eth.* 1094b28ff., and *De Part. An.* 639a1-5.

Similarly the property of having external angles equal in sum to four right angles belongs not only to triangles, but to all rectilinear figures;[61] hence the demonstration fits them all as rectilinear. Each definition invariably brings with it a specific property and character in which all things that fall under the definition participate—as, for example, the definition of triangle, or of rectilinear figure, or of figure in general.

In the second place, we must ask whether the mathematician's demonstration corresponds to the nature of his subject-matter, that is, whether he has given necessary and irrefutable reasons, and not arguments that are merely persuasive and full of probabilities. It is equally mistaken, says Aristotle, to demand demonstration from a rhetorician and to accept persuasive reasoning from a mathematician.[62] Every man who knows his science or his art should make his arguments appropriate to the things with which he is dealing. So Plato in the *Timaeus* rightly demands probable reasoning from the student of nature, since he is working on a subject-matter that is not precise, but incontrovertible and unshakeable arguments from one who is discoursing about intelligibles and about stable being.[63] Differences in subject-matter at once produce differences in the sciences and the arts that are concerned with them. Thus some things are unchangeable, others changing; some simpler, others more composite; some intelligible, others sensible. Even in mathematics we cannot demand the same degree of accuracy in all parts. If one part applies to sensibles and another is an investigation of intelligible matters, they will not be equally precise, but the latter more so than the former. This is why arithmetic, we say, is more exact than harmonics. Nor should we in general require that mathematics and the other sciences use the same demonstrations. For the differences between the subjects dealt with are not inconsiderable.

34

---

[61] 33.16 This proposition does not occur in Euclid, but Aristotle refers to it (*Post. Anal.* 85b38, 99a19); thus it probably appeared in a textbook used in the Academy. Heath (I, 340) judges it to be Pythagorean. Proclus refers to it again at 73.2 and demonstrates it at 382.24ff.

[62] 34.1 *Nic. Eth.* 1094b25-27.   [63] 34.7 *Tim.* 29bc.

In the third place, we assert that he who is to judge properly of mathematical arguments must also have studied the nature of sameness and otherness, of essential and accidental predication, of proportion, and all such matters. For it is with respect to these that almost every mistake is made by those who suppose they are giving a mathematical proof when they are not really doing so, as when they demonstrate by assuming the same to be different for each species of the subject or the different to be the same, or when they take an accidental attribute as essential or an essential one as accidental, asserting, for example, that the circle is more beautiful than the straight line or the equilateral triangle than the isosceles; for such distinctions are not the business of the mathematician.

35

In the fourth place, since mathematics occupies a middle position between the intelligible and the sense worlds and exhibits within itself many likenesses of divine things and also many paradigms of physical relations, we must observe the threefold character of its demonstrations, some being more intuitive, some more discursive, and some approaching the nature of opinion. Proofs must vary with the problems handled and be differentiated according to the kinds of being concerned, since mathematics is a texture of all these strands and adapts its discourse to the whole range of things. But enough of these matters.

**Chap. XII
The Pythagorean Classification of the Mathematical Sciences**

We must next distinguish the species of mathematical science and determine what and how many they are; for after its generic and all-inclusive form it is necessary to consider the specific differences between the particular sciences. The Pythagoreans considered all mathematical science to be divided into four parts: one half they marked off as concerned with quantity (ποσόν), the other half with magnitude (πηλίκον); and each of these they posited as twofold. A quantity can be considered in regard to its character by itself or in its relation to another quantity, magnitudes as either stationary or in motion. Arithmetic, then, studies quantity as such, music the relations between quantities,

36

geometry magnitude at rest, spherics[64] magnitude inherently moving. The Pythagoreans consider quantity and magnitude not in their generality, however, but only as finite in each case. For they say that the sciences study the finite in abstraction from infinite quantities and magnitudes, since it is impossible to comprehend infinity in either of them. Since this assertion is made by men who have reached the summit of wisdom, it is not for us to demand that we be taught about quantity in sense objects or magnitude that appears in bodies. To examine these matters is, I think, the province of the science of nature, not that of mathematics itself.

Now since, as the *Timaeus* has taught us,[65] the Demiurge took in hand the unity and diversity in the universe, and the mixture of sameness and otherness, to fill up the nature of soul, and constructed her out of these kinds together with rest and motion, let us say that it is by virtue of her otherness, that is, the plurality and diversity of the ratios in her, that the understanding, when she has been constituted and has noted that she is both one and many, projects numbers and the knowledge of numbers which is arithmetic; and by virtue of the unity of plurality in her and the community of bond that binds her together, she projects music. Hence arithmetic is elder than music, since the soul was first divided by the Demiurge and then bound together by ratios in the fashion explained by Plato. Again, her activities being firmly rooted in her constitution, she produces geometry out of her own nature, that is, the one essential figure and the creative principles of all the figures, while by virtue of the motion in her she produces spherics. For she herself revolves in circles but abides always the same by virtue of the causes of the circle, namely, the straight line and the circumference. Hence also geometry comes into being before spherics, as rest precedes motion.

The soul produces these sciences by looking not to her

37

[64] 36.2 I.e. astronomy. See Heath I, 11f.

[65] 36.17 *Tim.* 35a. At 36.22 read with Hultsch ἑαυτὴν for ἑαυτὸ in Friedlein.

infinite capacity for developing forms, but rather to the species within the compass of the Limit. For this reason, they say, these sciences exclude infinity from plurality and magnitude and concern themselves straightway with the Limited. For Nous set in the soul the principles of all things, including those of plurality and magnitude. Since she is through and through homogeneous with herself, one and undivided, and on the other hand differentiated and expressive of the ordered world of forms, she possesses from the intelligible world a share of both the primal Limit and the Unlimited. But she thinks the Unlimited in accordance with the Limit and gives birth to forms of life and ideas of all sorts through the Unlimited in her. Her thinking, however, constitutes these sciences not after the fashion of the Unlimited that belongs to life, but in accordance with the Limit inherent in these sciences; for they bear the likeness of Nous, not of life. This, then, is the doctrine of the Pythagoreans and their fourfold division of the mathematical sciences.

38

**Chap. XIII
Geminus'
Classifica-
tion of the
Mathemati-
cal Sciences**

But others, like Geminus,[66] think that mathematics should be divided differently; they think of one part as concerned with intelligibles only and of another as working with perceptibles and in contact with them. By intelligibles, of course, they mean those objects that the soul arouses by herself and contemplates in separation from embodied forms. Of the mathematics that deals with intelligibles they posit arithmetic and geometry as the two primary and most authentic parts, while the mathematics that attends to sensibles contains six sciences: mechanics, astronomy, optics, geodesy, canonics, and calculation. Tactics they do not think it proper to call a part of mathematics, as others do, though they admit that it sometimes uses calculation, as in the enumera-

[66] 38.4 Geminus was apparently a Stoic philosopher from the island of Rhodes and a pupil of Posidonius. He wrote a comprehensive mathematical work, probably between 73 and 67 B.C., to which Proclus refers no less than twenty times in this commentary. This work has unfortunately been lost, but another treatise, on astronomy, is still extant (see Gow, 287). But on the uncertainties regarding his date and birthplace, and even his name, see Heath II, 222f.

39

tion of military forces,[67] and sometimes geodesy, as in the division and measurement of encampments. Much less do they think of history and medicine as parts of mathematics, even though writers of history often bring in mathematical theorems in describing the lie of certain regions or in calculating the size, breadth, or perimeters of cities,[68] and physicians often clarify their own doctrines by such methods, for the utility of astronomy to medicine is made clear by Hippocrates and all who speak of seasons and places. So also the master of tactics will use the theorems of mathematics, even though he is not a mathematician, if he should ever want to lay out a circular camp to make his army appear as small as possible, or a square or pentagonal or some other form of camp to make it appear very large.

These, then, are the species of general mathematics. Geometry in its turn is divided into plane geometry and stereometry. There is no special branch of study devoted to points and lines, inasmuch as no figure can be constructed from them without planes or solids; and it is always the function of geometry, whether plane or solid, either to construct figures or to compound or divide figures already constructed. In the same way arithmetic is divided into the study of linear numbers, plane numbers, and solid numbers; for it examines number as such and its various kinds as they proceed from the number one, investigating the generation of plane numbers, both similar and dissimilar, and progressions to the third dimension. Geodesy[69] and calculation are analogous to these sciences,[70] since they discourse not about intelligible but about sensible numbers and figures. For it is not the function of geodesy to measure cylinders or cones, but heaps of earth considered as cones and wells considered as cylinders; and it does not use intelligible straight lines, but sensible

[67] 38.16   Reading with Barocius λόχων instead of λόγων in Friedlein. See Tannery, IX, 126.

[68] 38.22   Omitting with Barocius the dittograph καὶ διαμέτρους ἢ περιβόλους in Friedlein. See Tannery, *loc. cit.*

[69] 39.20   Mensuration in general, not merely land-measuring. Heath, I, 16.

[70] 39.21   I.e. to geometry and arithmetic.

40

ones, sometimes more precise ones, such as rays of sunlight, sometimes coarser ones, such as a rope or a carpenter's rule. Nor does the student of calculation consider the properties of number as such, but of numbers as present in sensible objects; and hence he gives them names from the things being numbered, calling them sheep numbers or cup numbers.[71] He does not assert, as does the arithmetician, that something is least; nevertheless with respect to any given class he assumes a least, for when he is counting a group of men, one man is his unit. Again optics and canonics[72] are offshoots of geometry and arithmetic. The former science uses visual lines and the angles made by them; it is divided into a part specifically called optics, which explains the illusory appearances presented by objects seen at a distance, such as the converging of parallel lines or the rounded appearance of square towers, and general catoptrics,[73] which is concerned with the various ways in which light is reflected. The latter is closely bound up with the art of representation and studies what is called "scene-painting,"[74] showing how objects can be represented by images that will not seem disproportionate or shapeless when seen at a distance or on an elevation. The science of canonics deals with the perceptible ratios between notes of the musical scales and discovers the divisions of the mono-

41

chord,[75] everywhere relying on sense-perception and, as Plato says, "putting the ear ahead of the mind."[76]

In addition to these there is the science called mechanics, a part of the study of perceptible and embodied forms. Under it comes the art of making useful engines of war, like the machines that Archimedes[77] is credited with devising for defense

[71] 40.5 Cf. the scholium to Plato's *Charm.* 165e and also *Laws* 819bc.

[72] 40.9 The mathematical theory of music.

[73] 40.16 See note at 19.27.

[74] 40.19 σκηνογραφική, i.e. applied perspective.

[75] 40.23 κανών, whence the name "canonics."

[76] 41.1 *Rep.* 531ab.

[77] 41.6 Archimedes of Syracuse, who was killed during the capture of Syracuse by the Romans in 212 B.C. His mathematical achievements are numerous and outstanding. He investigated the squaring of the circle and of other curvilinear plane figures, and the computing

against the besiegers of Syracuse, and also the art of wonder-working, which invents figures moved sometimes by wind, like those written about by Ctesibius and Heron,[78] sometimes by weights, whose imbalance and balance respectively are responsible for movement and rest, as the *Timaeus* shows,[79] and sometimes by cords and ropes in imitation of the tendons and movements of living beings. Under mechanics also falls the science of equilibrium in general and the study of the so-called center of gravity, as well as the art of making spheres imitating the revolutions of the heavens, such as was cultivated by Archimedes, and in general every art concerned with the moving of material things. There remains astronomy, which inquires into the cosmic motions, the sizes and shapes of the heavenly bodies, their illuminations and distances from the earth, and all such matters. This art draws heavily on sense-perception and coincides in large measure with the science of nature. The parts of astronomy are gnomonics,[80] which occupies itself with marking the divisions of time by the placing of sun-dials; meteorology, which determines the different risings of the heavenly bodies and their distances from one another and teaches many and varied details of

42

---

of the area of curved surfaces and of the volume of the sphere, cone, and cylinder. In mechanics he developed the theory of the lever and of the center of gravity and invented the whole science of hydrostatics. The mechanical inventions that Proclus mentions here and at 63.19ff, were regarded by Archimedes as merely incidental and relatively unimportant. His works *On the Sphere and Cylinder, Measurement of a Circle, On Plane Equilibriums*, and several others are still extant. See Heath II, 17-103, and Gow, 221-244. An English translation of his works was published by Heath in 1897.

[78] 41.10 Heron of Alexandria should probably be placed in the third century of our era, though the evidence is controversial; see Heath II, 300-306. Besides the references in Proclus, there is other evidence that he wrote a commentary on Euclid's *Elements* (see Heath, *Euclid* I, 21-24). Ctesibius seems to belong to an earlier period, possibly to the second century B.C. There is, however, another tradition that he was the teacher of Heron (see Heath II, 298), which would put him much later.

[79] 41.12 *Tim.* 57dff.

[80] 41.25 From γνώμων, "sun-dial."

astronomical theory; and dioptrics,[81] which fixes the positions of the sun, moon, and stars by means of special instruments. Such are the traditions we have received from the writings of the ancients regarding the divisions of mathematical science.

Chap. XIV
In What
Way Dia-
lectic is the
Capstone of
the Mathe-
matical
Sciences

Leaving these matters, let us look back and consider what Plato meant in the *Republic* when he declared dialectic to be the capstone of the mathematical sciences,[82] and what is the unifying bond among them reported by the author of the *Epinomis*.[83] Our answer is that, as Nous is set over understanding and dispenses principles to it from above, perfecting it out of its own riches, so in the same way dialectic, the purest part of philosophy, hovers attentively over mathematics, encompasses its whole development, and of itself contributes to the special sciences their various perfecting, critical, and intellective powers—the procedures, I mean, of analysis, division, definition, and demonstration. Being thus endowed

43

and led towards perfection, mathematics reaches some of its results by analysis, others by synthesis, expounds some matters by division, others by definition, and some of its discoveries binds fast by demonstration, adapting these methods to its subjects and employing each of them for gaining insight into mediating ideas. Thus its analyses are under the control of dialectic, and its definitions, divisions, and demonstrations are of the same family and unfold in conformity with the way of mathematical understanding. It is reasonable, then, to say that dialectic is the capstone of the mathematical sciences. It brings to perfection all the intellectual insight they contain, making what is exact in them more irrefutable, confirming the stability of what they have established and referring what is

[81] 42.4 From διόπτρα, an optical instrument for measuring angles or altitudes. Reading τὰς ἐποχάς, with Tannery IX, 126; the ἐ in Friedlein's MSS apparently originated as a marginal correction for the α in the erroneous ἀποχάs.

[82] 42.11 *Rep.* 534e.

[83] 42.12 Reading with Barocius ὃν ὃ for ὃν in Friedlein. This is a reference to the δεσμός of the mathematical sciences in *Epin.* 991e. See Novotny's note on this passage in his *Commentary on the Epinomis*, Prague, 1960, 222-223.

pure and incorporeal in them to the simplicity and immateriality of Nous, making precise their primary starting-points through definitions and explicating the distinctions of genera and species within their subject-matters, teaching the use of synthesis to bring out the consequences that follow from principles and of analysis to lead up to the first principles and starting-points.

As for the unifying bond of the mathematical sciences, we should not suppose it to be proportion, as Eratosthenes[84] says. For though proportion is said to be, and is, one of the features common to all mathematics, there are many other characteristics that are all-pervading, so to speak, and intrinsic to the common nature of mathematics. We should prefer to say that the immediate bond of union between them is that single and entire science of mathematics which contains in itself in simpler form the principles of all the particular sciences, that science which teaches their common nature as well as their differences and what traits are the same in them all and what belong to more or fewer of them. Those who study mathematics in the proper way advance to this science from the particular ones. But even higher than it, dialectic could be said to be the bond of union among the mathematical sciences or—to repeat Plato's designation in the *Republic*—their capstone. For this perfects general mathematics and sends it up towards Nous by means of its peculiar powers, showing that it is truly a science and rendering it steadfast and irrefutable. Yet highest in rank among the unifying bonds is that very Nous which contains in itself all dialectical resources in undifferentiated fashion, combining their variety in simplicity, their partiality in completeness of insight, their plurality in unity. Nous, then, wraps up the developments of the dialectical methods, binds together from above all the discursiveness of mathematical reasoning, and is the perfect

44

[84] 43.23 Eratosthenes of Cyrene, librarian at Alexandria in the second half of the third century B.C. He was a man of varied scientific attainments, to whom Archimedes dedicated two of his works. He is best remembered for having calculated the circumference and diameter of the earth, with surprisingly accurate results. See Heath II, 104-109; Van der Waerden, 228-234; Gow, 242-246.

terminus of the upward journey and of the activity of knowing. So much for these questions.

As for the name itself that is applied to mathematics and mathematical studies, from what source could we say the ancients got it for these sciences, and what relevant meaning could it have? In my opinion, such a designation for the science of dianoetic reasoning did not come about by accident, as most names do. According to the tradition, the Pythagoreans recognized that everything we call learning is remembering, not something placed in the mind from without, like the images of sense pictured in the imagination, nor transitory, like the judgments of opinion. Though awakened by sense-perception, learning has its source within us, in our understanding's attending to itself. They realized too that, although evidences of such memories can be cited from many areas, it is especially from mathematics that they come, as Plato also remarks. "If you take a person to a diagram," he says, "then you can show most clearly that learning is recollection."[85] This is why Socrates in the *Meno* uses this kind of argument to prove that learning is nothing but the mind's remembering its own ideas.[86] The explanation is that what remembers is the understanding. This part of the soul has its essence in these mathematical ideas,[87] and it has a prior knowledge of them, even when it is not using them; it possesses them all in an essential, though latent, fashion and brings each of them to light when it is set free of the hindrances that arise from sensation. For our sense-perceptions engage the mind with divisible things, the imagination fills it with moving shapes, and desires divert it to the life of feeling. Every divisible thing is an obstacle to our returning upon ourselves, every formed thing disturbs our formless knowledge, and every feeling is an impediment to passionless activity. Consequently when we remove these hindrances we are able to know by understanding itself the ideas that it has, and then we become knowers in actuality, that is, pro-

45

Chap. XV
The Origin
of the Name
Mathematics

46

---

[85] 45.17 *Phaedo* 73b.   [86] 45.21 *Meno* 82bff.
[87] 45.23 See below (51ff.) for an exposition of this doctrine in its wider context.

ducers of genuine knowledge. But so long as we remain in bondage, with the eye of the mind closed, we shall never attain the perfection to which we are adapted.

This, then, is what learning (μάθησις) is, recollection of the eternal ideas in the soul; and this is why the study that especially brings us the recollection of these ideas is called the science concerned with learning (μαθηματική). Its name thus makes clear what sort of function this science performs. It arouses our innate knowledge, awakens our intellect, purges our understanding, brings to light the concepts that belong essentially to us, takes away the forgetfulness and ignorance that we have from birth, sets us free from the bonds of unreason; and all this by the favor of the god[88] who is truly the patron of this science, who brings our intellectual endowments to light, fills everything with divine reason, moves our souls towards Nous, awakens us as it were from our heavy slumber, through our searching turns us back upon ourselves, through our birthpangs perfects us, and through the discovery of pure Nous leads us to the blessed life. And so, dedicating this composition to him, we proceed to delineate the theory of the science of mathematics.[89]

47

[88] 46.25 This god is probably Hermes, identified by the Greeks with the Egyptian Thoth or Theuth, the inventor of writing and of all the sciences and arts dependent on it (cf. Plato, *Phaedr.* 274c-275b; *Phil.* 18b). For the Hellenistic conception of the god Hermes-Thoth and the origin of his appellation "thrice-great" see Festugière, *La Révélation d'Hermès Trismégiste* I, 67-74. Cf. the reference at 155.24 to the "triadic god" whose appellation has been conferred by "the wise men most familiar with theological mysteries."

[89] 47.8 From these words one would infer that this first prologue was intended to serve as an introduction to a general treatise on mathematics. The beginning of the second prologue gives a different statement of the author's subject, not general mathematics but geometry in particular and more particularly Euclid's *Elements*. Is it possible that this first prologue was intended to be what these concluding words suggest and that it later became detached from the larger project and prefixed to the more specific prologue to the commentary on Euclid? See Supplementary Note at the end of this volume.

Book II
Chap. I
Geometry a
Part of Gen-
eral Mathe-
matics; Its
Subject-
matter

I N THE preceding discourse we have examined the common characters pervading all mathematical science, following Plato's lead and also using thoughts collected from
other sources that are relevant to the present study. It follows next to speak of geometry itself and of the treatise on
the *Elements* that lies before us and for whose sake the whole
of this work has been undertaken.

That geometry is a part of general mathematics and occupies a place second to arithmetic, which completes and defines
it (for everything that is expressible[1] and knowable in geometry is determined by arithmetical ratios), has been asserted by
the ancients and needs no lengthy argument here. It would
be reasonable to begin our exposition of geometry with an
examination of its subject-matter, to see what rank it holds in
the scale of things and the kind of being it has. For when we
have examined this carefully, the power and utility of the
science that knows it will be evident, as will the good that it
confers upon those who learn it.

It is obviously difficult to decide in what class of things to
put the subject-matter of geometry without missing the truth
about it. If we regard the figures that the geometer talks about
as belonging to the sense world and inseparable from matter,
how can we any longer say that geometry emancipates us from
sensible things, converts us to the realm of bodiless existence,
habituates us to the sight of intelligibles, and prepares us for
activity in accordance with Nous? And where among sensible
things have we seen the point without parts, the line without
breadth, the surface without thickness, the equality of the
lines from the center, or in general any of the polygonal and
polyhedral figures about which geometry teaches us? And

49

---

[1] 48.12 ῥητόν, i.e. "commensurable" (see note at 6.21). This passage leaves out what Proclus elsewhere says is the characteristic feature
of geometry as distinct from arithmetic, that it deals with incommensurable magnitudes and infinite divisibility (see 60.7ff.).

how can the propositions of this science remain irrefutable when the figures and forms of sensible things are only more or less what they are, moving and changing in every way and full of all the indeterminateness of matter, when equality is composed of its opposite inequality, and indivisibles parade as divisible and separated?

But if the objects of geometry are outside matter, its ideas pure and separate from sense objects, then none of them will have any parts or body or magnitude. For ideas can have magnitude, bulk, and extension in general only through the matter which is their receptacle, a receptacle that accommodates indivisibles as divisible, unextended things as extended, and motionless things as moving. How, then, can we still bisect the straight line or the triangle or the circle? Or speak of the difference between angles, or of increase and decrease of figures such as triangles or squares? How can we talk of contact between circles and straight lines? All these things indicate that the subject-matter of geometry is divisible and not composed of partless ideas. Besides difficulties of this sort, we must recall that Plato calls geometrical forms understandables[2] and asserts that they separate us from sensible things and incite us to turn from sensation to Nous—the ideas[3] of the understanding being, as I said, indivisible and unextended, in keeping with the peculiar character of the soul.

If we must formulate a theory in agreement both with the facts themselves and with this teaching of Plato's, let us proceed by making the following distinctions. Every universal, that is, every One that includes a Many, either appears in the particulars[4] and has its existence in them and is inseparable from them, holding its place in their ranks, moving as they move and remaining motionless when they are stationary; or

50

---

[2] 50.11 διανοητά. See note at 10.27.

[3] 50.14 λόγων. See note at 12.14.

[4] 50.20 Omitting with Schönberger πέφυκεν ἢ φαίνεται, which are included, though with a question mark, in Friedlein's text. They do not fit into the grammatical structure of the sentence, and I surmise that they were once questions written in the margin regarding the meaning of φαντάζεσθαι and were later ineptly incorporated in the text.

51 exists prior to the Many and produces plurality by offering its appearances to the many instances, itself ranged indivisibly above them but enabling these derivatives to share in its nature in a variety of ways; or is formed from particulars by reflection and has existence as an after-effect, a later-born addition to the Many. According to these three modes of being, I think we shall find that some universals are prior to their instances, some are in their instances, and some are constituted by virtue of being related to them as their predicate.

Of these three kinds of universal forms—briefly stated, the universal shared in by its particulars, the universal in its particulars, and the universal that supplements the particulars —let us note that there are differences in the underlying matter. If we assume two classes of things that participate in the universal, namely, sense objects and objects that have existence in the imagination (for matter likewise is twofold, as Aristotle somewhere says:[5] the matter of things tied to sensation and the matter of imagined objects), we shall admit that the corresponding universals are of two kinds: one perceptible, since it is participated in by sense objects, and the other imaginary, as existing in the plurality of pictures in the imagination. For imagination, both by virtue of its formative activity and because it has existence with and in the body,

52 always produces individual pictures that have divisible extension and shape, and everything that it knows has this kind of existence. For this reason a certain person[6] has ventured to call it "passive Nous." Yet if it is Nous, how could it be other than impassive ($\dot{a}\pi a\theta\eta s$) and immaterial? And if feeling ($\pi\dot{a}\theta os$) accompanies its activity, has it any longer a right to be called Nous? For impassivity belongs to Nous and intel-

---

[5] 51.17 Punctuating, with Barocius and Schönberger, to close the parenthesis with $\phi\eta\sigma\iota$, not with $\kappa a\theta\dot{o}\lambda ov$ as Friedlein does. Aristotle distinguishes (*Met.* 1036a9-12) between $\ddot{v}\lambda\eta$ $a\iota\sigma\theta\eta\tau\dot{\eta}$ and $\ddot{v}\lambda\eta$ $vo\eta\tau\dot{\eta}$; but Proclus' $\ddot{v}\lambda\eta$ $\phi av\tau a\sigma\tau\hat{\omega}v$ is justified, since Aristotle elsewhere (*De An.* 433a10) assumes that $\phi av\tau a\sigma\iota a$ is a form of $vo\eta\sigma\iota s$.

[6] 52.3 The reference is to Aristotle; cf. *De An.* 430a24. For a similar interpretation of Aristotle see Proclus' *Commentary on the Timaeus* I, 244.20, and III, 158.9, Diehl.

lectual nature, whereas whatever can be affected ($\pi\alpha\theta\eta\tau\iota\kappa\acute{o}\nu$) is far removed from that highest being. But I think he intended rather to express the middle position it occupies between the highest and the lowest types of knowledge and so called it at the same time "nous," because it resembles the highest, and "passive," because of its kinship with the lowest. For the knowing which is not of shapes and figures has its intelligible objects in itself, and its activity is concerned with these, its own contents; it is itself one with the things it knows, free of any impression or affection coming from elsewhere. But the lowest forms of knowledge work through the sense organs; they are more like affections, receiving their opinions from without and changing as their objects change. Such is what sense-perception is, the result of "violent affections," as Plato says.[7] By contrast the imagination, occupying the central position in the scale of knowing, is moved by itself to put forth what it knows, but because it is not outside the body, when it draws its objects out of the undivided center of its life, it expresses them in the medium of division, extension, and figure. For this reason everything that it thinks is a picture or a shape of its thought. It thinks the circle as extended, and although this circle is free of external matter, it possesses an intelligible matter provided by the imagination itself. This is why there is more than one circle in the imagination, as there is more than one circle in the sense world; for with extension there appear also differences in size and number among circles and triangles.

If, then, in sensible circles there is a universal that makes each of them a circle and all of them similar to one another because conformed to a single idea, yet differing in size and in their underlying subjects, so likewise in imaginary circles there is a common element in which they participate by virtue of which they all have the same form. They differ only on one point, their imagined sizes.[8] For if you imagine several concentric circles, they will all have their existence in a single

53

[7] 52.20   *Tim.* 42a.

[8] 53.12   Proclus is thinking of concentric circles, as the following sentence shows.

immaterial substratum and their life inseparable from a simple body that surpasses indivisible being only by being extended; but they will differ from one another in that some will be larger, some smaller, some encircling, and some encircled.

Let us, then, think of the universal in its instances as of two sorts, the universal in sense objects and the universal in objects of imagination, and likewise the idea of the circle—or of the triangle or of figure itself—as of two kinds, one presiding over intelligible matter, the other over perceptible. Prior to both, as we have seen, are the idea in the understanding and the idea in nature, the former the support of imagined circles and of the single form in them, the latter of perceived circles, such as the circles in the heavens and all circles generated by nature. As the idea in the understanding is undivided, so also is the idea in nature. For extended things exist without extension in the realm of immaterial causes, and divided things without division and magnitudes without magnitude, just as in the opposite direction indivisible things are divided and objects without magnitude have magnitude in the region of material causes. For this reason the circle in the understanding is one and simple and unextended, and magnitude itself is without magnitude there, and figure without shape;[9] for such objects in the understanding are ideas devoid of matter. But the circle in imagination is divisible, formed, extended—not one only, but one and many, and not a form only, but a form in instances—whereas the circle in sensible things is inferior in precision, infected with straightness, and falls short of the purity of immaterial circles.

When, therefore, geometry says something about the circle or its diameter, or about its accidental characteristics, such as tangents to it or segments of it and the like, let us not say that it is instructing us either about the circles in the sense world, for it attempts to abstract from them, or about the form in the understanding. For the circle [in the understanding] is one, yet geometry speaks of many circles, setting them forth individually and studying the identical features in all of them; and

54

---

⁹ 54.8 Punctuating with Barocius and Schönberger to close the parenthesis with τοιαῦτα, not with ἀσχημάτιστον as in Friedlein.

that circle [in the understanding] is indivisible, yet the circle in geometry is divisible. Nevertheless we must grant the geometer that he is investigating the universal, only this universal is obviously the universal present in the imagined circles. Thus while he sees one circle [the circle in imagination], he is studying another, the circle in the understanding, yet he makes his demonstrations about the former. For the understanding contains the ideas but, being unable to see them when they are wrapped up, unfolds and exposes them and presents them to the imagination sitting in the vestibule; and in imagination, or with its aid, it explicates its knowledge of them, happy in their separation from sensible things and finding in the matter of imagination a medium apt for receiving its forms.[10]

55

Thus thinking in geometry occurs with the aid of the imagination. Its syntheses and divisions of the figures are imaginary; and its knowing, though on the way to understandable being, still does not reach it, since the understanding is looking at things outside itself. At the same time the understanding sees them by virtue of what it has within; and though employing projections of its ideas, it is moved by itself to make them external. But if it should ever be able to roll up its extensions and figures and view their plurality as a unity without figure, then in turning back to itself it would obtain a superior vision of the partless, unextended, and essential geometrical ideas that constitute its equipment. This achievement would itself

---

[10] 55.6   The brackets in this passage indicate words not in Proclus which I have inserted for the purpose of clarifying the pronouns in his text. I.M. raises the pertinent question whether for Proclus mathematical reasoning is about universals or about the pictures in the imagination. I should reply that it is certainly about universals, but about universals grasped by means of pictures in the imagination. Obviously no picture in the imagination is a universal; but such pictures enable us to understand the unpicturable universal in its variety and complexity. This view of universals is not foreign to Plato, but it may well be so to Aristotle; his polemic against the Platonic Ideas suggests that it was. But whether or not Aristotle understood Plato's conception of the universal, there is no doubt that Proclus interpreted it in this way and that this is what he regarded as the eventual object of mathematical reasoning. (On the identity of I.M. see the Preface.)

be the perfect culmination of geometrical inquiry, truly a gift of Hermes, leading geometry out of Calypso's arms,[11] so to speak, to more perfect intellectual insight and emancipating it from the pictures projected in imagination.[12]

56

Every true geometer should cultivate such efforts and make it his goal to arouse himself to move from imagination to pure and unalloyed understanding, thus rescuing himself from extension and "passive nous" for the dianoetic activity that will enable him to see all things without parts or intervals—the circle, the diameter, the polygons in the circle, all in all and each separately. This is why even in our imagination we show circles as inscribed in polygons and polygons as inscribed in circles, in imitation of the proof that the partless ideas exist in and through one another. And this is why we use diagrams to illustrate the structure and construction of figures, their divisions, positions, and juxtapositions. We invoke the imagination and the intervals that it furnishes, since the form itself is without motion or genesis, indivisible and free of all underlying matter, though the elements latent in the form are produced distinctly and individually on the screen of imagination. What projects the images is the understanding; the source of what is projected is the form in the understanding; and what they are projected in is this "passive nous" that unfolds in revolution about the partlessness of genuine Nous, setting a distance between itself and that indivisible source of pure thought, shaping itself after the unshaped forms, and becoming all the things that constitute the understanding and the unitary ideas in us.

So much for what we have to say about the matter of geometry. We are not unaware of what the philosopher Porphyry[13] in his *Miscellaneous Inquiries* and most of the

[11] 55.21 Cf. *Odyssey* V, 55-147, where Hermes conveys to the nymph Calypso the gods' command that she release Odysseus and send him on his way homeward.

[12] 55.23 Nicolai Hartmann (*Des Proclus Diadochus Philosophische Anfangsgründe der Mathematik*, Giessen, 1909, 35) sees Proclus in this passage anticipating Descartes' analytic geometry.

[13] 56.24 Porphyry of Tyre was a devoted disciple of Plotinus (see 21.21). He wrote a *Life of Plotinus* and numerous commentaries on

57 Platonists have set forth, but we believe that what we have said is more in agreement with the principles of geometry and with Plato's declaration that the objects of geometry are understandables. These [principles and Plato's declaration] are in harmony with each other because, although the causes of the geometrical forms in accordance with which the understanding projects its demonstrations about them exist previously in the understanding, the several figures that are divided and compounded are projections in the imagination.

Chap. II
The Objects
and Methods
of Geometri-
cal Science

Let us next speak of the science itself that investigates these forms. Magnitudes, figures and their boundaries, and the ratios that are found in them, as well as their properties, their various positions and motions—these are what geometry studies, proceeding from the partless point down to solid bodies, whose many species and differences it explores, then following the reverse path from the more complex objects to the simpler ones and their principles. It makes use of synthesis and analysis, always starting from hypotheses and first principles that it obtains from the science above it and employing all the procedures of dialectic—definition and division for establishing first principles and articulating species and genera, and demonstrations and analyses in dealing with the consequences that follow from first principles, in order to show the more complex matters both as proceeding from the simpler and also conversely as leading back to them. It treats in one part the definitions of its objects, in another the axioms and the postulates that are the starting-points of its demonstrations,

58 and in another the demonstrations of the properties that belong essentially to its objects. Each science has its own class of things that concern it and whose properties it proposes to investigate, and also its own peculiar principles that it uses in demonstration; and the essential properties likewise differ in the various sciences. The axioms are common to all sci-

---

Plato, Aristotle, and other philosophers, most of which have disappeared. Among them was a commentary on Euclid, to which we have references later in Proclus' text. He is best known now for his *Introduction to the Categories of Aristotle*, which became an important medieval textbook in logic and which is extant.

ences, although each uses them in the fashion appropriate to its own subject-matter; but the genus studied and its essential properties are peculiar to each science.

Among the objects of geometrical inquiry are triangles, squares, circles, figures, and magnitudes in general and their boundaries; others are properties inherent in them, their parts, ratios, and contacts, their equalities, excesses, and deficiencies when laid alongside one another, and all such matters; still others are the postulates and axioms through which all these are demonstrated—for example, that it be permitted to draw a straight line from any point to any other, and that if equals be taken from equals the results are equal, and their consequences. Hence not every problem or question is a geometrical one, but only those that arise out of the principles of geometry; and anyone who is refuted on these principles would be refuted as a geometer; arguments not based on them are not geometrical, but ungeometrical. The latter are of two kinds: either they proceed from premises altogether unlike those of geometry, such as a question in music, which we say is ungeometrical because it arises from hypotheses quite different from the principles of geometry; or they use geometrical principles but in a perverse sense, as when it is asserted that parallel lines meet. Hence geometry also furnishes criteria whereby we can discriminate between statements that follow from its principles and those that depart from them. The various tropes for refuting fallacies when they occur have this function.

Geometrical principles yield consequences different from those that follow from arithmetical ones. And why speak of the[14] . . . ? They are far inferior to these. For one science is more accurate than another, as Aristotle says;[15] that is, a science that starts from simpler principles than one whose starting-point is more complex, or one that states why a fact

[14] 59.9 Between περὶ τῶν and πάμπολυ a few words have been lost, but the sequence of thought is clear. The same disturbing influence is seen in lines 11-12, which express the exact opposite of what Proclus must have written. I have followed Barocius in my translation here.
[15] 59.11 *Post. Anal.* 87a31-37.

is so than one which says that it is so, or a science concerned with intelligibles than one that applies to objects in the sense world. According to these criteria of exactness, arithmetic is more precise than geometry, for its principles are simpler. A unit has no position, but a point has; and geometry includes among its principles the point with position, while arithmetic posits the unit. Likewise geometry is superior to spherics and arithmetic to music, for in general they furnish the principles of the theorems subordinate to them. And geometry is superior to mechanics and optics, for the latter discourse about objects in the sense world.

60

The principles of arithmetic and geometry, then, differ from those of the other sciences, yet their own hypotheses are distinct from each other, in the sense mentioned above; nevertheless they have a certain community with one another, so that some theorems demonstrated are common to the two sciences, while others are peculiar to the one or the other. The statement that every ratio is expressible belongs to arithmetic only and not at all to geometry, for geometry contains inexpressible ratios.[16] Likewise the principle that the gnomons into which a square can be divided have a lower limit in magnitude is peculiar to arithmetic;[17] in geometry a least magnitude has no place at all. Peculiar to geometry are the propositions regarding position (for numbers do not have position), the propositions about contacts (for contacts occur only when there are continuous magnitudes), and the propositions about irrationals (for the irrational has a place only where infinite divisibility is possible). Common to both sciences are the theorems regarding sections (such as Euclid presents in his second book), with the exception of the division of a line in extreme and mean ratio. Of these common theorems some[18] have come to arithmetic from geometry, others from arithmetic to geometry, while others are equally

---

[16] 60.9 On ἄρρητοι λόγοι see note at 48.12.

[17] 60.11 On the Pythagorean use of the gnomon in figured numbers see Heath I, 76-84, and *Euclid* I, 370f.

[18] 60.20 Reading with Barocius τὰ for τὸ in Friedlein.

at home in both because derived by them from general mathematics. The principles governing alternation, conversion, composition, and division of ratios are thus shared by both. The theory of commensurable magnitudes is developed primarily by arithmetic and then by geometry in imitation of it. This is why both sciences[19] define commensurable magnitudes as those which have to one another the ratio of a number to a number, and this implies that commensurability exists primarily in numbers. For where there is number there also is commensurability, and where commensurability there also number. The properties of the triangle and the square are studied primarily by geometry, but arithmetic borrows them and uses them analogically, for figures are contained in numbers as in their causes. Thus in seeking the causes of certain results we turn to numbers, both when we see precisely the same properties, such as that every polygon can be divided into triangles, and when we are content with approximations, as when, in geometry we have found a square double a given square but do not have it in numbers, we say that a square number is the double of another square number when it is short by one, like the square of seven, which is one less than the double of the square of five.[20]

We have carried rather far this exposition of the community between the principles of these two sciences and their differences. For the geometer must understand what common first principles are required for their common theorems and what are the principles from which their special theorems are derived, so that he may distinguish between geometrical matters and those that do not belong to geometry, assigning some to one science, some to the other.

[19] 61.1 Reading with Grynaeus and Schönberger τούτων instead of τούτῳ in Friedlein.

[20] 61.17 In *Rep.* 546c Plato refers to "rational and irrational (ἄρρητοι) diameters of five" (i.e. diagonals of a square having five for its side) and says that the square on the rational diameter is less by one than that on the irrational diameter. The rational diameter is therefore 7, the irrational $\sqrt{50}$.

Let us now turn back for another look at the science of geometry as a whole, to see what its starting-point is and how far it ranges from it, so as to get a view of the ordered cosmos of its ideas. Let us note that it is coextensive with all existing things, applies its reasonings to them all, and includes all their kinds in itself. At the upper and most intellectual height it looks around upon the region of genuine being, teaching us through images the special properties of the divine orders and the powers of the intellectual forms, for it contains even the ideas of these beings within its range of vision. Here it shows us what figures are appropriate to the gods, which ones belong to primary beings and which ones to the substance of souls. In the middle regions of knowledge it unfolds and develops the ideas that are in the understanding; it investigates their variety, exhibiting their modes of existence and their properties, their similarities and differences; and the forms of figures shaped from them in imagination it comprehends within fixed boundaries and refers back to the essential being of the ideas. At the third level of mental exploration it examines nature, that is, the species of elementary perceptible bodies and the powers associated with them, and explains how their causes are contained in advance in its own ideas. It contains likenesses of all intelligible kinds and paradigms of sensible ones; but the forms of the understanding constitute its essence, and through this middle region it ranges upwards and downwards to everything that is or comes to be. Always philosophizing about being in the manner of geometry, it has not only ideas but pictures of all the virtues—intellectual, moral, and physical—and presents in due order all the forms of political constitution, showing from its own nature the variety of the revolutions they undergo.[21]

In these areas its activity is immaterial and theoretical; but when it touches on the material world it delivers out of itself a variety of sciences—such as geodesy, mechanics, and optics—by which it benefits the life of mortals. Through these sciences

[21] 63.5   This is probably a reference to Plato's analogy between the state and the individual soul, which provides the premise for his theory of the successive stages of political degeneration in the *Republic*.

it has devised instruments of war and defenses for our cities, made familiar the succession of the seasons and the lie of various regions, taught how to measure distances by land or sea, constructed balances and scales for determining arithmetical equality when a city needs it, invented models for exhibiting the order of the whole heaven, and many things incredible to men it has unveiled and made credible to all. Recall what Hieron of Syracuse is said to have remarked about Archimedes, who had built a three-masted vessel which Hieron had ordered made for sending to King Ptolemy of Egypt. When all the Syracusans together were unable to launch it and Archimedes made it possible for Hieron alone to move it down to the shore, he exclaimed, in his amazement: "From this day forth we must believe everything that Archimedes says." Tradition has it that Gelon made the same remark when, without destroying the crown that had been made, Archimedes discovered the weight of each of its component materials. Many of our predecessors have recorded such things in praise of mathematics, and for this reason we have presented here only a few of the many facts we might have cited to show the range and utility of geometrical knowledge.

64

Chap. IV
The Origin
and Development of
Geometry

Next we must speak of the development of this science during the present era. The inspired Aristotle[22] has said that the same beliefs have often recurred to men at certain regular periods in the world's history; the sciences did not arise for the first time among us or among the men of whom we know, but at countless other cycles in the past they have appeared and vanished and will do so in the future. But limiting our investigation to the origin of the arts and sciences in the present age, we say, as have most writers of history,[23] that

[22] 64.9 ὁ δαιμόνιος Ἀριστοτέλης; so also at 76.8, 116.24, 284.23. See *De Caelo* 270b19, *Pol.* 1329b25; also Plato, *Tim.* 22-23, *Critias* 109d, *Laws* 677b.

[23] 64.19 Herodotus II, 109; Diod. Sic. I, lxix, 5; lxxxi, 1-2; Strabo XVII, Chap. 3. Aristotle (*Met.* 981b23) credits Egypt with being the birthplace of geometry but assigns a different cause, viz. the leisure enjoyed by the priestly class. On Egyptian geometry see Heath I, 121-128; Van der Waerden, 15-36; Gow, 124-133.

geometry was first discovered among the Egyptians and originated in the remeasuring of their lands. This was necessary
for them because the Nile overflows and obliterates the boundary lines between their properties. It is not surprising that
the discovery of this and the other sciences had its origin in
necessity, since everything in the world of generation proceeds from imperfection to perfection. Thus they would naturally pass from sense-perception to calculation and from
calculation to reason. Just as among the Phoenicians the
necessities of trade and exchange gave the impetus to the
accurate study of number, so also among the Egyptians the
invention of geometry came about from the cause mentioned.

65

Thales,[24] who had travelled to Egypt, was the first to introduce this science into Greece. He made many discoveries
himself and taught the principles for many others to his successors, attacking some problems in a general way and others
more empirically. Next after him Mamercus,[25] brother of the
poet Stesichorus, is remembered as having applied himself to
the study of geometry; and Hippias of Elis[26] records that he
acquired a reputation in it. Following upon these men, Pythag-

[24] 65.7   The following account of the development of geometry
among the Greeks appears to be based on a history composed by
Eudemus of Rhodes, a pupil of Aristotle, which Proclus had at his
disposal but which has since been lost. For the evidence see Heath I,
118-120; and for Proclus' use of this source, *Euclid* I, 35-38. Thales of
Miletus lived in the early sixth century B.C. and was universally
counted as one of the Seven Sages. Since he wrote nothing, our knowledge of his geometrical discoveries is dependent on the traditions about
him recorded by later writers. Some of this evidence comes from
Eudemus, through Proclus; see 157.11, 250.20, 299.4, 352.15. For
estimates of his achievements see Heath I, 130-137; Gow, 138-145;
Van der Waerden, 85-90.

[25] 65.12   Of Mamercus nothing is known beyond this mention;
even his name is uncertain, for the MSS of Proclus contain variant
readings "Ameristus" and "Mamertius." Stesichorus belongs to the
late seventh and early sixth centuries.

[26] 65.14   Hippias of Elis, the famous Sophist of the fifth century
B.C., the inventor of a curve known as the quadratrix which, originally
intended for the solution of the problem of trisecting any angle, also
served (as the name implies) for squaring the circle (272.7, 356.11).
See Heath I, 23, 182; Gow, 162-164; Van der Waerden, 146.

oras[27] transformed mathematical philosophy into a scheme of liberal education, surveying its principles from the highest downwards and investigating its theorems in an immaterial[28] and intellectual manner. He it was who discovered the doctrine of proportionals[29] and the structure of the cosmic figures.[30] After him Anaxagoras of Clazomenae[31] applied himself to many questions in geometry, and so did Oenopides of Chios,[32] who was a little younger than Anaxagoras. Both these men are mentioned by Plato in the *Erastae*[33] as having got a

66

[27] 65.16 Pythagoras of Samos, the founder of a school of philosophy and mathematics in southern Italy in the fifth century B.C. Pythagoras' own contributions are difficult to identify, since he left no writings, and all the discoveries of the school are credited to him (see note at 22.14 above). But he influenced Plato profoundly and through him all later Greek science and philosophy down to the time of Proclus. For a survey of Pythagorean achievements in arithmetic see Heath I, 65-117; and for Pythagorean geometry, 141-169. See also Van der Waerden, 92-105, and Gow, 147-158.

[28] 65.18 ἀύλως, i.e. in abstraction from sensible things, but surely not "without concrete representation," as Van der Waerden (90) translates it.

[29] 65.19 Reading ἀναλόγων for ἀλόγων in Friedlein. See Heath I, 84f.

[30] 65.20 I.e. the five regular solids. For the controversies concerning the contribution of Pythagoras or the early Pythagoreans to the development of the theory of the regular solids see Heath I, 158-162, and Kurt von Fritz, in *RE, s.v.* "Theaitetos."

[31] 65.21 Anaxagoras of Clazomenae lived at Athens during the first half of the fifth century B.C., where his ideas made a great stir and eventually brought about his indictment for impiety ánd his withdrawal to Lampsacus. We know practically nothing of his achievements in geometry, though the fragments of his book *On Nature* show that he was a theoretical scientist of extraordinary ability. He used the idea of infinite divisibility in his cosmology and was the first to give the true explanation of lunar and solar eclipses.

[32] 66.2 Two propositions in Euclid's first book are attributed by Proclus to Oenopides of Chios, viz. XII (283.7) and XXIII (333.5). These are very simple problems, and it is likely that his importance in the history of geometry is due rather to improvements in method that he instituted, such as the rule limiting construction to the use of the ruler and compass (see Heath I, 175). Von Fritz (in *RE, s.v.* "Oinopides") attributes to him also the recognition of the problem as a kind of theoretical inquiry distinct from the theorem (see note on Zenodotus at 80.15).

[33] 66.3 *Erastae* 132a.

reputation in mathematics. Following them Hippocrates of Chios,[34] who invented the method of squaring lunules, and Theodorus of Cyrene[35] became eminent in geometry. For Hippocrates wrote a book on elements, the first of whom we have any record who did so.

Plato, who appeared after them, greatly advanced mathematics in general and geometry in particular because of his zeal for these studies. It is well known that his writings are thickly sprinkled with mathematical terms and that he everywhere tries to arouse admiration for mathematics among students of philosophy. At this time also lived Leodamas of Thasos,[36] Archytas of Tarentum, and Theaetetus of Athens,[37] by whom the theorems were increased in number and brought into a more scientific arrangement. Younger than Leodamas

[34] 66.4   Hippocrates of Chios, a contemporary and fellow-citizen of Oenopides. Proclus tells us later (213.3-11) that he reduced the problem of duplicating the cube to that of finding two mean proportionals. Besides this achievement he is credited with having effected the quadrature of lunes and with having proved that the areas of circles are proportional to the squares on their diameters. See Heath I, 182-209; Gow, 164-172; Van der Waerden, 131-136. Part of the actual text of Hippocrates' quadrature of lunes is preserved in Simplicius' *Commentary on Aristotle's Physics, CAG* IX, 60.22-68.32, Diels.

[35] 66.6   Theodorus of Cyrene is said to have been the teacher of Plato (Diog. Laert. III, 6). The *Theaetetus* of Plato has him present in Athens during the last period of Socrates' life. This dialogue in fact contains all the ancient evidence about him that is available; but for its significance see von Fritz, in *RE, s.v.* "Theodoros."

[36] 66.15   Proclus tells us later (211.19-23) that Plato is said to have taught Leodamas the method of analysis. Cf. Diog. Laert. III, 24. Archytas was an older contemporary and friend of Plato (cf. Plato's *Epistle VII* 338c, 350a), eminent as statesman and philosopher and as author of the first treatise on mechanics based on mathematical principles. He solved the problem of finding two mean proportionals by a remarkable construction in three dimensions (Heath I, 213-214, 246-249). "It is perhaps worth pointing out that Van der Waerden (153) attributes Book VIII of the *Elements* to Archytas." (I.M.)

[37] 66.16   Theaetetus was one of the two greatest mathematicians of the fourth century B.C. He laid the foundations of the theory of irrationals as we find it in Euclid's tenth book and distinguished their main varieties; he also contributed substantially to the theory of the five regular solids developed in Euclid's thirteenth book. See Heath I, 209-212, and von Fritz, in *RE, s.v.* "Theaitetos."

were Neoclides and his pupil Leon,[38] who added many discoveries to those of their predecessors, so that Leon was able to compile a book of elements more carefully designed to take account of the number of propositions that had been proved and of their utility. He also discovered *diorismi*, whose purpose is to determine when a problem under investigation is capable of solution and when it is not. Eudoxus of Cnidus,[39] a little later than Leon and a member of Plato's group, was the first to increase the number of the so-called general theorems;[40] to the three proportionals already known he added three more and multiplied the number of propositions concerning the "section"[41] which had their origin in Plato, employing the method of analysis for their solution. Amyclas of Heracleia,[42] one of Plato's followers, Menaechmus,[43] a student of Eudoxus who also was associated with Plato, and

67

---

[38] 66.19   Of these men we know nothing more than is here stated (Heath I, 319).

[39] 67.2   Eudoxus ranks with Theaetetus among the greatest mathematicians of the fourth century. In astronomy he was the author of the theory of concentric spheres for explaining the motions of the heavenly bodies. His great contributions to geometry were the new theory of proportion expounded in Euclid V and VI, the method of exhaustion for measuring and comparing the areas and volumes of curvilinear plane and solid surfaces, and the solution of the problem of doubling the cube. See Heath I, 322-334; Gow, 183-185; Van der Waerden, 179-190.

[40] 67.4   It is a disputed question what these καθόλου θεωρήματα are. Theorems true of everything falling under the concept of magnitude? Or axioms, such as those underlying the reasoning of Euclid V and VI? "I am inclined to think that the words refer primarily to the contents of Book V of the *Elements*, i.e. the Eudoxian theory of proportion." (I.M.) See Heath I, 323f., and Van der Waerden, 183.

[41] 67.6   Does this refer to the sectioning of solids by planes or the sectioning of a straight line in extreme and mean ratio? See Heath I, 324f.

[42] 67.8   Amyclas is otherwise unknown.

[43] 67.9   Menaechmus, as a pupil of Eudoxus and of Plato, must have lived in the fourth century. From the saying of Eratosthenes quoted at 111.22f. it is generally inferred that Menaechmus discovered the conic sections. Proclus' other references to him indicate that he wrote also on the methodology of mathematics; cf. 72.24ff., 78.9ff., 254.4. See Heath I, 251-255; Van der Waerden, 190f.; Gow, 185-187.

his brother Dinostratus[44] made the whole of geometry still more perfect. Theudius of Magnesia[45] had a reputation for excellence in mathematics as in the rest of philosophy, for he produced an admirable arrangement of the elements and made many partial theorems more general.[46] There was also Athenaeus of Cyzicus,[47] who lived about this time and became eminent in other branches of mathematics and most of all in geometry. These men lived together in the Academy, making their inquiries in common. Hermotimus of Colophon pursued further the investigations already begun by Eudoxus and Theaetetus, discovered many propositions in the *Elements,* and wrote some things about locus-theorems. Philippus of Mende,[48] a pupil whom Plato had encouraged to study mathematics, also carried on his investigations according to Plato's instructions and set himself to study all the problems that he thought would contribute to Plato's philosophy.

All those who have written histories bring to this point their account of the development of this science. Not long after these men came Euclid, who brought together the *Elements,* systematizing many of the theorems of Eudoxus, perfecting many of those of Theaetetus, and putting in irrefutable demonstrable form propositions that had been rather loosely established by his predecessors. He lived in the time of Ptolemy the First, for Archimedes, who lived after the time of the first Ptolemy, mentions Euclid. It is also reported that

68

[44] 67.11   Dinostratus applied Hippias' quadratrix to the squaring of the circle. Heath I, 225-230; Van der Waerden, 191f.

[45] 67.12   From the fact that Theudius was a member of the Academy in Plato's time it has been inferred that the propositions in elementary geometry cited by Aristotle come from his *Elements* (Heath I, 321); but see von Fritz, in *RE, s.v.* "Theudius."

[46] 67.15   Reading with Barocius μερικῶν instead of ὁρικῶν. See von Fritz, *loc.cit.*

[47] 67.16   Athenaeus of Cyzicus is otherwise unknown; and so also is Hermotimus of Colophon, mentioned in line 20 below.

[48] 67.23   Philippus of Mende is undoubtedly the same as the Philippus of Opus who edited and published Plato's *Laws* and who is said to have been the author of the *Epinomis* (cf. 42.12). He wrote numerous works, chiefly on astronomy, but also some mathematical treatises whose titles are preserved. See von Fritz, in *RE, s.v.* "Philippos."

Ptolemy once asked Euclid if there was not a shorter road to geometry than through the *Elements*, and Euclid replied that there was no royal road to geometry. He was therefore later than Plato's group but earlier than Eratosthenes[49] and Archimedes, for these two men were contemporaries, as Eratosthenes somewhere says. Euclid belonged to the persuasion of Plato and was at home in this philosophy; and this is why he thought the goal of the *Elements* as a whole to be the construction of the so-called Platonic figures.

There are many other mathematical writings of Euclid, full of remarkable precision and scientific insight. Such are his *Optics*, his *Catoptrics*, his *Elements of Music*, and his little book on *Divisions*. But we should especially admire him for the work on the elements of geometry because of its arrangement and the choice of theorems and problems that are worked out for the instruction of beginners. He did not bring in everything that he could have collected, but only what could serve as an introduction. He also included reasonings of all sorts, both proofs founded on causes and proofs based on signs,[50] but all of them impeccable, exact, and appropriate to science. Besides these the book contains all the dialectical methods: the method of division for finding kinds, definitions for making statements of essential properties, demonstrations for proceeding from premises to conclusions, and analysis for passing in the reverse direction from conclusions to principles. The various forms of conversion, both the simple and the more complex, can be accurately learned in this treatise.[51] One sees when conclusion and hypothesis can be interchanged as wholes, when the whole with a part and a part with the whole are interchangeable, and when only a part with a part. We mark also the coherence of its results, the economy and orderliness in its arrangement of primary and corollary propositions, and the cogency with which all the several parts are presented. Indeed, if you add or take away any detail whatever, are you not inadvertently leaving the way of science and

69
Chap. V
Euclid's
Mathematical Works

---

[49] 68.18   On Eratosthenes and Archimedes see 41.6 and 43.23.
[50] 69.12   On this distinction see 206.15.
[51] 69.22   On geometrical conversion see 252-254, 409.1-6.

70

being led down the opposite path of error and ignorance? Since there are many matters that seem to be dependent on truth and to follow from scientific principles but really lead away from them and deceive the more superficial students, he has given us methods for clear-sighted detection of such errors; and if we are in possession of these methods, we can train beginners in this science for the discovery of paralogisms and also protect ourselves from being led astray. The work in which he teaches us this apparatus he entitled *Fallacies*. It enumerates in order the various methods of refutation[52] and for each of them provides exercise for our understandings by a variety of theorems, setting the true beside the false and adapting his refutations of error to the seductions we may encounter. This book is cathartic and gymnastic, while the *Elements* contains an impeccable and complete exposition of the science itself of geometrical matters.

Chap. VI
The Purpose
of the
Elements

If now anyone should ask what the aim of this treatise is, I should reply by distinguishing between its purpose as judged by the matters investigated and its purpose with reference to the learner. Looking at its subject-matter, we assert that the whole of the geometer's discourse is obviously concerned with the cosmic figures. It starts from the simple figures and ends with the complexities involved in the structure of the cosmic bodies, establishing each of the figures separately but showing for all of them how they are inscribed in the sphere and the

71

ratios that they have with respect to one another. Hence some[53] have thought it proper to interpret with reference to the cosmos the purposes of individual books and have inscribed above each of them the utility it has for a knowledge of the universe. Of the purpose of the work with reference to the student we shall say that it is to lay before him an elementary exposition ($\sigma\tau o\iota\chi\epsilon\iota\omega\sigma\iota\varsigma$, as it is called) and a method of perfecting ($\tau\epsilon\lambda\epsilon\iota\omega\sigma\iota\varsigma$) his understanding for the whole of geometry. If we start from the elements, we shall be able to understand the other parts of this science; without the elements we cannot grasp its complexity, and the learning of the rest will

[52] 70.11 $\tau\rho\delta\pi o\iota$. See 59.5.
[53] 71.3 Proclus is presumably referring to editors of the *Elements*.

be beyond us. The theorems that are simplest and most fundamental and nearest to first principles are assembled here in a
suitable order, and the demonstrations of other propositions
take them as the most clearly known and proceed from them.
In this way also Archimedes in his book on *Sphere and Cylinder* and likewise Apollonius[54] and all other geometers appear
to use the theorems demonstrated in this very work as generally accepted starting-points. This, then, is its aim: both to
furnish the learner with an introduction to the science as a
whole and to present the construction of the several cosmic
figures.

But—to inquire briefly about its title—what is the meaning
of this very word στοιχείωσις and of the word στοιχεῖον from
which it is derived? Some theorems we are accustomed to call
"elements" (στοιχεῖα), others "elementary" (στοιχειώδη), and
others do not qualify for either designation. We call "elements" those theorems whose understanding leads to the
knowledge of the rest and by which the difficulties in them are
resolved. As in written language there are certain primal
elements, simple and indivisible, to which we give the name
στοιχεῖα[55] and out of which every word is constructed, and
every sentence, so also in geometry as a whole there are
certain primary theorems that have the rank of starting-points
for the theorems that follow, being implicated in them all and
providing demonstrations for many conjunctions of qualities;
and these we call "elements." "Elementary" propositions are
those that are simple and elegant and have a variety of applications but do not rank as elements because the knowledge of
them is not pertinent to the whole of the science: for example,

72

Chap. VII
The Mean-
ing of "Ele-
ment"

[54] 71.19 Apollonius of Perga, in Pamphylia, belongs to the latter
half of the third century B.C. His monumental treatise on *Conics* is
one of the most imposing productions of ancient mathematics and
earned for him in antiquity the title of the "great geometer." It con-
sisted of eight books, of which the first four survive in Greek (ed.
Heiberg, 1891-1893), and the next three in an Arabic version, the
eighth having completely disappeared. See Heath II, 126-196; Van
der Waerden, 237-263; Gow, 246-264. Almost nothing is known of
his life.

[55] 72.8 One of the many uses of the word στοιχεῖα was to designate the letters of the alphabet.

the theorem that the perpendiculars from the vertices of a triangle to the sides meet in a common point. Propositions whose understanding is not relevant to a multitude of others or which do not exhibit any grace or elegance—these do not have the force of elementary propositions. The term "element," however, can be used in two senses, as Menaechmus tells us. For what proves is called an element of what is proved by it; thus in Euclid the first theorem is an element of the second, and the fourth of the fifth. In this sense many propositions can be called elements of one another, when they can be established reciprocally. From the proposition that the exterior angles of a rectilinear figure are equal to four right angles we can prove the number of right angles to which the interior angles of the figure are equal, and vice versa. An element so regarded is a kind of lemma.[56] But in another sense "element" means a simpler part into which a compound can be analyzed. In this sense not everything can be called an element of anything [that follows from it], but only the more primary members of an argument leading to a conclusion, as postulates are elements of theorems. This is the sense of "element" that determines the arrangement of the elements in Euclid's work, some of them being elements of plane geometry, and some elements of stereometry. This also is the meaning the word has in numerous compositions in arithmetic and astronomy entitled "elementary treatises" (στοιχειώσεις).

It is a difficult task in any science to select and arrange properly the elements out of which all other matters are produced and into which they can be resolved. Of those who have attempted it[57] some have brought together more theorems, some less; some have used rather short demonstrations, others have extended their treatment to great lengths; some have avoided the reduction to impossibility, others proportion; some have devised defenses in advance against attacks upon the starting-points; and in general many ways of constructing elementary expositions have been individually invented. Such

73

[56] 73.4   For Proclus' explanation of lemma see 211.1ff.
[57] 73.18   *Sc.* for geometry.

a treatise ought to be free of everything superfluous, for that
is a hindrance to learning; the selections chosen must all be
coherent and conducive to the end proposed, in order to be of
the greatest usefulness for knowledge; it must devote great attention both to clarity and to conciseness, for what lacks these
qualities confuses our understanding; it ought to aim at the
comprehension of its theorems in a general form, for dividing
one's subject too minutely and teaching it by bits make knowledge of it difficult to attain. Judged by all these criteria, you
will find Euclid's introduction superior to others. Its usefulness contributes to the study of the primary figures;[58] its
method of proceeding from simpler to more complex matters
and its laying the foundations of the science on the "common
notions"[59] produce clarity and articulateness; and by moving
towards the questions under investigation by way of primary
and basic theorems, it makes the demonstration general.
The matters that appear to be omitted either can be
learned through the same methods as those it employs, like
the construction of the scalene and the isosceles triangles; or
they are unsuitable for a selection of elements because they
lead to great and unlimited complexity, such as the material
that Apollonius has elaborated at considerable length about
unordered irrationals;[60] or they can be constructed from traditional premises, such as the many species of angles and
lines. These matters are passed over in this work, and though
they may receive rather fuller treatment in others, they can
be learned from simple premises. So much we thought it desirable to record about the general nature of this elementary
introduction.

[58] 74.13   I surmise that Proclus' text has lost something here and
that what he wrote is that the usefulness of the book for the understanding of the ἀρχικὰ σχήματα contributes to the understanding of the
κοσμικὰ σχήματα, as he says at 83.1f.

[59] 74.15   Κοιναὶ ἔννοιαι occurs frequently in Proclus but is nowhere
defined as a technical term. Cf. also κοιναὶ ἐπίνοιαι (188.12).

[60] 74.23   The Greek text of this book has been lost, but ver Eecke
(ad loc.) notes an attempted restoration of it from an Arabic manuscript by Woepke, in Mémoires présentés à l'Académie des Sciences
xiv, 658-720.

Chap. VIII
The Arrangement
of the
Propositions
in the *Elements*

The general arrangement of its propositions we should explain somewhat, as follows. Since this science of geometry is based, we say, on hypothesis and proves its later propositions from determinate first principles—for there is only one unhypothetical science, the other sciences receiving their first principles from it—he who prepares an introduction to geometry should present separately the principles of the science and the conclusions that follow from the principles, giving no argument for the principles but only for the theorems that are derived from them. For no science demonstrates its own first principles or presents a reason for them; rather each holds them as self-evident, that is, as more evident than their consequences. The science knows them through themselves, and the later propositions through them. This is the way the natural scientist proceeds, positing the existence of motion and producing his ideas from a definite first principle. The same is true of the physician and of the expert in any other science or art. Whoever throws into the same pot his principles and their consequences disarranges his understanding completely by mixing up things that do not belong together.[61] For a principle and what follows from it are by nature different from each other.

76

First of all then, to repeat what I said, it was incumbent on him to set apart the principles from their consequences; and this is just what Euclid does in practically every book, besides setting forth at the outset of his whole treatise the common principles of the science. Next he divides them into hypotheses, postulates, and axioms,[62] for these are all different from each other. Axiom, postulate, and hypothesis are not the same thing, as the inspired Aristotle somewhere says.[63] When a proposition that is to be accepted into the rank of first principles is something both known to the learner and credible in itself, such a proposition is an axiom: for example, that things

---

[61] 75.22 An echo of Plato, *Phaedo* 101e.

[62] 76.6 Note that Proclus describes Euclid as dividing the κοιναί ἀρχαί of geometry into ὑποθέσεις, αἰτήματα, and ἀξιώματα instead of the ὅροι, αἰτήματα, and κοιναί ἔννοιαι of our Euclid text.

[63] 76.8 *Post. Anal.* 76a31-77a4.

equal to the same thing are equal to each other. When the student does not have a self-evident notion of the assertion proposed but nevertheless posits it and thus concedes the point to his teacher, such an assertion is a hypothesis. That a circle is a figure of such-and-such a sort we do not know by a common notion in advance of being taught, but upon hearing it we accept it without a demonstration. Whenever, on the other hand, the statement is unknown and nevertheless is taken as true without the student's conceding it, then, he says, we call it a postulate: for example, that all right angles are equal. This characteristic of postulates is evidenced by the strenuous efforts that have been made to establish one of them,[64] as though nobody could concede it without more ado. In this way axiom, postulate, and hypothesis are distinguished according to Aristotle's teaching. Often, however, they are all called hypotheses, just as the Stoics call every simple statement an axiom,[65] so that according to them even hypotheses are axioms, whereas according to others axioms are hypotheses.

Again the propositions that follow from the first principles he divides into problems and theorems, the former including the construction of figures, the division of them into sections, subtractions from and additions to them, and in general the characters that result from such procedures, and the latter concerned with demonstrating inherent properties belonging to each figure. Just as the productive sciences have some theory in them, so the theoretical ones take on problems in a way analogous to production. Some of the ancients, however, such as the followers of Speusippus and Amphinomus,[66] insisted on calling all propositions "theorems," consider-

[64] 76.21 This appears to be a reference to Post. V and to the attempts made in antiquity to demonstrate it. See 191.23ff. and note.

[65] 77.3 See Diog. Laert. VII, 65; von Arnim II, 62-72; and Benson Mates, *Stoic Logic*, Berkeley, 1961, 18.

[66] 77.16 Speusippus was Plato's nephew and his successor as head of the Academy. Nothing otherwise is known of Amphinomus, who is referred to later at 202.11, 220.9, and 254.4. These references confirm the implication of this passage that he was a contemporary of Speusippus.

ing "theorems" to be a more appropriate designation than "problems" for the objects of the theoretical sciences, especially since these sciences deal with eternal things. There is no coming to be among eternals, and hence a problem has no place here, proposing as it does to bring into being or to make something not previously existing—such as to construct an equilateral triangle, or to describe a square when a straight line is given, or to place a straight line through a given point. Thus it is better, according to them, to say that all these objects exist[67] and that we look on our construction of them not as making, but as understanding them, taking eternal things as if they were in the process of coming to be. Hence we can say that all propositions have a theoretical and not a practical import. Others, on the contrary, such as the mathematicians of the school of Menaechmus, thought it correct to say that all inquiries are problems but that problems are twofold in character: sometimes their aim is to provide something sought for, and at other times to see, with respect to a determinate object, what or of what sort it is, or what quality it has, or what relations it bears to something else. Both parties are right. The school of Speusippus are right because the problems of geometry are of a different sort from those of mechanics, for example, since the latter are concerned with perceptible objects that come to be and undergo all sorts of change. Likewise the followers of Menaechmus are right because the discovery of theorems does not occur without recourse to matter, that is, intelligible matter. In going forth into this matter and shaping it, our ideas are plausibly said to resemble acts of production; for the movement of our thought in projecting its own ideas is a production, we have said, of the figures in our imagination and of their properties. But it is in imagination that the constructions, sectionings, superpositions, comparisons, additions, and subtractions take place, whereas the contents of our understanding all stand fixed, without any generation or change.

There are, then, both geometrical problems and geometrical theorems. But because theory is the predominant element in

78

79

<hr>

[67] 78.4   For ταὐτά in Friedlein read ταῦτα, with Tannery IX, 126.

geometry, as making is in mechanics, every problem has also some theory in it; but the reverse is not true, for demonstrations in general are the product of theory. All the propositions in geometry after the first principles are obtained by demonstration, so that "theorem" is the more general term. And not all theorems require the assistance of problems: there are some which contain in themselves the demonstration of what is sought. Those who distinguish theorem from problem say that every problem admits the possibility of antithetical predicates in its matter—the attribute sought as well as its opposite—whereas a theorem admits only a given attribute, not its antithesis also. (By "matter" here I mean the genus of the thing being studied, such as triangle, square, or circle; by "attribute" I mean something that is by itself accidental, such as "equal," "divided into segments," "in such-and-such a position," or something similar.) When, therefore, we propose to inscribe an equilateral triangle in a circle, we call it a problem, for it is possible to inscribe a triangle that is not equilateral; or again to construct an equilateral triangle on a given finite line is a problem, for it is possible to construct one that is not equilateral. But when a man sets out to prove that the angles at the base of an isosceles triangle are equal, we should say he is proposing a theorem, for it is not possible that the angles at the base of an isosceles triangle should not be equal. Thus if anyone were to set it up as a problem to inscribe a right angle in a semicircle, he would be regarded as being ignorant of geometry, for any angle inscribed in a semicircle is a right angle. In general, then, all cases in which the property is universal, that is, coextensive with the whole of the matter, must be called theorems; but whenever the character is not universal, that is, does not belong to the whole genus of the subject, then it must be called a problem. The proposal to bisect a given finite line is a problem, for it can also be divided into unequal segments; or to bisect a rectilinear angle, for it can be divided unequally; likewise to describe a square from a given line, for we could construct a figure that is not a square. All such questions belong to the class of problems.

80

On the other hand, the followers of Zenodotus,[68] who belonged to the succession of Oenopides, although he was a pupil of Andron, used to distinguish theorem from problem in the sense that a theorem seeks to know what character is attributed to the matter it is investigating, whereas a problem asks under what conditions something exists. Hence the followers of Posidonius[69] likewise distinguished between a proposition that inquires whether or not something exists and one[70] that seeks to know what or of what sort it is, maintaining that the theoretical proposition ought to be stated in declarative form (for example, "In every triangle the sum of two of its sides is greater than the third," or "The angles at the base of an isosceles are equal"), whereas the problematic proposition should be stated as a question (for example, "Is it possible to construct a triangle on this straight line?"). For there is a difference, they said, between simply inquiring in general whether it is possible to erect a perpendicular to this line at this point and investigating the nature of the perpendicular.

It is clear from these considerations that there is a distinction between a problem and a theorem. That Euclid's *Elements* contains both problems and theorems will be evident from the individual propositions and from his practice of placing at the end of his demonstrations sometimes "This is what was to be done" and at other times "This is what was to be proved." The latter is the mark of a theorem, although, as we said, demonstration also occurs in problems; nevertheless

81

---

[68] 80.15 Zenodotus and Andron are otherwise unknown; but this passage is an important part of the evidence that von Fritz presents for the significant contributions to methodology made by Oenopides; see note at 66.2 above.

[69] 80.21 Posidonius of Apamea was head of a school of Stoicism at Rhodes in the late second century B.C. For his contributions to mathematical geography and astronomy see Heath II, 219-222. Proclus appeals to him frequently; cf. 143.8, 170.13, 176.6, 200.2, 216.20, 217.24.

[70] 80.22 Omitting πρόβλημα in Friedlein, since to take it with the immediately following πρότασιν would violate the distinction that Proclus is expounding. It appears to be another marginal note that has got into the text and at a most inappropriate place. See Tannery IX, 126.

sometimes the demonstration is used for the sake of the construction—that is, we bring it in to prove that what was proposed has been done—whereas on other occasions it deserves attention on its own account because it is able to set forth the nature of the object investigated. You will find that Euclid sometimes interweaves theorems with problems, using them alternately, as in the first book; but sometimes one or the other predominates. The fourth book consists entirely of problems, the fifth book of theorems. So much for these matters.

Next we must define the aim of the first book and set forth its several divisions, and then we shall be able to begin the examination of the Definitions. What this book proposes to do is to present the principles of the study of rectilinear figures. Although the circle is naturally superior to the straight line and the study of it a higher form of being and knowledge, yet instruction in the nature of straight lines is more suitable for us who are less than perfect intelligences and are striving to convert our understanding from sensible to intelligible objects. Rectilinear figures are akin to sensibles, but the circle to intelligibles; for what is simple, uniform, and determinate accords with the nature of being, whereas to be diversified and to possess indefinitely more containing sides is a characteristic of sense objects. In this book, therefore, are presented the first and most fundamental rectilinear figures, the triangle and the parallelogram. For these are the genera that include the causal principles of the elements, the isosceles and scalene triangles and their compounds, the equilateral triangle and the square, from which the figures of the four elements are constructed.[71] We shall therefore discover how to construct the equilateral triangle and the square, the one on a given straight line, the other from a given line.[72] The equi-

[71] 82.20 Στοιχεῖα here means the four primary bodies, the elements of the physical world. How these are related to the equilateral triangle and the square is expounded in Plato's *Tim.* 53c-55c.

[72] 82.22 This distinction between the *constructing* of an equilateral triangle *on* a finite line and the *describing* of a square *from* a given line seems to have been traditional among Greek geometers, though the reason for it is hard to see. Proclus observes it consistently (cf.

83

Chap. X
The Divi-
sions of
Book I

lateral triangle[73] is the proximate cause of three of the elements—fire, air, water—and the square the cause of earth.
Consequently the aim of the first book is dependent on the
entire treatise and contributes to the full understanding of
the cosmic elements. Furthermore, it introduces the learner to
the science of rectilinear figures by revealing their first principles and establishing them with precision.

The book is divided into three major parts. The first reveals
the construction of triangles and the special properties of their
angles and their sides, comparing triangles with one another
as well as studying each by itself. Thus it takes a single triangle
and examines now the angles from the standpoint of the sides
and now the sides from the standpoint of the angles, with respect to their equality or inequality; and then, assuming two
triangles, it investigates the same properties[74] in various ways.
The second part[75] develops the theory of parallelograms, beginning with the special characteristics of parallel lines and
the method of constructing the parallelogram and then demonstrating the properties of parallelograms. The third part[76]
reveals the kinship between triangles and parallelograms both
in their properties and in their relations to one another. Thus
it proves that triangles or parallelograms on the same or equal
bases have identical properties;[77] it shows [what is the relation

---

78.2 above) and apparently sees some profound significance in it (see
note at 423.20).

[73] 82.23 After ἰσόπλευρον τρίγωνον the text of Grynaeus skips without a break to ἵνα γὰρ τὸ τριχῇ διαστάν at 86.16. This same gap occurs
in several other MSS. Evidently the codex from which they are all
derived had lost some of its pages. See C. Wachsmuth in *Rheinisches
Museum* XXIX, 1874, 317; and the note at 416.14.

[74] 83.14 I.e. equality and inequality.

[75] 83.15 XXVII to XXXIV.

[76] 83.19 XXXV to the end.

[77] 83.24 I.e. are shown to be equal; the qualifying phrase "with
respect to their equality or inequality" (83.13f.) appears to govern the
whole passage. Proclus' statement is unusually loose, since triangles
on the same or equal bases are equal only when they lie between the
same parallels, a condition that must be taken as understood here.
See XXXV-XL.

84

Chap. XI
A Warning
to the
Reader

between] a triangle and a parallelogram on the same base,[78] how to construct a parallelogram equal to a triangle,[79] and finally, with respect to the squares on the sides of a right-angled triangle, what is the relation of the square on the side that subtends the right angle to the squares on the two sides that contain it.[80] Something like this may be said to be the purpose of the first book of the *Elements* and the division of its contents.

As we begin our examination of details, we warn those who may encounter this book not to expect of us a discussion of matters that have been dealt with over and over by our predecessors, such as lemmas, cases, and the like. We are surfeited with those topics and shall touch on them but sparingly. But whatever matters contain more substantial science and contribute to philosophy as a whole, these we shall make it our chief concern to mention, emulating the Pythagoreans whose byword and proverb was "a figure and a stepping-stone, not a figure and three obols."[81] By this they meant that we must cultivate that science of geometry which with each theorem lays the basis for a step upward and draws the soul to the higher world, instead of letting it descend among sensibles to satisfy the common needs of mortals and, in aiming at these, neglect to turn away hence.

[78] 84.1 This must refer to XLI, but some words have been lost or else Proclus expressed himself most elliptically. I have filled out his text with a clause identical to one which occurs four lines later.
[79] 84.1 XLIV.
[80] 84.6 XLVII.
[81] 84.17 Σχᾶμα καὶ βᾶμα, ἀλλ' οὐ σχᾶμα καὶ τριώβολον. Taylor (I, 113) notes: "I do not find this aenigma among the Pythagoric symbols that are extant; so that it is probably no where mentioned but in the present work."

DEFINITIONS

I. *A point is what has no parts.*

IN ADVANCING from the more composite to the simpler figures, the geometer proceeds from the three-dimensional solid to the plane that bounds it, from the plane to the line that is its boundary, and from the line to the point devoid of all extension. This has often been said and is evident to everyone. But since these limits, because of their simplicity, are often thought to be more august than the complex natures they delimit, and yet often resemble accidents in having their existence in the objects bounded by them, we must decide under which of these two classes of beings they are to be considered.

I begin, then, by remarking that in immaterial things, which subsist as ideas separate from matter and as forms grounded in themselves alone, the substance of the simpler is always more primary than that of the more composite. For this reason both in Nous and in the intermediate orders of souls[1]— that is, those natures that directly breathe life into bodies— the limiting factors have an essential priority over the things that are limited, as being less divisible, more uniform, and more sovereign; for among immaterial forms unity is more perfect than plurality, the partless more perfect than what proceeds in any way from it, and what bounds more perfect than what gets its limit from something other than itself. On

---

[1] 86.1   On the intermediate position of souls and their life-giving function see Proclus' *Elements of Theology*, Props. 188-190. For the basic principles governing the hierarchy of the intelligible world see especially the first six propositions in the *Elements*. To give references for all the details of Proclus' cosmology, here and at later points in this commentary, would extend these notes beyond all convenient bounds. The reader who wishes to study them further is advised to consult Dodds' commentary on the *Elements*, the detailed exposition in Rosán, or the more summary accounts in Thomas Whittaker, *The Neo-Platonists*, 2nd edn., Cambridge, 1918, 157-180; and in Friedrich Ueberweg, *Die Philosophie des Altertums*, ed. Karl Praechter, Basel, 1953, 625-631.

the other hand, the forms which, requiring matter, have their foundation in what is outside themselves, and have departed from their own nature to be dispersed among their several subjects and possess only an imported unity, have been allotted more complex ideas rather than simpler ones. Thus in the objects that appear in imagination and in the matter of imagined shapes, as well as in perceptible things that are generated by nature, the ideas of the bounded objects have priority, the ideas of their boundaries being subsequent and, as it were, adventitious. In order that an object in three dimensions may not stretch to an infinite size in our thought or perception, it is limited on all sides by planes; and so that the plane may not slip away into boundlessness, the line comes to be in it to contain and define it; and the point does the same thing for the line, these simples existing for the compounds.

This also is clear, moreover, that in the forms separable from matter the ideas of the boundaries exist in themselves and not in the things bounded, and it is because they remain precisely what they are that they become agents for bringing to existence the entities dependent upon them. But in the forms inseparable from matter the limits surrender themselves to the things they limit; they establish themselves in them, becoming, as it were, parts of them and being filled with their inferior characters. This is why in this region the partless partakes of divisible existence and the breadthless of breadth; and the limiting elements are no longer able to preserve their simplicity and purity, for they are altered by having come to be in a substratum that is other than themselves. Matter muddies their precision; the idea of the plane gives the plane depth, that of the line blurs its one-dimensional nature and becomes generally divisible, and the idea of the point ends by becoming bodily in character and extensible together with the things that it bounds. For all ideas when they flow into matter —the ideas of the understanding into intelligible and those of nature into perceptible matter—are filled with their substrates: they forsake their native simplicity for alien combinations and extensions.

87

But if all things in Nous and in the soul are without parts or intervals, how does it happen that in the realm of matter some of them are subject to division preeminently and others because of the nature of matter? Is it not that among immaterial forms there is a gradation of rank between primary, intermediate, and later forms? Some forms are more uniform, others more inclined to plurality; some hold their powers together in concentration, others endeavor to scatter theirs; some sit close to the Limit, others incline towards the Unlimited. For although they all partake of these two principles, yet some are more the offspring of one and have a larger share of its nature, while others are similarly related to the other. This is why in that higher region the point is completely without parts and yet, although its being is determined by the Limit, it secretly contains the potentiality of the Unlimited, by virtue of which it generates all intervals;[2] and the procession of all the intervals does not exhaust its infinite capacity. Body on the other hand—that is, the idea of body—has a greater share of the nature of the Unlimited, wherefore it belongs among the things bounded from without and divisible to infinity in all directions. The forms intermediate between these two, according to their distance from one or the other extreme, belong respectively to the class of things inclined more towards the Limit and to those that enjoy boundlessness. Consequently these forms both bound and are bounded: they bound insofar as, owing their existence to the Limit, they are able to impose limits on other things; and they are bounded insofar as, by their participation in the Unlimited, they need to be limited by other things.

The point, then, being a limit, preserves its character when things participate in it. But since it also secretly possesses the nature of the Unlimited and strives to be everywhere in the things that it bounds, it is present in them an infinite number of times; and since in the intelligible world the Unlimited is a power generative of extended bodies, so it is potentially such

88

---

[2] 88.5 Removing Friedlein's period after διαστήματα and ending the sentence with δύναμιν in line 7; the following τὸ δὲ is obviously the correlate to τὸ μὲν in line 2.

in the things that share in it. For among the higher realities—the intelligibles—the Unlimited is the first creative cause and generative power of all things, but in enmattered forms it is imperfect and only potentially everything. In sum, those forms that by their simplicity and absence of parts occupy the highest station among first principles preserve their specific natures when things share in them, but they do so in a lesser degree than the more composite ideas. For matter is able to share in the composite ideas more clearly; it is to them that it is adapted, rather than to the simplest principles of being. For this reason, although traces of the most exalted principles descend into matter, yet the characters that it receives from principles of the second and third ranks are much more clearly evident. Hence it partakes more of the principle of body than of the principle of the plane, of the plane more than of the form of the line, and of the line more than of the point that bounds and holds them together. For the idea of the point is the first member of this entire series; it unifies all things that are divided, it contains and bounds their processions, it brings them all on the stage and encompasses them about. This is why even in sensible likenesses, although different things have different boundaries, the point is the limit of them all.

We should not suppose, as the members of the Stoic school did, that these limits—I mean the limits found in body—exist merely as the product of reflection. To be reminded that natures of this sort with their creative presiding ideas exist in things we need only look at the whole of the cosmos—at its circular revolutions and the centers of these circuits, that is, the axes that penetrate them all. For the centers actually exist as holding together the spheres, unifying their extensions and constraining and compressing their forces about the centers. The axes wrap the spheres about them and, while themselves remaining fixed, carry them around in revolutions about themselves. And the poles of the spheres, which limit the axes and from their positions control all the circuits—are they not clear evidence that points have creative and controlling powers capable of making a whole of the disparate parts, providing them with their unity and their never-ceasing motion? This is why Plato declares that the substance of these axes is as hard as adamant, thus indicating their irreversible, everlasting, steadfast, unchangeable being. The whole "spindle," he says, moves about these axes, celebrating their unity in a dance.[3] Other doctrines of a more secret kind assert that the Demiurge who presides over the cosmos rides[4] upon the poles and through his divine love turns the whole towards himself. The Pythagoreans claimed that the pole should be called "the seal of Rhea," as the place through which the life-giving goddess dispenses her mysterious and effective power to the All; and the center they said is "the guardhouse of Zas," since Zeus set his creative watch in the bosom of the cosmos and established it securely there about the middle. For if the center remains fixed, the All likewise maintains its orderly arrangement unperturbed and its revolution unending, and all things preserve their stations unchanged. The gods that guard the poles have been assigned the function of assembling the separate and unifying the manifold members of the whole, while those appointed to the axes keep the circuits in everlasting revolution around and around. And if I may add my own conceit, the centers and the poles of all the spheres symbolize the wry-necked gods[5] by imitating the mysterious union and synthesis which they effect; the axes represent the mainstays of all the cosmic orders, since they hold together the unities and revolutions in the visible cosmos, as the intelligible centers hold together the cosmos

91

---

[3] 90.11   This passage contains reminiscences both of *Rep.* 616cff. and of *Tim.* 40c.

[4] 90.13   ἐποχούμενον. This is a clear reference to the ὄχημα, or vehicle, in which every soul, divine or human, is, so to speak, embodied according to Neoplatonic thought. For this doctrine see Proclus' *Elements of Theology*, Props. 196, 204-211, and Dodds' Appendix II, 313-321. Cf. 138.8.

[5] 91.3   τῶν ἰυγγικῶν θεῶν, Chaldaic divinities (see Kroll, 39-42, 73f.), apparently represented in the form of the bird called ἰυγξ (see Aristotle's description in *Hist. An.* 504a12-19). They are alluded to elsewhere by Proclus (e.g. *Commentary on the Republic* II, 213.1, Kroll; *Commentary on the Cratylus*, 33.15, Pasquali), but these passages throw no more light on their nature and function than does the present one. Barocius translates *conciliantium deorum*, and ver Eecke *des dieux conciliateurs*.

of the intelligibles; and the very spheres are likenesses of the perfecting divinities, joining end to beginning and surpassing all other figures in simplicity, uniformity, and perfection.

We have expanded somewhat largely on these matters in order to show that points, and limits in general, have power in the cosmos and that they have the premier rank in the All by virtue of carrying the likenesses of the first and most sovereign causes. For the centers and poles of the cosmos are not limits such as exist in limited things; rather they have an actual foundation and a self-sufficient being and power that extend throughout the whole of the divisible world. Most people, observing that limits exist imperfectly in limited things, have a confused conception of their being. Some say that they are only abstracted by reflection from sensible things, others that they have no existence apart from our thoughts. But the forms of all of them do exist in the intelligible world, they exist in the orders of soul, they exist in nature and, last of all, in bodies. Let us then note how they have their being in each class of things corresponding to the position of that class. All limits exist preeminently in Nous, but partless and without differentiation of kind, so that they all subsist covertly and indivisibly in a single form under the idea of the point. Likewise they all exist in souls, but under the form of the line; that is the reason why Timaeus constructed the soul out of straight lines and circles,[6] for every circle is only a line. And they all exist in the things of nature, but under the idea of the plane. This is why Plato thought it proper to exhibit the ideas constitutive of natural bodies with the help of planes;[7] the analysis of bodies into plane surfaces brings us to the proximate cause of their appearances. Finally, all the limits are in bodies, since all the forms exist in them, but in a material fashion in accordance with the divisible nature of bodies.

Consequently all the limits are everywhere, and each comes to light in its proper place, their appearances varying according to the power that prevails in them. As to the point, it is everywhere indivisible and distinguished by its simplicity from

92

[6] 92.7   On the straight lines in the soul see note at 16.22 above.
[7] 92.11   I.e. in *Tim.* 53c-55c.

divisible things; but as it descends in the scale of being, even the point takes on the character distinctive of divisibles. Sometimes it has its seat altogether above them in keeping with the superiority of its cause, sometimes it is ranged beside them, and sometimes it takes up temporary residence among them and, drinking as it were from the partibility of inferior beings, relaxes its own partlessness. Just as the unit in one of its aspects is generative of numbers and in another aspect serves as the matter underlying numbers, and in neither case is number itself but a principle of number in one or the other of these ways, so likewise the point is sometimes the constitutive principle of magnitudes and at other times a principle in a different sense, but not as the generative cause.

93

But is the point the only thing that is without parts, or is not this a characteristic also of the instant in time and of unity among numbers? The answer is that the philosopher, whose field of inquiry is the universe of beings, should examine everything that is in any way divisible as well as the natures of the indivisibles that are sovereign over them, whereas the scientist in a special area—conducting his inquiry from certain limited starting-points to which alone he refers his results, without attending to the procession of beings in the cosmos—has the responsibility of examining and expounding only that indivisible nature which[8] is appropriate to his first principles. It is his responsibility to see that simplicity which is primary in the objects that he studies. In geometrical matter, then, the point alone is without parts, and in arithmetic the unit; and the definition of the point, though it may be imperfect from another point of view, is perfect as far as the science before us is concerned. The physician says that the elements of bodies are fire, water, and the like, and he carries his analysis of bodies only thus far; but the physicist proceeds to other elements simpler than these. The former defines as element what is simple to sense-perception, the other what is simple in thought; and each of them is right with regard to his own science. We must not therefore con-

[8] 93.15 Reading with Barocius ἤ for ῇ and, in the next line, ὁρᾶν for ὁρᾷ in Friedlein.

94   sider the definition of point mistaken, nor judge that it is imperfect; for with respect to the subject-matter of geometry and the starting-points of this science, it is adequately given. It all but clearly says that "what is without parts is a point for my purposes and a principle for me; and the simplest object is none other than this." In such fashion must we understand the statement of our geometer.

By denying parts to it, then, Euclid signifies to us that the point is the first principle of the entire subject under examination. Negative definitions are appropriate to first principles, as Parmenides teaches us in setting forth the first and ultimate cause by means of negations alone. For every first principle is constituted by a different essence from that of the things dependent on it, and to deny the latter makes evident to us the peculiar property of the principle. For that which is their cause, but not any one of the things of which it is the cause, becomes in a sense knowable through this method of exposition.

But someone may object: How can the geometer contemplate a partless something, a point, within the imagination if the imagination always apprehends things as shaped and divisible? For not only the ideas in the understanding, but also the impressions of intellectual and divine forms, are accepted by the imagination in accordance with its peculiar nature, which furnishes forms to the formless and figures to what is without figure. To this difficulty we reply that the imagination in its activity is not divisible only, neither is it indivisible.

95   Rather it moves from the undivided to the divided, from the unformed to what is formed. For if the imagination were divisible only, it would be unable to preserve in itself the various impressions of the objects that come to it, since the later ones would obscure those that preceded them—just as no body can at the same time and in the same place have a series of shapes, for the earlier ones are erased by the later. And if it were indivisible, the imagination would not be inferior to the understanding or to the soul, which views everything as undivided; nor could it exercise form-giving functions. It is necessary therefore that its activity should

start from what is partless within it, proceed therefrom to project each knowable object that has come to it in concentrated form, and end by giving each object form, shape, and extension. If, then, it has a nature of this kind, the character of indivisibility is in a certain sense within it, and it is primarily by virtue of this character that we must say it contains the being of the point; and by virtue of the same character the form of line also exists wrapped up within it. Possessing this double character of indivisibility and divisibility, the imagination contains the point in undivided and intervals in divided fashion.

Since the Pythagoreans, however, define the point as a unit that has position, we ought to inquire what they mean by saying this.[9] That numbers are purer and more immaterial than magnitudes and that the starting-point of numbers is simpler than that of magnitudes are clear to everyone. But when they speak of the unit as not[10] having position, I think they are indicating that unity and number—that is, abstract number[11]—have their existence in thought; and that is why each number, such as five or seven, appears to every mind as one and not many, and as free of any extraneous figure or form. By contrast the point is projected in imagination and comes to be, as it were, in a place and embodied in intelligible matter. Hence the unit is without position, since it is immaterial and outside all extension and place; but the point has position because it occurs in the bosom of imagination and is therefore enmattered. Owing to its affinity with the principles, the unit is simpler than the point;[12] for the point, by

96

[9] 95.23   The purpose of the following paragraph appears to be to dispute the Aristotelian interpretation (*Met.* 1080b16-20 and *passim*) that the Pythagoreans considered numbers to have magnitude.

[10] 95.26   It is certain that μὴ has dropped out between θέσιν and ἔχουσαν in 96.1, otherwise the contrast between τὴν μὲν μονάδα and τὸ δὲ σημεῖον (96.6) is not expressed.

[11] 96.3   μοναδικός. For the meaning given this term in the translation see note at 95.23.

[12] 96.12   στιγμή. This is the older word for "point," supplanted in Euclid and his successors by σημεῖον. Proclus ordinarily uses the Euclidean term, as in the preceding sentence; but here and at 59.18 (both times in a historical context) he uses the earlier Pythagorean

having position, goes beyond the unit. And additional determinants in the bodiless concepts effect a lessening of being in the things that accept them.

II. *A line is length without breadth.*[13]

The line is second in order as the first and simplest extension, what our geometer calls "length," adding "without breadth" because the line also has the relation of a principle to the surface. He taught us what the point is through negations only, since it is the principle of all magnitudes; but the line he explains partly by affirmation and partly by negation. The line is length, and in this respect it goes beyond the undividedness of the point; yet it is without breadth, since it is devoid of the other dimensions. For everything that is without breadth is also without depth, but the converse is not true. Thus in denying breadth of it he has also taken away depth, and this is why he does not add "without depth," since this is implied in the absence of breadth.

The line has also been defined in other ways. Some define it as the "flowing of a point,"[14] others as "magnitude extended in one direction." The latter definition indicates perfectly the nature of the line, but that which calls it the flowing of a point appears to explain it in terms of its generative cause and sets before us not line in general, but the material[15] line. This line owes its being to the point, which, though without parts, is the cause of the existence of all divisible things; and the "flowing" indicates the forthgoing of the point and its genera-

97

term. Aristotle uses both, but στιγμή more often. Plato uses neither; and Aristotle reports that he rejected the concept as a geometrical fiction (*Met.* 992a20-22). See Ross's note on this passage (*Aristotle's Metaphysics*, Oxford, 1958, I, 203-207) and Heath, *Euclid* I, 155f.

[13] 96.16 Aristotle (*Topics* 143b11) cites this definition of the line. It was therefore current before Euclid's time and perhaps should be attributed to Plato and his school.

[14] 97.7 Referred to by Aristotle in *De An.* 409a4. The definition of line as "magnitude extended in one direction" is essentially Aristotle's, as Heath notes (*Euclid* I, 158); see *Met.* 1020a11-12.

[15] 97.11 The text reads ἄυλον, "immaterial," but this must be a slip, since the line described in the immediate sequel is the material line, and I have translated it accordingly, as does ver Eecke.

tive power that extends to every dimension without diminution and, remaining itself the same, provides existence to all divisible things.

All these things are known to everyone. But let us recall the more Pythagorean doctrine that posits the point as analogous to the monad, the line to the dyad, the surface to the triad, and the solid to the tetrad. On the other hand, considering them as extended, we shall find that the line is one-dimensional, the surface two-dimensional, and the solid three-dimensional; hence Aristotle says[16] that body comes to completion with the number three. It is no wonder that the point, because of its partlessness, has been primarily associated with the monad; but of the things that come after the point, although they correspond to the numbers that arise from the monad and keep the same relation[17] to the point that the numbers have to the monad, yet each participates in what is immediately before it and has the same value in relation to its next and successor as its antecedent has to it. That is, the line has the rank of two with respect to the point, but of one with respect to the surface; the surface has the rank of three with respect to the point and the line, but of two with respect to the solid; and so body is tetradic with respect to the point and triadic with respect to the line.[18] Both of these orderings have their justification, but that of the Pythagoreans is closer to first principles, for it starts from the top and follows the nature of things. The point is twofold, because it exists either by itself or in the line. As a limit only, and one, possessing neither wholeness nor parts, it is a likeness of the very summit of being and so is ranked as analogous to unity. For unity is primarily there where ancestral unity dwells, as the Oracle says.[19] Since the line is the first thing to have parts and to be a

98

[16] 97.25 *De Caelo* 268a8.

[17] 98.1 Reading with Barocius τοῦτον for τούτων in Friedlein.

[18] 98.7f. The text here is puzzling, but I have not ventured to emend it.

[19] 98.18 τὸ λόγιον. Kroll investigated the contents, origin, and date of this Oracle to which Proclus and other late Neoplatonists refer and brought together the fragments that can be rescued from these citations. Proclus' interest in these Oracles is well attested by his

whole, and since it is both monadic because unidimensional and dyadic because of its forthgoing—for if it is an infinite line it partakes of the indefinite dyad, and if it is finite it requires two limits, a whence and a whither[20]—for these reasons it is an imitation of wholeness and of that grade of being which is extended oneness and generates duality. For this it is that produces transformation into length,[21] that is, into divisible extendedness in one dimension together with participation in duality. The surface is both triad and dyad; being the receptacle of the primary figures as well as the first nature that takes on form and shape, it resembles both the triad that primarily bounds all beings and also in a way the dyad which divides this triadic nature. But the solid, extended in three directions and defined by the tetrad that comprehends all ratios in itself,[22] carries our thoughts to that intelligible cosmos which by the aid of the tetradic property—that is, the feminine and generative power—produces the separation of the orders of bodily things and the division of the universe into three.

These matters could be worked out further. Because the line is second and owes its existence to the first change from partlessness, the Pythagorean doctrine properly calls it dyadic. That the point comes after the monad and the line after the

99

---

biographer Marinus; see the Introduction, "Proclus: His Life and Writings." Both Porphyry and Iamblichus before him had ‧ written commentaries on the Chaldaean Oracles. Those who would like to pursue this inquiry further should begin with Dodds' essay in the *Journal of Roman Studies* XXXVII, 1947, reprinted as Appendix II in his *The Greeks and the Irrational* (Berkeley, 1951), where they will find abundant references to the recent literature. For the Oracle cited here see Kroll, 15. A new edition of the Chaldaean Oracles by Edouard des Places is promised for the current year.

[20] 98.22   I can make nothing of πρὸς τὸν ἀπ' αὐτῆς in Friedlein, and his conjecture πρῶτον ἐπ' αὐτῆς is little more intelligible. Either Barocius did not have these words in his text, or if he did, he chose not to translate them. I follow his example.

[21] 99.2   Reading with Grynaeus ἔκστασιν for ἔκτασιν in Friedlein.

[22] 99.10   The first solid number, according to Pythagorean lore, is four—three dots making the base and one the apex of a triangular pyramid. The numbers 1, 2, 3, 4 yield the most consonant intervals in the musical scale. See Van der Waerden, 95.

dyad and the surface after the triad the *Parmenides* indicates when it denies first plurality of the One and then wholeness;[23] if plurality comes before wholeness, so also number comes before the continuous, the dyad before the line, and the monad before the point. For it is fitting to describe as "not many" the monad that generates plurality [and as "neither whole nor part" the point that brings the whole into being].[24] For of the whole it is said that it has parts.

So much can be said about the line on more speculative grounds. But we should also accept what the followers of Apollonius say, namely, that we have the idea of the line when we ask only for a measurement of length, as of a road or a wall. For breadth does not enter into our consideration, since we reckon only the distance in one direction. Similarly when we measure a plot of land we look only at the surface, and when we are measuring a well, the three-dimensional cavity; in this case we consider all the dimensions together and declare that such-and-such is the volume of the well, according to its length, breadth, and depth. And we can get a visual perception of the line if we look at the middle division separating lighted from shaded areas, whether on the moon or on the earth. For the part that lies between them is unextended in breadth, but it has length, since it is stretched out all along the light and the shadow.

III. *The limits of a line are points.*

Every compound gets its boundary from the simple, and every divisible thing from the indivisible. The principles of mathematics provide images of these truths; for when Euclid says that the line is limited by points, he is clearly making the line as such unlimited, as not having any limit because of its own forthgoing. So just as the dyad is bounded by the monad and, when controlled by it, sets a term to its own unchecked boldness, so also the line is bounded by points. And being dual in nature, when it participates in the point, which contains the

---

[23] 99.22 *Parm.* 137c.

[24] 100.2 I have accepted with modifications Friedlein's suggestions for filling up this lacuna.

idea of unity, it does so in the fashion of a dyad. Now in imagined and perceived objects the very points that are in the line limit it, but in the region of immaterial forms the partless idea of the point has prior existence. As it goes forth from that region, this very first of all ideas expands itself, moves, and flows towards infinity and, imitating the indefinite dyad, is mastered by its own principle, unified by it, and constrained on all sides. Thus it is at once unlimited and limited—in its own forthgoing unlimited, but limited by virtue of its participation in its limitlike cause. For as it goes forth, it is held by itself within the compass of that cause and is bounded by its unifying power. Hence also in [sensible] likenesses the points

102

that constitute the extremity and the beginning of a line are said to bound it. In that upper realm, then, the limit transcends what is limited, but here it is twofold, for it exists in the limited thing itself. This affords a remarkable illustration of the principle that the forms existing in themselves are causally prior to the things that participate in them but, in giving themselves to their participants, take on an existence after their kind, becoming plural and divisible as their subjects are, and enjoying their diversity.

This further fact we must also anticipate about the line: our geometer makes a threefold use of it. Thus he takes it as limited at both ends, as in the problem "Upon a given finite line to construct an equilateral triangle";[25] then as unlimited in one direction and limited in the other, as in the problem "To construct a triangle from three straight lines that are equal to three given straight lines"[26] (for in the construction he says "given a straight line limited in one direction but unlimited in the other"); [and, lastly, as unlimited in both directions, as in the problem "To a given infinite line to draw a perpendicular to it from a point lying outside the line"].[27] Thus line is understood in three senses by our geometer.

Besides these matters, this point also is worthy of attention and should not be passed over: in what sense are points said

[25] 102.15   In I.   [26] 102.18   As in XXII.
[27] 102.20   I have adopted and translated Barocius' plausible filling of the lacuna here. The proposition used as illustration is XII.

103

to be limits of a line, and of what sort of line are they limits? They cannot be limits of the infinite line, nor of every finite line. For there is a line which is finite but does not have points as its limits. The circle is such a line, bending back upon itself and making no use of limits as does the straight line. Such also is the ellipse.[28] Perhaps, then, we should consider the line only insofar as it is a line.[29] For we can conceive a segment of a circumference bounded by points and a part of an ellipse likewise having points as its boundaries; but the circle and the ellipse have another property by virtue of which they are not only lines, but also productive of figures. If, then, we consider both of them as lines, they have points as limits; but if they are thought of as producing the sorts of figures mentioned, then they bend back on themselves. And if you think of them as they are being drawn, you will find where they are bounded by points; but taking them as already drawn, with their beginnings and their ends joined together, you can no longer see their extremities.

IV. *A straight line is a line which lies evenly with the points on itself.*

Plato assumes that the two simplest and most fundamental species of line are the straight and the circular and makes

104

all other kinds mixtures of these two, both those called spiral, whether lying in planes or about solids, and the curved lines that are produced by the sections of solids. According to Plato, the point, if we may say so, appears to bear the likeness of the One, for the One also is without parts, as he has shown in the *Parmenides*. Since there are three hypostases[30] below the One—namely, the Limit, the Unlimited, and the Mixed— it is through them that the species of lines, angles, and figures come to be. Corresponding to the Limit are, in surfaces, the circular line, the angle bounded by circular lines, and the

[28] 103.6   θυρεός, "shield."

[29] 103.7   This is a strange phrase, for it contradicts the statement (101.7) that the line "as such" is unlimited. The point of this paragraph is only to show that, *if* a line has limits, those limits are points. See Heath, *Euclid* I, 165.

[30] 104.9   See notes at 3.5 and 5.18.

circle;[31] and, in solids, the sphere. To the Unlimited corresponds the straight line in all these groups, for it is found in them all, presenting its characteristic appearance on each occasion. And the mixtures in all of them correspond to the principle of the Mixed. For there are mixed lines, such as spirals; mixed angles, such as the semicircular and the horned angles;[32] mixed figures, such as sections of plane figures and arches;[33] and mixed solids, such as cones, cylinders, and the like. Hence the Limit, the Unlimited, and the Mixed are present in all of them. Aristotle's opinion is the same as Plato's; for every line, he says, is either straight, or circular, or a mixture of the two.[34] For this reason there are three species of motion—motion in a straight line, motion in a circle, and mixed motions.

Some dispute this classification, denying that there are only two simple lines and saying that there is also a third, namely, the cylindrical helix, which is traced by a point[35] moving uniformly along a straight line that is moving around the surface of a cylinder. This moving point generates a helix any part of which coincides homoeomerously with any other, as Apollonius has shown in his treatise *On the Cochlias*. This characteristic belongs to this helix alone. For the segments of a spiral in a plane are dissimilar, as are those of the spirals about a cone or sphere; the cylindrical spiral alone is homoeomeric, like the straight line and the circle. Are there not, then, three simple lines, instead of two only?

To this difficulty we shall reply by saying that this helix is indeed homoeomeric, as Apollonius has shown, but is by no means simple. For to have similar parts and to be simple are not the same thing. Among natural bodies gold and silver

105

---

[31] 104.13 These are chosen as examples of lines, angles, and figures respectively, as is implied by κατὰ πάντα ταῦτα in line 14.

[32] 104.18 The former is the angle made by the diameter and the circumference of a circle, the latter that made by the circumference and a tangent to the circle; cf. 127.14. On the horned angle see Heath I, 178, 382, and *Euclid* II, 39ff.

[33] 104.19 ἀψῖδες, i.e. arcs other than semicircles.

[34] 104.24 For Plato see *Parm.* 145b; for Aristotle *De Caelo* 268b17f. and *Phys.* 261b29.

[35] 105.3 Reading σημεῖον for σημείου in Friedlein.

consist of similar parts but are not for that reason simple. The very mode of generating the cylindrical helix shows that it is a mixture of simple lines, for it is produced by the movement of a straight line about the axis of a cylinder and by the movement of a point along this line. It owes its existence, then, to two [dissimilar][36] simple motions, so that it is to be classed among the mixed, not the simple lines. For what comes to be out of diverse principles is not simple, but mixed; and Geminus has rightly declared that, although a simple line

106   can be produced by a plurality of motions, not every such line is mixed, but only one that arises from dissimilar motions. Imagine a square undergoing two motions of equal velocity, one lengthwise and the other sidewise; a diagonal motion in a straight line will result. But this does not make the line a mixed one, for it is not brought into being by a line different from itself and moving simply, as was the case with the cylindrical helix mentioned. Nor is it true that a circular line comes about by mixture if one imagines a straight line moving in a right angle[37] and describing a circle with its middle point; for when a straight line is moving thus, its extremities, moving nonuniformly,[38] describe straight lines, whereas the middle point, moving nonuniformly, describes a circle, and the other points ellipses.[39] So a circular line is generated as a result of nonuniform motion of the middle point, under the condition given that the line is moving not naturally, but with its extremities on the sides of a right angle.[40] But enough of these matters.

[36] 105.23   Consistency with 105.25 and 106.3 requires that ἀνόμοιοι be inserted here.

[37] 106.10   I.e. with its extremities on the two sides of a right angle.

[38] 106.13   Reading ἀνομάλως for ὁμαλῶς here, after Tannery, who has analyzed this example in II, 36.

[39] 106.14f.   Presumably this illustration is taken from Geminus, who may have provided the demonstration for the interesting theorem employed. For this demonstration see ver Eecke, 96, note 4.

[40] 106.19   The point of this illustration is not clear. It would appear that the circle thus resulting is a mixed line, like the ellipse. Either Proclus has incorrectly transcribed what he found in Geminus, or Geminus was himself not clear about the distinction between simple and mixed lines. Elsewhere in Proclus' citation of Geminus' doctrine

One might think that, although both the straight line and the circle are simple lines, the straight line is the simpler. For it contains not even any dissimilarity in thought, whereas concavity and convexity in the circle indicate difference; and the straight line does not suggest the circle, whereas the circular line does bring to mind the idea of the straight line, if not through its mode of generation, at least by its relation to a center. What, then, if someone should say that the circle needs the straight line for its existence? For if one end of a finite line remains stationary and the other moves, it will describe a circle whose center is the stationary extremity of the straight line. Should we not reply that what describes the circle is not the line, but the point that moves about the stationary point? The line only defines its distance from the center, whereas what produces the circle is the point in circular movement. But enough of this.

It appears that the circular line belongs with the Limit and has the relation to other lines that the Limit has to all things; for of the simple lines the circular alone is limited and makes a figure, whereas the straight line belongs with the Unlimited and hence can be projected indefinitely without end. So as all other things arise from the Limit and the Unlimited, likewise the whole class of mixed lines, both those in planes and those about solids, come from the circle and the straight line. For this reason the soul contains in advance the straight and the circular in her essential nature, so that she may supervise the whole array of unlimiteds as well as all the limited beings[41] in the cosmos, providing for their forthgoing by the straight line and for their reversion by the circle, leading them to plurality by the one and collecting them all into unity by the other. And not only the soul, but also he who constituted the soul and furnished her with these two powers possesses in himself the primordial causes. For "holding in advance the

107

108

---

of mixed lines there is unclearness, if not confusion. See notes below at 111.9 and 113.23.

[41] 107.23 Reading with Grynaeus and apparently Barocius περα-τοειδῆ instead of περιττοειδῆ in Friedlein.

beginning, middles, and ends of all things," says Plato, "he bounds straight lines as he moves around by nature."[42] He goes forth to all things with his providential activity while he is turned upon himself, "abiding in his accustomed nature," as the *Timaeus* says.[43] The straight line is a symbol of the inflexible, unvarying, incorruptible, unremitting, and all-powerful providence that is present to all things; and the circle and circular movement symbolize the activity that returns to itself, concentrates on itself, and controls everything in accord with a single intelligible Limit. The demiurgic Nous has therefore set up these two principles in himself, the straight and the circular, and produced out of himself two monads,[44] the one acting in a circular fashion to perfect all intelligible essences, the other moving in a straight line to bring all perceptible things to birth. Since the soul is intermediate between sensibles and intelligibles, she moves in circular fashion insofar as she is allied to intelligible nature but, insofar as she presides over sensibles, exercises her providence in a straight line. So much regarding the similarity of these concepts to the order of being.

109

Euclid gives the definition of the straight line that we have set forth above, making clear by it that the straight line alone covers a distance equal to that between the points that lie on it. For the interval between any two points is the length of the line that these points define, and this is what is meant by "lying evenly with the points on itself."[45] If you take two points on a circle or any other kind of line, the length of the

---

[42] 108.7 *Laws* 716a. Burnet's text of the *Laws* has εὐθείᾳ instead of εὐθείας in Proclus (line 6) which gives a different meaning.

[43] 108.10 *Tim.* 42e.

[44] 108.18 A monad, in Proclus' thought, is an originative principle of a series of beings, as the number one is of the series of numbers. See his *Elements of Theology*, Prop. 21.

[45] 109.13 On the many attempts to explain the meaning of this phrase see Heath, *Euclid* I, 166ff., and Tannery II, 540-544. Proclus' attempt to make it the equivalent of Archimedes' "definition"—the shortest distance between two points—is hardly successful. Tannery sees the expression as having its origin in the language of the everyday craftsman, such as the carpenter evening a piece of wood, or the stonemason levelling a surface by a stretched cord or dressing a wall by means of a plumb line.

line between the two points taken is greater than the distance between them. This seems to be a characteristic of every line except the straight. Hence it accords with a common notion that those who go in a straight line travel only the distance they need to cover, as men say, whereas those who do not go in a straight line travel farther than is necessary.

Plato, however, defines the straight line as that whose middle intercepts the view of the extremes.[46] This is a necessary property of things lying on a straight line but need not be true of things on a circle or any other extension. This is why astronomers say that the sun is in eclipse at the time when both it and the moon are on a straight line with our eye; for then our view of it is intercepted by the moon which has come between it and us. Perhaps this property of the straight line affords a proof that in the realm of being, as things go forth from their causes, the middle orders of things have the property of differentiating the separateness[47] of the extremes from the community of nature that unites them, just as in the process of reversion they draw back to their first causes the entities that lie at intervals from themselves.

But Archimedes defined the straight line as the shortest of all lines having the same extremities.[48] Because, as Euclid's definition says, it lies evenly with the points on itself, it is the shortest of all lines having the same extremities; for if there were a shorter line, this one would not lie evenly with its own extremities. In fact all other definitions of the straight line fall back upon the same notion—that it is a line stretched to the utmost, that one part of it does not lie in a lower and another in a higher plane, that all of its parts coincide similarly with all others, that it is a line that remains fixed if its

110

---

[46] 109.22 *Parm.* 137e. This is the only pre-Euclidean definition of straight line that we hear of. Aristotle quotes it in *Topics* 148b27. Heath (*Euclid* I, 168) suggests that Euclid's definition is based on Plato's, changed to eliminate any implied appeal to vision, which, as a physical fact, could not properly find a place in a purely geometrical definition.

[47] 110.7 Reading with Barocius and Grynaeus ἀποστάσεως instead of ὑποστάσεως in Friedlein.

[48] 110.12 In his *On the Sphere and Cylinder* I, *ad init.*

end points remain fixed, that it cannot make a figure with another line of the same nature.[49] All these definitions express the property which the straight line has by virtue of being simple and exhibiting the single shortest route from one extremity to the other. So much for definitions of the straight line.

111   Geminus divides lines first into incomposite and composite, calling a composite line one that is broken and forms an angle.[50] Incomposite[51] lines he then divides into those that make figures and those that extend indefinitely. By those that make figures he means the circular, the elliptical, and the cissoidal;[52] and by those that do not, the section of a right-angled cone, the section of an obtuse-angled cone,[53] the conchoid, the straight line, and all such. Again, following another method, he divides incomposite lines into simple and mixed.[54] And of simple lines some, such as the circular, make figures;

[49] 110.23   The first, third, and fourth of these alternative "definitions" are found in Heron, and the second and fifth are found in Euclid himself (I. 4; XI. 1), though not as definitions (Heath, *Euclid* I, 168).

[50] 111.3   Proclus gives us two classifications of lines from Geminus. The second appears at 111.9-112.18, to which a later passage (176.27-177.23) adds some interesting details. On these classifications see Heath, *Euclid* I, 160ff.

[51] 111.4   σύνθετον is clearly a mistake for ἀσύνθετον here.

[52] 111.6   The cissoid is the curve invented by Diocles (late second or early first century B.C.) for solving the problem of doubling the cube. On its construction see Heath I, 264-266, and *Euclid* I, 164.

[53] 111.8   I.e. the parabola and the hyperbola respectively. Before Apollonius the conic sections were thought of as made by a plane at right angles to one of the sides of the cone; hence a right-angled cone yielded what was after Apollonius called a "parabola," an obtuse-angled cone a "hyperbola," and an acute-angled cone an "ellipse." But the older theory and its terminology persisted after Apollonius' time, as Proclus' language shows. See note to 420.23 below. On the construction of the conchoids see Heath, *Euclid* I, 160f.

[54] 111.9ff.   Tannery (II, 37) has pointed out that the classification of lines and figures into simple and mixed, insisted on by Geminus, does not occur in Pappus. Since this threefold classification of lines into straight, circular, and mixed is traced by Proclus back to Plato (103.21-104.5), its origin, Tannery thinks, should be sought in an attempt to return to Plato's ideas. This was an unfortunate encroachment by philosophy on the domain of mathematics, for it could not result in any rational classification either of lines or of figures.

others, like the straight line, are unbounded. Of mixed lines some lie in planes, others are on solids: of those in planes some return upon themselves, like the cissoid, and others extend indefinitely; and of those in solids some come to our attention through the sections of solids, and others lie around the solids. For the helix on a sphere or a cone is around a solid, but a conic or spiric section arises from such-and-such a section of a solid. Some of these sections, in particular the conic, were discovered by Menaechmus—and Eratosthenes refers to this when he says, "Don't produce the conic section triads of Menaechmus"[55]—others by Perseus,[56] who composed an epigram on his discovery:

112
>Having discovered three spirals on five sections
>Perseus honored the gods with this dedication.

The three conic sections are the parabola, the hyperbola, and the ellipse. Of the spiric sections one is interlaced like a horse's hobble,[57] another is broad in the middle and thins out at the sides, and another is elongated and has a narrow middle portion but broadens out at the two ends.[58] The other mixtures are limitless in number, for the number of solid figures is infinite and their sections are of numerous kinds. For if the straight line in circular motion generates a surface, so also do conic sections, and conchoids, and circular lines themselves;[59] and as these resulting solids are cut in all sorts of ways, they reveal the varied species of lines.

Of the lines that are on the surface of bodies some are homoeomeric, like the spiral around a cylinder, but all the others are anhomoeomeric. Hence from these distinctions it may be gathered that the only three lines that are homo-

[55] 111.23 See Eutocius' *Commentary on Archimedes* (Heiberg, *Archimedes* III, 112.20).

[56] 111.24 Perseus is known to us only from this and another reference in Proclus (356.12). On the spiric curves that he is said to have discovered and investigated see Heath II, 203-206.

[57] 112.5 I.e. like the figure 8.

[58] 112.8 On the corresponding spiric surfaces see 119.9-17.

[59] 112.14 The translation of the preceding obscure sentence is based on the interpretation of Tannery II, 24.

eomeric are the straight line, the circle, and the cylindrical helix. Two of them lie in a plane and are simple; one is mixed and lies around a solid. This has been clearly shown by Geminus, who had previously demonstrated[60] that the two lines drawn from a point to a homoeomeric line and making equal angles with it are themselves equal. Ambitious students should go to his writings for the proofs, since he also shows how spirals, conchoids, and cissoids are generated.

We have given the names and classifications of these lines[61] in order to encourage the able student to inquire into them; we consider it superfluous in the present work, however, to make a precise inquiry into each of them, since our geometer has revealed here only the simple and fundamental lines, the straight line in the definition before us and the circumference in his account of the circle (for there he will tell us that the line that bounds a circle is a circumference). Nowhere does he mention mixed lines. And yet he knows mixed angles, such as the semicircular and the horned angles; mixed plane figures, such as segments and sections; and mixed solids, such as cones and cylinders. Of each of the others he has given the three kinds,[62] but of the line only these two, the straight and the circular, considering that in a treatise on the elements he should bring in only simple species; and all the other lines are too complex.[63] Let us also, then, following our geometer, end our classification of lines with these simple ones.

V. *A surface is what has length and breadth only.*

Next in rank after the point and the line comes the surface, which is extended in two ways, length and breadth, but remains without depth, thus possessing a simpler nature than what is extended in three ways. Hence our geometer adds

[60] 112.23   Despite Barocius and Grynaeus, Friedlein's προαποδείξας, *not* προσαποδείξας, is the correct reading, as is shown by 251.8.

[61] 113.7   Reading with Barocius ἱστορήσαμεν for ἱστορήσομεν in Friedlein.

[62] 113.20   The species corresponding to the Limit, the Unlimited, and the Mixed.

[63] 113.23   Perhaps another reason is that Euclid did not consider mixed lines of mathematical importance. See note at 111.9.

"only" after the mention of the two dimensions, implying that there is no third dimension in the surface. This word is indeed equivalent to the denial of depth; and our author's purpose here also is to indicate both the superior simplicity of the surface as compared to the solid by using a negation (or an addition equivalent to a negation) and its subordination to the beings above it by using affirmative characterizations. Others have defined it as the limit of a body, saying in a sense the same thing, for the bounding element falls short of what it bounds by one dimension. Others define it as magnitude extended in two ways, and others still differently, but meaning the same thing however they frame their definitions.

We have the notion of surface, it is said, when we measure pieces of land and determine their boundaries according to length and breadth; and we get some perception of it when we look at shadows. These are without depth, since they cannot go under the ground, and have only breadth and length. The Pythagoreans used to say that the surface is related to the triad, because all the figures on it have the triad as their first cause. For the circle, which is the principle of all curvilinear figures, carries a hidden trinity in its center, diameter, and circumference; and the triangle is the premier of all rectilinear figures, as everyone can see, because it is determined by the number three and formed by it.

115

VI. *The limits of a surface are lines.*

If we take these propositions as likenesses, we can understand that every being simpler than what immediately follows it supplies a boundary and limit to its successor. Soul bounds and perfects the activity of nature, nature does likewise for the motion in bodies, and prior to both of them Nous measures the revolutions of soul and the One measures the life of Nous itself, for the One is the measure of all things. So also in geometry the solid is bounded by the surface, the surface in turn by the line, and the line by the point, for the point is the limit of them all. In the realm of immaterial forms and partless ideas the line, being uniform in its forthgoing, bounds and contains the varied activity of the surface and immediately

116

unifies its boundlessness, while in the realm of their [sensible] likenesses the limiting factor belongs to the very thing that is limited and in this way furnishes it with its boundary.

If someone here also[64] should ask how lines can be the limits of the surface in general but not of all finite surfaces (for the surface of the sphere is limited, not however by lines but by itself), we should reply that, if we take surface as extended in two dimensions, we shall find that it is bounded by lines according to its length and its breadth; but if we consider the spherical surface as itself shaped and invested with an additional quality, we take it as having joined its end to its beginning and made of its two extremities one, a unity existing in potentiality only, not in actuality.[65]

VII. *A plane surface is a surface which lies evenly with the straight lines on itself.*

The older philosophers did not think to posit the plane (ἐπίπεδον) as a species of surface (ἐπιφάνεια) but took the two terms as equivalent for expressing magnitude in two dimensions. Thus the divine Plato said that geometry is the study of planes (ἐπίπεδα)[66] and contrasted it with stereometry as if he thought surface and plane were the same thing. Likewise also the inspired Aristotle. But Euclid and his successors make the surface the genus and the plane a species of it, as the straight line is a species of line. This is why, by analogy with the straight line, he defines the plane separately from the surface. For the straight line, he says, is equal to the interval that lies between its points, and the plane likewise occupies

117

[64] 116.4 Recall the similar difficulty raised at 102.23 with respect to points as limits of lines. Here Proclus is in effect contending only that, if a surface has limits, these limits are lines.

[65] 116.14 I suspect that Proclus has inadvertently interchanged "potentiality" and "actuality" here, and so apparently does ver Eecke. But Barocius and Schönberger adhere to the text, as I do.

[66] 116.21 Plato, *Rep.* 528d. Aristotle uses both ἐπιφάνεια and ἐπίπεδον for "surface" (e.g. *Cat.* 5a3f.). Plato, however, at least in the dialogues, does not use the former term at all in the sense of "surface." But Diogenes Laertius (III, 24) says that Plato was the first philosopher to give a name to the plane surface (ἐπίπεδον ἐπιφάνεια). This attribution may well be derived from some tradition about his oral teachings. See Heath, *Euclid* I, 169.

a place equal to that between two straight lines lying on it. This is what is meant by "lying evenly with the straight lines on itself." Others, meaning the same thing, have said that the plane is a surface stretched to the utmost, and others that it is a surface such that a straight line is congruent with all its parts. And some would say that it is the least of all the surfaces that have the same boundaries, or that it is one whose middle parts intercept the view of the extremities; and one could transfer all the definitions of the straight line to the plane surface by merely substituting the genus surface.[67] For these characters—the straight, the circular, and the mixed —extend all the way from lines to solids, as we have said; and they exist both in planes and in solids in an analogous fashion.[68] Hence the *Parmenides* says that every figure is straight, circular, or mixed.[69] If, then, you wish to consider straightness in surfaces, take the plane, which the straight line fits on in all ways; or if circularity, take the spherical surface; or if the mixture of the two, take the cylindrical, or the conical, or some similar surface.

But we must realize, says Geminus, that the mode of mixture is different in a mixed line so-called and in a mixed plane. Mixture in lines does not come about through composition, nor through blending. For example, the helix, a mixed line, does not have one part straight and another curved, like a mixture resulting from composition; nor when cut does the helix yield any trace of its simple constituents, as things that are blended do; its terms have been destroyed and fused together. So Theodorus the mathematician[70] is wrong in

118

[67] 117.13   On alternative definitions of plane surface in antiquity see Heath I, 171f.

[68] 117.17   This language is strange, but what Proclus evidently means by a "straight solid" is a solid which is bounded only by planes.

[69] 117.18   *Parm.* 145b; but note that this passage says nothing about solids.

[70] 118.7   "Theodorus the mathematician" is certainly not the Theodorus of Cyrene mentioned above (66.6), but probably the Theodorus of Soli, in Cilicia, who is cited by Plutarch on certain mathematical difficulties in the *Timaeus*. See Plutarch, *De Defectu Oraculorum* XXXII, 426F; *De Animae Procreatione in Timaeo* XX, 1022D; and Tannery II, 37.

assuming that mixture in lines is a blending. In the case of planes, however, mixture arises neither by composition nor by fusion, but rather by a kind of blending. If we think of a circle lying in a plane with a point above it and from the point project a straight line to the circumference of the circle and set the line in revolution, we shall produce a conical surface, which is mixed. Now if we cut it, we can resolve it back into simple surfaces; by making a section from the apex to the base we get a triangle, but by cutting it parallel to the base we get a section that is a circular plane. Yet in the case of lines the appearance does not show that mixture is the result of blending, for it does not take us back to the simple nature of the elements, whereas surfaces when cut reveal immediately from what sort of lines they have been produced. Thus the mode of mixture in lines is different, as has been said, from that in surfaces.

Just as with lines there were certain ones, as we saw, that are simple, namely, the straight and the circular, of which most people have a conception without being taught, though the species of mixed lines required a more technical understanding, so also we have at once a notion of the most elementary kinds of surfaces, the plane and the spherical, though it is only through science and scientific reasoning that we discover the variety of surfaces that arise by mixture. What is remarkable about them is that from the circle there can often be generated a mixed surface. This is what we say happens in the case of spiric surfaces,[71] for they are thought of as generated by the revolution of a circle standing upright and turning about a fixed point that is not the center of the circle.[72] Thus three kinds of spiric surface are generated, for the center[73] lies either on the circumference, or inside, or outside the circle. If the center is on the circumference, the continuous spiric surface is generated; if within the circle, the

119

[71] 119.9   Called "tores" in modern mathematics. On this passage see Heath II, 203-206, and *Euclid* I, 162f.

[72] 119.12   I.e. about an axis that does not pass through the center of the circle.

[73] 119.13   The axis of revolution.

interlaced spiric surface; and if outside, the open spiric surface. And the spiric sections are three in number corresponding to these different kinds of surface.[74] But every spiric surface is mixed, although the motion that generates it is one and circular. Mixed surfaces are also generated, however, not only from simple lines moving in the manner described, but also from mixed lines. The conic lines,[75] though three in number, make four mixed surfaces, called conoids. By the revolution of the parabola about its axis the right-angled conoid is generated; by the revolution of the ellipse are produced the so-called spheroids[76]—the elongated spheroid if the revolution is about the major axis, the flattened sort if about the minor axis; and from the revolution of the hyperbola comes another conoid.[77]

It should be realized that sometimes we get the idea of the surfaces from the lines and sometimes learn the lines from the surfaces. For example, from the conic and spiric surfaces we come to think also of conic and spiric lines. And we must anticipate this further point of difference between lines and surfaces: there are three homoeomeric lines, as was said before, but only two homoeomeric surfaces, the plane and the spherical—not the cylindrical, for not all the parts of a cylindrical surface coincide with one another.

We have said enough about the differences between surfaces. Our geometer has chosen one of them, the plane, for definition as the subject in which he will study the figures and their properties, for his inquiry can proceed more easily with this than with any other surface. On the plane it is possible to think of straight lines, circles, spirals, and the sections and contacts of straight lines and circles, as well as to make ap-

120

---

[74] 119.17 Proclus is in error here; all three spiric sections (described in 112.4-8) arise from plane sections of the open spiric surface. See Heath, *Euclid* I, 163, and Tannery II, 24-28.

[75] 119.21 Reading with Barocius κωνικαὶ for κινητικαὶ in Friedlein.

[76] 119.25 I.e. ellipsoids.

[77] 120.2 The obtuse-angled conoid, according to Barocius' text. Grynaeus has the unintelligible ἐλυκωνοειδές. The ἄλλο in Friedlein has evidently replaced a word which was illegible or unintelligible to some copyist.

plication of areas and construct the various kinds of angles. Not all these matters can be investigated on any other surface. For how could we understand a straight line or a rectilinear angle on a spherical surface? Or how on a conical or cylindrical surface could we study the sections of circles or straight lines? It is reasonable, then, that he should define this particular surface and build his entire treatise upon it. For this reason also he gives his work the subtitle "plane geometry." And thus we must think of the plane as projected and lying before our eyes and the understanding as writing everything upon it, the imagination becoming something like a plane mirror to which the ideas of the understanding send down impressions of themselves.

VIII. *A plane angle is the inclination to one another of two lines in a plane which meet one another and do not lie in a straight line.*

Some of the ancients put the angle in the category of relation, calling it the inclination either of lines or of planes to one another; others place it under quality, saying that, like straight and curved, it is a certain character of a surface or a solid; others refer it to quantity, asserting that it is either a surface or a solid quantity. For the angle on a surface is divided by a line, that in solids by a surface, and what is divided by them, they say, can only be a magnitude; and it is not linear magnitude, for a line is divided by a point. So it remains that it is either a surface or a solid quantity.

But if it is a magnitude and all finite homogeneous magnitudes have a ratio to one another, then all homogeneous angles, at least those in planes, will have a ratio to one another, so that a horned angle will have a ratio to a rectilinear. But all quantities that have a ratio to one another can exceed one another by being multiplied; a horned angle, then, may exceed a rectilinear, which is impossible, for it has been proved that a horned angle is less than any rectilinear angle.[78]

And if it is only a quality, like heat or coldness, how can it be divided into equal parts? For equality and inequality belong no less to angles than to magnitudes, and divisibility

[78] 122.7 For the demonstration see Euclid III. 16.

in general is an intrinsic property of angles and magnitudes alike. But if the things to which these properties intrinsically belong are quantities and not qualities, then it is clear that angles are not qualities. Of quality the relevant modifications are more and less, not equal and unequal. We should then have to refrain from saying that angles are unequal, one greater than another, and instead call them unlike, one of them more an angle than another. Anyone can see that this is alien to the nature of mathematics; for an identical definition applies to all angles, and one is not more an angle than another.

As to the third possibility, if the angle is an inclination and in general belongs to the class of relations, it will follow that, when the inclination is one, there is one angle and not more. For if the angle is nothing other than a relation between lines or between planes, how could there be one relation but many angles? If you imagine a cone cut by a triangle from apex to base, you will see one inclination at the apex of the half-cone, that of the sides of the triangle, but two separate angles, one the angle on the plane of the triangle, the other on the mixed surface of the cone; and both of these angles are contained by the above-mentioned two lines. The relation of these lines, then, did not make the angle. And yet it is necessary that we call the angle either a quality, or a quantity, or a relation. Figures are qualities, the ratios between them are relations, and so we must refer the angle also to some one of these three genera.

Such are the difficulties. Now Euclid says the angle is an inclination, whereas Apollonius calls it the contracting of a surface at a point under a broken line, or of a solid under a broken surface (this seems to be the way in which he defines angle in general).[79] But let us follow our "head"[80] and say

123

---

[79] 123.18 Reading οὕτως for οὗτος in Friedlein. Ver Eecke suggests that Proclus is referring to a definition occurring in a work of Apollonius which had probably disappeared from circulation.

[80] 123.19 This is clearly a reference to Syrianus of Alexandria, who was head of the Academy during Proclus' early years in it (see the Introduction, "Proclus: His Life and Writings"). There are a score or more such references to Syrianus in the *Commentary on the Timaeus*.

that the angle as such is none of the things mentioned but exists as a combination of all these categories, and this is why it presents a difficulty to those who are inclined to make it any one of them. The angle is not the only thing that has this character. The triangle also has it: it partakes of quantity (a triangle is said to be equal or unequal to another, and quantity is, as it were, matter for these properties); and quality also belongs to it by virtue of its shape. This is why triangles can be called similar and equal, deriving one attribute from one category, the other from another. So the angle surely needs the underlying quantity implied in its size, it needs the quality by which it has something like a special shape and character of existence, and it needs also the relation of the lines that bound it or of the planes that enclose it. The angle is something that results from all of these, and is not just one of them. It is divisible and receptive of equality and inequality by virtue of the quantity in it, but it is not compelled to accept classification among homogeneous magnitudes because it also has a distinguishing quality that often makes angles incomparable with one another. Nor is the angle made one if the inclination is one, since the quantity between the inclined sides completes its being. Now if we observe these distinctions, we shall be able to solve the difficulties. We shall find the peculiar property of the angle is not that it is, as Apollonius said, a contracting of a surface or a solid, although this contributes to its nature, but rather that contracted surface itself at the given point, contained by the inclined lines or by a single line bent upon itself, or that contracted solid itself that underlies the planes inclined towards one another. Thus one may define it as a qualified quantity, constituted by such-and-such a relation,[81] and not quantity as such, nor quality nor relation alone.

So much needed to be said about the nature of angles to obtain a preliminary understanding of the whole genus before we distinguish its species.[82]

124

125

[81] 125.1 *Sc.* of lines or planes.

[82] 125.6 The promise in this apparently introductory sentence is not fulfilled until 126.7. I agree with Schönberger that the passage from 125.6 to 126.6 has been misplaced; its opening sentence suggests

Such are the three opinions about the angle. Eudemus the Peripatetic,[83] who wrote a book on the angle, declared it to be a quality. Looking at the way in which it is produced, he says that it is nothing other than the fracture of a line; and if straightness is a quality, so also is fracture; therefore, since the angle has its origin in quality, it is certainly quality. But Euclid, and all who claim it is inclination, classify it among relations. It is made a quantity by those who say the angle is "the first interval under the point." Plutarch[84] is one of these, and he insists that Apollonius held the same opinion. For there must be, he argues, some first interval under the inclination[85] of the containing lines or planes. Yet since the interval under the point is continuous, it is impossible to determine the first interval, for every interval is infinitely divisible. Besides, even if in some way we could determine the first interval and draw a straight line through it, we would produce a triangle, not a single angle. Carpus of Antioch[86] says the angle is a quantity, specifically, the distance between the containing lines or planes. Although this is a distance "in one sense" ($\dot{\epsilon}\phi'$ $\ddot{\epsilon}\nu$), the angle nevertheless is not a line, he contends, for

126

---

that it was designed to come at 123.14, after "to some one of these three genera."

[83] 125.7   Eudemus of Rhodes (called "the Peripatetic" here and at 379.2) was the author of the history of geometry that Proclus evidently used for the historical account beginning at 65.7. He is cited as Proclus' authority also at 299.3, 333.6, 352.14ff., 419.15. Eudemus appears to have derived his conception of the angle as quality from Aristotle: cf. *Cat.* 10a11 with *Phys.* 188a25; and Heath, *Euclid* I, 177f.

[84] 125.16   This is Plutarch of Athens, the teacher of Proclus and the immediate predecessor of Syrianus as head of the Academy. See the Introduction.

[85] 125.19   Reading with Barocius κλίσιν for κλάσιν in Friedlein.

[86] 125.25   Called later "Carpus the engineer" (241.19). He is included in Pappus' *Collection*, and Simplicius (in *CAG* VIII, 192.23, Kalbfleisch) cites Iamblichus as including him among the "Pythagoreans" who solved the problem of squaring the circle. Nothing is known, however, of the "curve of double motion" which he used in his solution (Heath I, 225). Tannery (II, 554) is inclined to think that he lived at the time of Heron, or a little later. On the date of Heron see note at 41.10 above. Proclus quotes a passage of considerable length from Carpus below (241.18ff.).

not everything extended "in one sense" is a line. But this is the height of paradox, if there is a magnitude other than a line that extends in only one sense. But enough of this.

Of angles we must remark that some exist on surfaces, others in solids; and of those on surfaces some are on simple, others on mixed surfaces. That is, there can be an angle on a cylindrical or a conical, as well as on a spherical surface or a plane. Of those on simple surfaces some are on spherical surfaces, and others have their existence on planes. For example, the zodiacal circle at the equinoctial intersection makes two angles at the tips of the intersecting circumferences. Such angles lie on spherical surfaces. Of those on planes some are contained by simple lines, others by mixed, and others by the two combined. In the ellipse, for example, an angle is formed between the axis and the boundary of the ellipse. One of these lines is mixed, the other simple. And if a circle cuts an ellipse, there will be an angle contained by its circumference and the boundary of the ellipse. Whenever cissoid lines converge to a single point, as do ivy leaves (indeed the cissoids get their name from this resemblance), they make an angle, one obviously contained between mixed lines; and whenever the hippopede, one of the spiric sections, makes an angle with another,[87] this angle also is contained within mixed lines. The angles formed by straight lines and by circumferences are contained by simple lines. Of these again some are contained by similar lines. For two circumferences intersecting or tangent to one another produce angles, and of three sorts: either biconvex, when the convex parts of the circumferences are outside;[88] or biconcave, when both concave segments are outside, angles that are called "scraper-like"; or a mixture of convex and concave, like the angles of lunes. And an angle may be contained also by a straight line and a circumference, and in either of two ways: either by a straight line and a convex circumference, such as the angle in a semicircle; or by a straight line and a concave circumference,

127

[87] 127.1   Omitting τὰ in Friedlein. On the hippopede see 112.5.
[88] 127.8   See the diagrams in the next note.

like the horned angle.[89] The angles formed by two straight
lines will all be called rectilinear, and they also are differentiated into three kinds.[90]

All these angles that are constructed on plane surfaces our
geometer defines in this treatise under the common designation of "plane angle," asserting that their genus is inclination
and their locus the plane (for angles have position), that they
are produced by two lines (and not by three or more, as are
solid angles) that come together not as parts of a straight
line extending in one direction, but at an inclination[91] to one
another and with an area contained between them. Now this
definition, in the first place, seems to deny that an angle can
be produced by a single line. Yet the cissoid, a single line,
makes an angle, and so does the hippopede. We call "cissoid"
the line as a whole, not its parts (for then we could say that it
is its parts converging on one another that make the angle);
likewise it is the whole spiric section, not its parts, that we
call the hippopede. Thus each of them, being one, makes an
angle with itself, not with another line. In the second place,
this definition appears to be mistaken in defining angle as
inclination; for how can there be two angles formed by one
inclination? How can we continue to speak of equal and
unequal angles? And there are all the other objections that are
customarily brought against this opinion. Thirdly, the condition "not lying on a straight line" is unnecessary with regard
to certain angles, such as those formed by circular lines. The
definition is complete without it, for the inclination of these

128

---

[89] 127.14 The following diagrams will illustrate the species of
angles mentioned. DCE is biconvex, ACB is biconcave, and ACD and
BCE are lunular angles. Of those formed by a circle and a straight
line CBD is the angle in a semicircle, and ABD is a horned angle.

[90] 127.16 Obviously right, obtuse, and acute. Cf. Deff. X-XII.
[91] 128.1 Reading with Barocius κλίσις for κλάσις in Friedlein.

lines to one another will make the angle, and it is at the outset impossible that circular lines should lie in a straight line.

So much we had to say about Euclid's definition, in part interpreting and in part exposing difficulties in it.

IX. *When the lines containing the angle are straight, the angle is called rectilinear.*

The angle is a symbol and a likeness, we say, of the coherence that obtains in the realm of divine things—of the orderliness that leads diverse things to unity, divided things to the indivisible, and plurality to conjunction and communion. For the angle functions as a bond between the several lines and planes, focussing magnitude upon the unextendedness of points and holding together every figure that is constructed by means of it. Hence the Oracles call these angular conjunctions the "bonds"[92] of the figures, because of their resemblance to the constraining unities and couplings in the divine world by which things separated are joined to one another. Plane angles typify the more immaterial, the simpler, and the more perfect modes of unification, whereas the angles in solid bodies represent those unifying processes that go forth even to the lowest realities and provide community for things sundered and a congenial ordering for things that are utterly disparate. Of plane angles some represent the primary and unmixed unifying agencies, others those that contain the infinity of their own progressions;[93] some the unifying[94] forces of the intelligible forms; others those of sensible ideas; and others the binding principles of intermediate things. Circular angles imitate the causes that enwrap intelligible diversity in a unity, for circular lines ever bending back upon themselves are images of Nous and intelligible forms; rectilinear angles represent the presiding causes in sensible things that bring about the interdependence of their ideas; and mixed angles show forth the causes that preserve the community between

92 129.7 συνοχηίδας. For the Oracles see note at 98.18 above; and Kroll, 73.11-12, 74.2.
93 129.17 A cryptic reference to the distinction between right angles and obtuse or acute angles. See 131.13-134.7.
94 129.19 Reading with Barocius ἐνοποιούσας for ἐνοποιοῦσιν in Friedlein.

sensible and intelligible forms in a single and unshakeable unity.

We must therefore look to these paradigms also when assigning the causes of particular things. Among the Pythagoreans we find some angles dedicated to certain gods, others to others.[95] Thus Philolaus makes the angle of a triangle sacred to some, and the angle of a square sacred to others, assigning different angles to different gods, or the same angle to more than one god and several angles to the same god, according to the various potencies in him. And I think the philosopher of Asine[96] has in mind these features of the demiurgic triangle, the primary cause of all the order among the elements, when he sets some gods at the sides and others at the angles, the former presiding over the forthgoing and potentiality of things, the latter over the unification of wholes and the reassembling into unity of the things that have issued forth. Thus do these features of the angle bring our thoughts around to the contemplation of being.

If the lines are here said to contain[97] the angle, this is not to be wondered at, for in this world unity and partlessness are introduced from without. And among the gods and in the realm of the truly real, the complete and indivisible Good has primacy over the things that are plural and separated.

131

X-XII. *When a straight line set up on a straight line makes the adjacent angles equal to one another, each of the equal angles is a right angle and the straight line standing on the other is called a perpendicular to that on which it stands. An obtuse angle is an angle greater than a right angle. An acute angle is an angle less than a right angle.*

These are the three kinds of angles that Socrates in the *Republic* says are accepted as hypotheses by geometers,[98] the

[95] 130.10   This passage is supplemented by what is said at 132.17 and 167.1. On Philolaus see note at 22.14.

[96] 130.15   Reading with Grynaeus and Barocius 'Ασιναῖος instead of Friedlein's emendation 'Αθηναῖος. This is clearly a reference to Theodorus of Asine, an immediate disciple of Porphyry and of Iamblichus, whom Proclus calls ὁ μέγας Θεόδωρος in his *Commentary on the Timaeus* (I, 213.3 and *passim*, Diehl).

[97] 130.23   See 318.13 and note.

[98] 131.11   *Rep.* 510c. On this use of the term "hypothesis" see note at 178.3.

angles produced when the rectilinear angle is divided into species—the right, the obtuse, and the acute. The first of them is distinguished by equality, sameness, and likeness; the others are characterized by relative greatness and smallness and in general by inequality, difference, and indefinite more-and-less. Most geometers are unable to give a reason for this classification but take it as a hypothesis that there are three angles; and if we demand an explanation, they deny that we have a right to ask it of them. But the Pythagoreans, who refer the solution of this triple distinction to first principles, have no difficulty in giving the causes of this difference among rectilinear angles. For one of their principles is constituted by the Limit which is the source of the definiteness and self-identity of all things that have come to completion, the cause also of equality and of everything in the better of the two columns of contraries;[99] their other principle is the Unlimited which produces progression to infinity, increase and diminution, inequality, and every sort of difference among the things it generates, and in general is the head of the inferior column. Therefore since rectilinear angles also come to be in accordance with these principles, it is reasonable that the idea which proceeds from the Limit should produce the one right angle, ruled by equality and similarity to every other right angle, always determinate and fixed in nature, not admitting of either growth or diminution; whereas the idea that comes from the Unlimited, being second in rank and dual in nature, reveals a pair of angles about the right angle characterized by inequality of greater and smaller, more and less, and subject to unlimited variation, the one through degrees of obtuseness, the other through degrees of acuteness.

132

For these reasons, therefore, they refer right angles to the immaculate essences in the divine orders and their more particular potencies, as causes of the undeviating providence that presides over secondary things—for what is upright, un-inclined to evil, and inflexible accords with the character of those high gods—whereas they say that obtuse and acute angles are left in the charge of the divinities that supervise

[99] 132.2   See note at 7.3.

133

134

the forthgoing of things and the change and variety of their powers. The obtuse angle is an image of the extension of the forms to everything, while the acute is a likeness of the cause that discriminates and activates all things. Furthermore, the rightness that preserves identity of being is like the essence in things themselves, whereas obtuseness and acuteness resemble their attributes, for they are receptive of the more-and-less and undergo indefinite change without end. Rightly, then, they exhort the soul to make her descent into the world of generation after the undeviating form of the right angle, inclining no more to one side than to the other, nor being affected more by some things than by others, for the possession of fellow-feeling drags her down into the error and indeterminacy of matter. The perpendicular thus is also a symbol of directness, purity, undefiled unswerving force,[100] and all such things, a symbol of divine and intelligent measure. For by perpendiculars we measure the altitude of figures, and it is by reference to the right angle that we define the other rectilinear angles, since they have no limiting principle in themselves; they are considered only as exceeding or falling short,[101] each of them being in itself indeterminate. Hence they say that virtue is like rightness, whereas vice is constituted after the fashion of the indeterminate obtuse and acute, possessing both excesses and deficiencies and showing by this more-and-less its own lack of measure. We shall therefore lay it down that the right among rectilinear angles is the image of perfection, undeviating energy, intelligent limit and boundary, and everything similar to them, and that the obtuse and acute angles are likenesses of indefinite change, irrelevant progression, differentiation, partition, and unlimitedness in general. So much for these matters.

To the definitions of the obtuse and acute angles[102] we must add the genus: each of them is rectilinear, one larger and the other smaller than a right angle. Not every angle smaller than a right angle is acute, for the horned angle is

[100] 133.14   Omitting Friedlein's comma after δυνάμεως.
[101] 133.19   Sc. of the right angle.
[102] 134.8   Sc. as given at 131.7-8.

smaller than any right angle—indeed smaller than any acute angle—but is not acute; and likewise the semicircular angle is smaller than any right angle but is not acute. The explanation is that these are mixed, not rectilinear angles. Clearly also many angles contained between circular lines appear to be greater than right angles, but they are not for that reason obtuse; for the obtuse angle must be rectilinear. I call attention to this and observe also that in defining a right angle our geometer takes a straight line standing on another straight line and making the adjacent angles equal to one another, whereas he explains the obtuse and the acute angles without assuming a straight line inclined towards one side, referring instead to the right angle; for the right angle is the measure of angles other than right, just as equality is the measure of unequal things, and the lines inclined towards either side are, as he saw, infinite in number, not one only as is the perpen-

135 dicular. Furthermore, his comment that the angles are equal "to one another" I regard as a mark of his geometrical precision. For it would be possible for these angles to be equal to other angles and not be right angles; hence it is because they are equal to one another[103] that they are necessarily right angles. The addition of "adjacent" does not seem to me irrelevant, as some have incorrectly supposed; it makes plain the definition of the right angle. The reason why each of the angles is a right angle is that they are adjacent and equal, since the lack of inclination of the upright straight line towards either side is the cause of the equality of both the angles and of the rightness of each. It is, then, not simply their equality to one another, but their being adjacent that, together with their being equal, is the cause of the rightness of the angles.

In addition to what has been said, I think it proper to recall here also the purpose of the author of the *Elements*, that he is discoursing about the figures constructed in a single plane. Hence this definition of the perpendicular is not applicable to all perpendiculars, but only to that which lies in the same

---

[103] 135.3 ἀλλήλαις is obviously misplaced; it is meaningless with ὀρθάς but needed with the following ἴσας.

plane; this was not the occasion for a definition of the so-called solid perpendicular. As he has defined the plane angle, so also he defines the plane perpendicular, since the solid perpendicular necessarily makes right angles not on one line only, but on all the lines that touch it and that lie in the plane; for this is its property.

136

XIII. *A boundary is what is the limit of something.*

The term "boundary" (ὅρος) should not be applied to every magnitude—the boundary of a line, for example, is rather a "limit" (πέρας)—but to plane areas and solids. In this work our author calls "boundary" the line that encloses an area; and it is a limit in this sense, not as the point is said to be the limit of a line, but as that which shuts in and closes off something from what lies around it.[104] The term has been at home in geometry from the beginning, for geometry is the art by which men were accustomed to measure lands and keep their boundary marks (ὅρους) distinct; and it is from this activity that they became aware of this science. Hence when the author of the *Elements* calls the outer enclosing line a boundary, he naturally calls it also a limit of areas. For everything enclosed is limited by its enclosing line. Of the circle, for example, the circumference is the boundary and limit, but the plane surface itself is an area; and similarly with other figures.

XIV. *A figure is that which is contained by any boundary or boundaries.*

Since figure has many meanings and is divided into different kinds, we must first look at these differences in order to arrive at figure as it is presented in this definition. A figure, then, is something that results from change, arising from an

---

[104] 136.8 Proclus wishes to emphasize, first, that ὅρος and πέρας are not synonymous—the latter being the more general term, for which reason it is used as genus in this definition—and, secondly, that ὅρος had from the beginning a special application to areas, as it has in the definition that immediately follows. Aristotle, however, uses the two terms as synonymous; see Heath, *Euclid* I, 182, who cites *De Gen. An.* 745a 6, 9.

137 effect produced in things that are struck, or divided, or decreased, or added to, or altered in form, or affected in any one of various other ways. There are figures produced by art (for example, by modelling or sculpturing), in accordance with the idea preexisting in the artist's mind, the art providing the form and the matter receiving therefrom its shape, beauty, and seemliness. More august and imposing figures than these are the works of nature's craftsmanship, some of them containing the constitutive proportions in the sublunary elements, others in the heavens defining the powers and motions of the heavenly bodies.[105] For the heavenly bodies, both in themselves and in their relations to one another, present a great and marvellous variety of figures, exhibiting now one and now another of the shapes that bear the likeness of intelligible forms; and they copy in their rhythmic choruses the bodiless and immaterial forces resident in the figures. Beyond these are the figures of souls, the purest and most perfect in beauty, full of life, by their self-motion preeminent over things that are moved by external causes, and by their immateriality and lack of extendedness superior to extended and embodied things. About them the *Timaeus* has instructed us in revealing the essentially demiurgic character of the figure that belongs to souls.[106] Even more divine than the figures of souls are the intelligible figures; they are in every way superior to divided

138 things, shining everywhere with indivisible and intelligible light, generating, effecting, perfecting all things, being present equally in all of them though themselves[107] steadfast and unmoved, bringing unity to the figures of souls and keeping the aberrations of sensible figures within appropriate bounds. And high above all these are the perfect, uniform, unknowable, and ineffable figures of the gods which, being mounted[108] on the intelligible figures, impose limits upon the whole universe of figures and hold everything together in their unifying boundaries. Their properties have been represented

---

[105] 137.12 Reading with Schönberger αὐτῶν for αὑτῶν in Friedlein.
[106] 137.24 *Tim.* 36b-d.
[107] 138.3 Reading with Schönberger αὐτοῖς for αὑτοῖς in Friedlein.
[108] 138.8 ἐποχούμενα. See note at 90.13.

for us by the theurgic art[109] in its statues of the gods, whom it clothes in the most varied figures. Some of them it portrays by means of mystic signs that express the unknowable divine potencies; others it represents through forms and shapes, making some standing, others sitting; some heart-shaped, some spherical, and some fashioned still otherwise; some simple, others composed of several shapes; some stern, others mild and expressing the benignity of the gods; and still others fearful in shape. To these figures it adjoins various symbols for different gods, as they are appropriate to the divinities represented.

Figure, then, begins above with the gods themselves and extends down to the lowest orders of beings, exhibiting even in them its derivation from the first of causes. For the perfect figures are necessarily prior to the imperfect, those grounded in themselves prior to those that exist in other beings, and those that preserve their nature undefiled to those that are stuffed with their own privations. Material figures partake of the unshapeliness of matter and lack the purity that they should have; the figures in the heavens are divisible and have their existence in other things; the figures of souls admit of differentiation and variety and every kind of development; the intelligible figures, together with unity, contain progression to plurality; and at the head of them all stand the very figures of the gods, independent, uniform, simple, generative, having all perfection in themselves and from themselves offering to all things the perfecting agency of the forms.

We cannot, then, allow what is usually said, that figures in the sense world are produced by additions or subtractions or alterations. For such incomplete processes could not contain the original and primary cause of their products. Nor would

[109] 138.10 θεουργία, a kind of sympathetic magic practised for religious purposes by some of the later Neoplatonists. By this "art" an image invested with the symbols of a god, and duly consecrated, is made capable of "participating in divinity," of moving and of speaking (Proclus, *Commentary on the Timaeus* III, 155.18ff., Diehl). Proclus' interest in theurgy is abundantly attested (see the Introduction); but this is the only occurrence of the term in this *Commentary*. For further details about theurgy in the Hellenistic age, see Dodds, *The Greeks and the Irrational*, Berkeley, 1951, Appendix II, "Theurgy," esp. 291ff.

we see the same figures produced by contrary causes: the same shape, for example, could come about either by addition or by subtraction. Rather we shall posit that the causes mentioned are subservient to others in the process of generation and affirm that the end is defined for them by other and precedent causes. Neither is it true, as some say, that immaterial figures lack reality and only material things exist; nor true what still others say, that they exist indeed apart from matter, but only in thought and by abstraction. For how could the

140

precision, beauty, and orderliness of these figures be preserved if they were merely abstractions?[110] Being of the same kind as sense objects, they fall far short of certainty, precision, and accuracy; and if they later take on accuracy, orderliness, and perfection, from what source will these characters be derived? From sense objects? But sense objects do not have them to contribute. Or from intelligibles? But intelligibles have them more perfectly. To say they come from nonbeing is the most impossible of all; for nature has in no wise produced imperfect beings and left the perfect ones nonexistent, and it is impious to suppose that our soul generates things more accurate, perfect, and better ordered than Nous and the gods.

Prior to sense objects, therefore, are the self-moving intelligible and divine ideas of the figures. Although we are stirred to activity by sense objects, we project the ideas within us, which are images of things other than themselves; and by their means we understand sensible things of which they are paradigms and intelligible and divine things of which they are likenesses. As these ideas within us unfold, they reveal the forms of the gods and the uniform boundaries of the universe by which the gods, without command, bring all things back to themselves and enclose them. The gods have a wondrous knowledge of the universe of figures and a potency capable of generating and supporting all secondary things; the figures in the realm of nature have the power of creating appearances, though they are devoid of knowledge and intelligent comprehension;

141

individual souls have immaterial thought and spontaneous knowledge, but not the generative and activating

---

[110] 140.1ff. With this argument cf. the more extended one at 12.10ff.

cause. Therefore just as nature stands creatively above the visible figures, so the soul, exercising her capacity to know, projects on the imagination, as on a mirror, the ideas of the figures; and the imagination, receiving in pictorial form these impressions of the ideas within the soul, by their means affords the soul an opportunity to turn inward from the pictures and attend to herself. It is as if a man looking at himself in a mirror and marvelling at the power of nature and at his own appearance should wish to look upon himself directly and possess such a power as would enable him to become at the same time the seer and the object seen. In the same way, when the soul is looking outside herself at the imagination, seeing the figures depicted there and being struck by their beauty and orderedness, she is admiring her own ideas from which they are derived; and though she adores their beauty, she dismisses it as something reflected and seeks her own beauty. She wants to penetrate within herself to see the circle and the triangle there, all things without parts and all in one another, to become one with what she sees and enfold their plurality, to behold the secret and ineffable figures in the inaccessible places[111] and shrines of the gods, to uncover the unadorned divine beauty and see the circle more partless than any center, the triangle without extension, and every other object of knowledge that has regained unity. Clearly, then, the self-moved figure is prior to what is moved by another; the partless is prior to the self-moved; and prior to the partless is the figure which is identical with unity. For all figures attain consummation in the henads,[112] the source from which they all entered into being.

142

[111] 141.23 Restoring Grynaeus' αἰτίοις for ἀγγελοις in Friedlein.

[112] 142.5 The doctrine of the divine henads is a development in Neoplatonic thought after Plotinus. The henads are unities beside the One and help to bridge the gulf between the One and reality. Syrianus and Proclus gave the doctrine a theological interpretation by identifying the henads with the gods of traditional Greek mythology. The doctrine "represents an attempt to account for the existence of individuality by importing plurality into the first hypostasis, yet in such a manner as to leave intact the perfect unity of the One." See Proclus' *Elements of Theology*, Props. 113-127, with the commentary of Dodds, 259. On the relation of the divine henads to the figures see 146.1ff.

But we have drawn out at great length these matters of Pythagorean doctrine. Our geometer, looking at the figure in imagination and primarily defining it (although his formula fits sensible things in a secondary way), says that figure is what is contained by a boundary or boundaries. He takes it at once as joined with matter and extended in imagination and rightly calls it limited and bounded. For everything that has matter, whether intelligible or sensible, has a boundary coming from outside itself. Figure is not itself a limit, but limited; it is not its own boundary (the bounding is other than what is bounded), nor is it in it but contained by it. Since it is born with quantity and subsists with it, quantity is its substratum, while the definition of that quantity is the figure, that is, its form and shape.[113] For figure limits it, giving it a character and such-and-such a boundary, either simple or composite. Since figure too, like the idea of the angle, exhibits in its own subdivisions the twofold progression of the Limit and the Unlimited, it applies the single boundary and the simple form to the things it bounds when it acts in accordance with the Limit and the many boundaries by virtue of the Unlimited. This is why everything figured has either one or more than one boundary.

143

Euclid, then, calling figure the figured and enmattered thing coexistent with quantity, naturally designates it as contained. Posidonius, however, defines figure as the containing limit, separating the idea of figure from quantity and making it the cause of definiteness, limitation, and inclusion; for the factor that encloses is other than what is enclosed, the limit other than what is limited. It seems that he is looking at the outer enclosing boundary, while Euclid is looking at the whole of the object. So one of them says the circle is a figure by virtue of the whole plane surface with its outer circuit, whereas the other says it is a figure by virtue of the circumference. The one shows that he is defining the figured and considering it together with its substratum, the other that he wishes to ex-

---

[113] 142.22 I assume that καὶ μορφή, if it belongs in the text at all, has been misplaced.

press the idea of figure itself as limiting and confining the quantity.

If a captious logician should criticize Euclid's definition because it defines the genus by means of the species (for what is included within one boundary and what is included within more than one are species of figures), we must reply to him that genera[114] already contain in themselves the characters of their species. Whenever the ancients wish to make clear the nature of a genus from the powers it contains,[115] they appear to proceed by way of the species, although in fact they are explaining the genus from itself and from the powers it contains. Thus this single definition of figure includes, by virtue of the Limit and the Unlimited in it, the differentiae of the many particular figures; and he who defines it thus is not out of order in including in his definition the differentiae of its powers.

But whence comes the idea of figure and from what sort of principles is it perfected? I answer, first, that it owes its being to the Limit and the Unlimited and the Mixture of the two. This is why it generates some kinds by virtue of the Limit, others by reason of the Unlimited, and others according to the Mixed. For circular figures it invokes the idea of the Limit, for rectilinear that of the Unlimited, and for figures derived from both the idea of the Mixed. In the second place, it is perfected by the kind of wholeness that discriminates unlike parts so that, when it applies wholeness to any form, it also divides the figure into the different forms that constitute it. The circle, for example, and every rectilinear figure can be divided into parts unlike their ideas, a matter which the author of the *Elements* himself takes up in his *Divisions*, where he divides a given figure sometimes into like and sometimes into unlike parts. Thirdly, it has the potency of thoroughgoing plurality, and through this it exhibits all kinds of shapes and generates multiform ideas of figures, unfolding unceasingly

144

145

[114] 143.25 Reading with Grynaeus and Schönberger καὶ τὰ γένη for κατὰ γένη in Friedlein.

[115] 144.3 Putting Friedlein's comma after instead of before οἱ παλαιοί.

until it has reached the end term and revealed every variety of kind. Just as in the upper world it is shown that the One coexists with being and being exists in the One, so also the idea of figure shows that circular lines are implicated in straight and straight in circular; that is, it projects its whole nature in characteristic fashion in each thing, and all of them are in all when the whole is simultaneously in all of them and in each separately. And this power it possesses from that higher ordering. In the fourth place, it receives from the first number the measures applicable to the procession of the kinds and thus constitutes all things according to numbers, some by simpler, others by more complex numbers. Triangles, squares, pentagons, and all polygonal figures issue forth in company with the inexhaustible variations in numbers. Through what cause this comes about most men do not know; but to those who know the place of number and of figure the explanation of the cause is transparent. In the fifth place, from another and secondary kind of wholeness, that which divides into similar parts, it is equipped to divide forms into parts like one another, whereby it resolves the triangle into triangles and the square into squares. This is just what we do, as I said, when we exercise ourselves with [sensible] images of them; but the procedure has its prototype in the first principles themselves.

146   When we attend to these explanations, we are able to interpret many characteristics of figures by tracing them back to the causes that are prior to them. The one universal figure has the rank we have assigned to it, receiving its perfecting power from all the causes mentioned. Thence it proceeds to the species of the gods,[116] distributing its shapes and acting differently towards the different gods, to some giving simple shapes, to others mixtures of the simple one, allotting to some the basic figures that are generated in plane surfaces, and to those that mount[117] upon solid masses the appropriate solid figures. All of them are in all things, for the forms of the gods are complete and endowed with all powers, yet each one has a

---

[116] 146.6   Omitting κατ' after καὶ in Friedlein.
[117] 146.10   See note at 90.13.

characteristic property assigned to it according to a specific principle. One of them, for example, contains all things after the manner of a circle, another in triangular fashion, and still another after the fashion of a square; and similarly for the solid figures.

XV, XVI. *A circle is a plane figure contained by one line such that all the straight lines falling upon it from one point among those lying within the figure are equal to one another. And the point is called the center of the circle.*

147

The first and simplest and most perfect of the figures is the circle. It is superior to all solid figures because its being is of a simpler order, and it surpasses other plane figures by reason of its homogeneity and self-identity. It corresponds to the Limit, the number one, and all the things in the column of the better.[118] Hence whether you analyze the cosmic or the supercosmic world, you will always find the circle in the class nearer the divine. If you divide the universe into the heavens and the world of generation, you will assign the circular form to the heavens and the straight line to the world of generation; for insofar as the circular form is found in the changes and figures of the world of generation, it is derived from above, from the heavenly order. It is because of the circular revolution of the heavens that generation returns in a circle upon itself and brings its unstable mutability into a definite cycle. If you divide bodiless things into soul and Nous, you will say that the circle has the character of Nous, the straight line that of soul. This is why the soul, as she reverts to Nous, is said to move in a circle.[119] Soul bears to Nous the relation that generation has to the heavens. For the heavens move in a circle (he remarks that the circle is an imitation of Nous),[120]

[118] 147.5 See note at 7.3.
[119] 147.18 *Sc.* by Plato in the *Laws* and the *Timaeus.*
[120] 147.20 I suspect that the words put in parentheses by Friedlein are what was formerly a marginal comment made by an editor. If this supposition is correct, the intended subject of φησὶν is Proclus, not Socrates, as Barocius supposes. There is obviously a lacuna after ψυχῆς in the following line; I have adopted the supplement suggested by Barocius.

but the development of the soul [is in a straight line], because her property is to come to be now in one and now in another form. Again, if you distinguish body and soul, you will put everything that is body on the side of the straight line and make everything psychical partake of the identity and homogeneity of the circle. For the former is composite and possesses varied powers, like the rectilinear figures; the other is simple and intelligent, moving and acting of its own accord, turned inwards, and occupied with itself. Hence the *Timaeus*, though it constructs the primary bodies in the universe by means of straight lines, gives them a circular revolution, that is, a form derived from the soul which uses the cosmos as its vehicle.[121]

From what has been said it is clear that the circle everywhere has primacy over the other figures. But we must also contemplate the entire series to which the circle gives rise. Beginning above and ending in the lowest depth of things, it perfects all of them according to their suitableness for participation in it. On the gods it confers the power of reverting to and being unified with their principles, of remaining in themselves without departing from their own blessedness. The highest unities among them it sets up as centers and aiming-points for the secondary divinities, fixing the plurality of the powers in them firmly about these centers and holding them together by the simplicity of these unities. To intelligent beings the circle gives the power of being continuously active in relation to themselves, enabling them to be filled with knowledge from their own store, to assemble the intelligibles in themselves and perfect their insights from within. For Nous always gives itself the object of its thought, and this object is, as it were, its center; Nous clings to it, loves it, and becomes one with it, converging upon it the whole of its intellectual powers. Souls are illuminated by autonomous life and motion, which enables them to revert to Nous and circle about it, enjoying self-renewal through the special periodic revolutions which unfold the partlessness of Nous. Here again

[121] 148.4 See *Tim.* 34c for the circular revolution and 53cff. for the construction of the primary bodies. On the vehicle see note at 90.13.

the ranks of the intelligibles, like centers, will have preeminence over souls, whose activity it is to revolve about them. For every soul is centered in her intelligent part, where she is truly and most fully one; but because of her plurality she traverses a circle in her desire to embrace the Nous within herself. On the heavenly bodies the circle confers their likeness to Nous, their homogeneity and uniformity, their function of enclosing the universe within limits, their fixed and measured revolutions, their eternal existence without beginning or end, and all such things. The sublunary elements owe to the circle the cycle of their changes, their likeness to the heavenly cycle, the presence of the ungenerated among things generated, of the stationary amidst changing things, and of the bounded amongst divisibles. All things exist eternally through the cycle of generation, and the equilibrium among them all is maintained by its balancing destruction; for if generation were not recurrent, the order of things and the whole cosmic scheme would soon have been dissolved. Animals and plants owe to the circle the likeness between parents and offspring. For animals and plants are born from seed and produce seed in their turn: generation becomes reciprocal, with a recurring cycle of growth from the immature to the fully grown and back again, so that decay accompanies generation. On things that we call "contrary to nature" the circle imposes order by limiting their boundlessness and regulating even them rightly by using the last traces of the powers resident in it. Hence such unnatural events recur at determinate intervals, and times of dearth as well as of fruitfulness are based on the revolutions of the circles, as the myth of Muses has it.[122] All evils may have been banished from the divine to this mortal region, yet even they are in revolution, as Socrates says,[123] and have a share of cyclical ordering. Hence nothing is unmixedly evil and abandoned by the gods; rather the providence that perfects all things brings even the boundless variety of evils under the limit and order appropriate to them.

150

[122] 150.5 *Rep.* 545eff.   [123] 150.8 *Theaet.* 176a.

Thus the circle regulates all things for us down to its humblest beneficiaries and has left nothing without a share in its bounty, as it dispenses beauty, homogeneity, shapeliness, and perfection. Even in numbers it controls the middle centers of the whole procession as it unfolds from unity to the decad. For the numbers five and six alone exhibit the cyclic power by turning back upon themselves in the terms that are derived from them, since when they are multiplied,[124] they end with themselves. Multiplication, as a reaching for plurality, is a likeness of procession, while their ending with their own forms is an image of reversion. The power of the circle brings about both these processes by arousing the generating causes of plurality from what is at rest as a center and by enveloping, after its acts of generation, the plurality of its products into their originating causes. Two numbers, therefore, at the center of the series possess this property: one of them heads the whole class of things capable of reversion, namely, the class of the male and the odd; the other summons the female and the even and whatever belongs to the generative series back to their native causes, in conformity with the power of the circle.

But let us make an end here of these matters and observe[125] that the mathematical account of the circle reaches the height of precision. The definition states that it is a figure (since obviously it is limited and contained on all sides by a single boundary, and hence does not belong to the nature of the Unlimited, but to the column of the Limit), and moreover a plane figure (for figures are found both in planes and in solids), and the first of plane figures (since not only does it surpass solids in simplicity, but it also has the place of primacy among plane figures, being contained within a single line, without variety in its boundaries, and is thus akin to unity and defined by unity), and a figure having all the lines equal that are drawn to this bounding line from one of the points

151

152

---

[124] 150.22 *Sc.* by themselves.

[125] 151.14 Reading with Grynaeus θεωρήσωμεν instead of θεωρήσομεν in Friedlein.

within it.[126] For of the figures bounded by a single line some have all the lines drawn from the middle equal and others do not. The ellipse, for example, is contained by a single line, yet not all the lines drawn to it from the center are equal, but two of them only; and the plane figure delimited by the curve of the cissoid has one surrounding line but no center from which all lines drawn to it are equal. Then, since the center in the circle is always one point (for more than one point cannot be the center), he adds that the lines to the boundary "from one point" are all equal; for though there are an infinite number of points within the circle, one only of this infinite number has the character of the center. And since this one point from which all the lines to the circumference are equal lies either within or outside the circle (for every circle has a pole from which all the lines to the circumference are equal), he adds further "of the points lying within the figure." It is not without reason that he takes into account the center only, ignoring the pole, because he wishes to restrict his consideration to what lies in a plane, and the pole is above the plane assumed. Of necessity, therefore, he adds at the end that this point which lies within the circle and from which all the lines drawn to the circumference are equal is the center of the circle. For there are only two such points, the pole and the center, but one is outside the plane, the other within it. If you imagine, for example, a gnomon standing at the center of the circle, then its extreme point is the upper pole, and all the lines drawn from it to the circumference of the circle are demonstrably equal to one another. Likewise in a cone the apex of the whole figure is the pole of the circle at its base.

153

Now that it has been made precise what is meant by a circle, its center, the circumference of the circle, and the figure as a whole, let us once more ascend from these details to the contemplation of their paradigms. Let us think of the center among them, with its unitary, indivisible, and steadfast

[126] 152.3 The absence of Proclus' usual lucidity and precision in this passage, coupled with its grammatical irregularities, suggests that it has been corrupted.

superiority in every instance; the distances from the center, as the ways in which this unity issues forth as far as possible into indefinite plurality; and the circumference of the circle, as the element through which, in the reversion to the center by the things that have gone forth from it, the many powers are wrapped about their own unitary source, all pressing towards it and desiring activity around it. As in the circle the center, the distances, and the outer circumference all exist at the same time, so also in the paradigms there are no parts that are earlier in time and others that come to be later, but all are together at once—rest, procession, and reversion. But the figures differ from the paradigms in that the latter are without parts or spatial intervals, whereas the figures are divided, the center being in one place, the lines from the center in another, and the circumference that bounds the circle in still another. But up there they are all in one. If you take what corresponds to the center, you will find everything in it; if you take the procession coming out of the center, you will find that this also contains everything; and likewise if you take the reversion. When you have seen that they are all of them in each other, and have discounted the imperfection implicit in their extendedness, and have banished from thought the spatial position around which they are distributed, you will discover the truly real circle itself—the circle which goes forth from itself, bounds itself, and acts in relation to itself; which is both one and many; which rests and goes forth and returns to itself; which has its most indivisible and unitary part firmly fixed, but is moving away from it in every direction by virtue of the straight line and the Unlimited that it contains, and yet of its own accord wraps itself back into unity, urged by its own similarity and self-identity towards the partless center of its own nature and the One that is hidden there. And once it has embraced this center, it becomes homogeneous with it and with its own plurality as it revolves about it. What turns back imitates what has remained fixed; and the circumference is like a separate center converging upon it, striving to be the center and become one with it and to bring the reversion back to the point from which the procession began.

154

Such is the character of the center everywhere. It ranks as a goal for the beings that have their existence around it and as the source of the multiple processions from it. This is what the mathematical center typifies, since it is the end point of all the lines that lead to the circumference and presents equality to them as the image of its own unity. It is thus that the Oracles define it: "The center, from which all the lines to the rim are equal . . . ."[127] But as "from which" indicates the source of the divergence of the lines, so "to which" indicates the center of the circumference; for the circumference in all its extent gathers itself towards the center.

If we must identify the first cause through which the circular figure is brought to light and perfected, I should say it is the very highest of the intelligibles. For the center resembles the principle of the Limit, while the lines from it, being indefinite in number and length, typify Boundlessness, so far as in them lies; and the line which bounds their indefinite extendedness and gathers it back to the center is like the hidden cosmic order they constitute, which Orpheus[128] describes as moving in a circle:

> The Boundless in a circle
> Was moving unweariedly.

For since it moves in an intelligent fashion about the intelligible and has that as the center of its motion, it is properly said to work cyclically. Hence from this proceeds the triadic god who comprehends in himself the primary cause of the procession of the rectilinear figures; and from this comes the appellation[129] placed upon this god by the wise men most familiar with theological mysteries. And the triangle is the first

155

156

[127] 155.5   Barocius' translation contains the beginning of the following line "and to which . . . ," which is missing in Friedlein. Something like this may have been in the original text to explain Proclus' reference to this phrase in the following line. But Taylor notes that πρὸς ὅ is wanting in all the published collections of the Babylonian Oracles. Kroll (65) cites the text without this supplement.

[128] 155.18   Otto Kern (ed.), *Orphicorum Fragmenta*, Berlin, 1922, No. 71a.

[129] 155.26   Τρισμέγιστος (?). See note at 46.25.

of the rectilinear figures. The figures, then, first come to light in the successive hierarchies of the gods, but they have their being in the preexisting hidden causes in the intelligible world.

XVII. *A diameter of the circle is a straight line drawn through the center and terminated in both directions by the circumference of the circle; and such a straight line also bisects the circle.*

The author of the *Elements* himself makes it clear that he is defining not every diameter, but the diameter of the circle. The square also has a diameter, and so does the parallelogram in general, and among solid figures the sphere. But in these cases such a line is also called a "diagonal," and in the case of the sphere an "axis" also, "diameter" alone being used for the circle. Even for the ellipse, the cylinder, and the cone we are accustomed to say "axis," "diameter" being peculiar to the circle. The genus of diameter is straight line. But there are many straight lines in the circle, as there are indefinitely many points;[130] and just as the center is one of those points, so only that line is called the diameter which goes through the center and which neither stops short of the circumference nor goes beyond the boundary of the circle but is terminated by the circumference in both directions. This shows how the diameter is drawn. What is added at the end, that it cuts the circle in half, indicates its peculiar effect upon the circle in comparison with all the other straight lines drawn through the center but not terminated by the circumference in both directions.

The famous Thales is said to have been the first to demonstrate that the circle is bisected by the diameter. The cause of this bisection is the undeviating course of the straight line through the center; for since it moves through the middle and throughout all parts of its identical movement refrains from swerving to either side, it cuts off equal lengths of the circumference on both sides. If you wish to demonstrate this mathematically, imagine the diameter drawn and one part of

157

---

[130] 156.21f. Omitting with Barocius the full stop after σημείων and μὲν οὖν after ἀπείρων.

the circle fitted upon the other. If it is not equal to the other, it will fall either inside or outside it, and in either case it will follow that a shorter line is equal to a longer. For all the lines from the center to the circumference are equal, and hence the line that extends beyond will be equal to the line that falls short, which is impossible. The one part, then, fits the other, so that they are equal. Consequently the diameter bisects the circle.[131]

158

But if from one diameter two semicircles are produced, and if an indefinite number of diameters can be drawn through the center, it will follow that the number of semicircles is twice infinity.[132] This difficulty is alleged by some persons against the indefinite divisibility of magnitudes. We reply that a magnitude is indefinitely divisible, but not into an infinite number of parts. The latter statement makes an infinite number actual, the former merely potential; the latter assigns existence to the infinite, the other only genesis.[133] With one diameter, then, two semicircles come into being, and the diameters will never be infinite in number, even though they can be taken indefinitely. So the number of semicircles will never be twice infinity; those that are produced at any time will be twice a finite number, for the diameters taken at any time are always finite in number. And why should not every magnitude have only a finite number of divisions,[134] seeing that number exists before magnitude and sets bounds to its sections, thus forestalling infinity by bounding at any time only what has come into being?

[131] 158.2 "Presumably Proclus gives Thales' proof. Euclid's incorporating the assertion in a definition where it certainly doesn't belong is probably due to a desire to avoid proofs in which a geometric object is moved. In the same way Euclid probably states Post. IV to avoid a proof in which motion is used, like the one given by Proclus at 188.20ff. See von Fritz, in *Archiv für Begriffsgeschichte* I, 76ff." (I.M.)

[132] 158.5 "This is the earliest example known to me of an argument relating to those paradoxical features of infinite collections which Cantor was the first to straighten out." (I.M.)

[133] 158.10 There is more about the infinite below, at 278f. and 284ff.

[134] 158.16 Reading οὐ μέλλει for οὐ μέλει in Friedlein.

XVIII, XIX.[135] *A semicircle is the figure contained by the diameter and the circumference cut off by it. And the center of the semicircle is the same as that of the circle.*

From the definition of the circle our author discovered the nature of the center, which differs from all the other points in the circle; and from the center he defined the diameter and distinguished it from the other straight lines drawn within the circle. And from the diameter he teaches us what the semicircle is, that it is contained by two boundaries—two boundaries always different from each other, a straight line and a circumference—and that the straight line is not any chance line, but the diameter of the circle. For a segment of a circle less than and a segment greater than a semicircle will each be contained by a straight line and a circumference; but such segments are not semicircles, because the division of the circle is not made through the center.

Now all figures of this sort are dyadic, as the circle is monadic, and are composed of unlike elements. For every figure contained by two boundaries is contained either by two circumferences, like the lunule; or by a straight line and a circumference, like the figures just mentioned; or by two mixed lines, as when two ellipses intersect (for their boundaries will cut off and enclose an area);[136] or by a mixed line and a circumference, as when a circle cuts an ellipse; or by a mixed and a straight line, such as the half of an ellipse. The semicircle, then, is formed of unlike parts, but of parts that are simple and that are joined to one another by juxtaposition. Quite properly, therefore, before the definition of triadic figures, the exposition proceeds, after the circle, to the figure formed by two elements. Two straight lines cannot enclose an area, but a straight line and a circumference can; and so can two circumferences, either making angles, as in the lunular figure, or forming a figure without angles, as do two con-

159

160

---

[135] 158.21 These two definitions are numbered together as Def. XVIII in Heiberg's text of Euclid.

[136] 159.18 Reading τὸ for τὰ in Friedlein, and in the following line χωρίον for the impossible σημεῖον.

centric circles.[137] The area cut off between them is contained by two circumferences, the one inside, the other outside; and they do not form an angle, for they do not cut one another, as in the lunule and in the biconvex figure.[138]

It is clear, furthermore, that the center of the semicircle is the same as the center of the circle. For the diameter with the center upon it completes the semicircle, and from this center all the lines to the circumference are equal; and the circumference is also part of the circle, and lines drawn to all parts of the circumference of the circle are equal when drawn from the center. Hence the center of the circle and the center of the semicircle are one. It should be noted that this alone of the figures, that is, of plane figures, has its center on its perimeter. So one can summarize and say that there are three positions for the center: either within the figure, as with the circle, or on the perimeter, as with the semicircle, or outside it, as with certain conic lines.

Therefore the semicircle has the same center as the circle. What does this characteristic of the semicircle indicate to us, and of what things is it a likeness? Does it not show that the things which have not fully departed from first principles but still have some share of them can be concentric with them and participate in the same causes? For the semicircle has two things in common with the circle, its diameter and its circumference. This is why they have a common center. And so perhaps the semicircle is to be compared with those beings of secondary rank below the simplest principles which still participate in them and because of their kinship with them, though it is imperfect and halfway, can nevertheless be traced back to being and to their primary cause.

161

[137] 160.5   These two possibilities are illustrated as follows in one of our early manuscripts:

[138] 160.9   See illustrative diagram above at 127.14.

XX-XXIII.[139] *Rectilinear figures are those which are contained by straight lines, trilateral figures being those contained by three, quadrilateral those contained by four, and multilateral those contained by more than four straight lines.*

After the figure with one boundary, which has the relation of a first principle to all the figures, and the semicircle with its two boundaries, our author presents the procession of rectilinear figures corresponding to the endless number series. This explains why he mentions the semicircle, because with respect to its boundaries it has something in common with the circle on the one hand and with rectilinear figures on the other, just as two is intermediate between unity and plurality.[140] Unity produces a greater quantity by addition than by multiplication, whereas number has the reverse effect, producing a greater quantity by multiplication than by addition; but the number two produces an equal quantity when multiplied by itself and when added to itself. So just as two is a mean between unity and plurality, so the semicircle has a community with rectilinear figures with regard to its base and with the circle with regard to its circumference.

The rectilinear figures come forth in orderly fashion according to the series of numbers from three to infinity. This is why the author of the *Elements* begins here. He defines "trilateral," "quadrilateral," and then figures called by a common name "multilateral." Trilateral figures are also multilateral, but they have a special as well as the common designation, whereas for the others, since we are unable to follow the endless procession of the numbers, we are content to use the common designation. Our author mentions only the trilateral and the quadrilateral, since three and four are the first of the numbers, three being unmixedly odd among the odd numbers and four the most even of even numbers. He brings in these two for the construction of the rectilinear figures in order to show their dependence on all the numbers,

162

[139] 161.13   Def. XIX in Heiberg's text.
[140] 161.25   Literally "between unity and number." The Pythagoreans defined number as a plurality of units.

both even and odd; moreover, because in the first book he is going to explain triangles and parallelograms as the most elementary figures, he naturally stops the particular enumeration with them and includes all the others under the common designation "multilateral." So much for this.

163
We must now begin afresh and say that of plane figures some are contained by simple lines, others by mixed, and others by both sorts. Of those contained by simple lines some are contained by similar lines, such as rectilinear figures, and some by dissimilar lines, such as semicircles and segments and arches that are less than semicircles. Of those contained by similar lines some are bounded by circular, others by straight lines; and of those contained within circular lines some are bounded by one, others by two, and others by several. The circle itself is bounded by one line; of those bounded by two, some are without angles, like the "crown" that is contained between the circumferences of two concentric circles, and others with angles, like the lunule; and of those bounded by more than two there are an indefinite number, for some figures are contained by three or four or more circumferences. If three circles are tangent, for example, they will cut off a three-sided area bounded by three circumferences; and if there are four tangent circles, there will be an area bounded by four circumferences; and so on in the same way. Of figures contained by straight lines some are bounded by three, others by more. It is not possible for an area to be contained by two straight lines, still less by one such; so every area that is contained within one or two boundaries belongs either to those bounded by mixed lines or to those bounded by circumferences. "Bounded by mixed lines" can
164
mean either that mixed lines contain the area, such as the area cut off by the curve of a cissoid, or that unlike boundaries contain it, as with the arch. For mixture may come about in two ways, either by juxtaposition or by fusion.[141] Not every

---

[141] 164.4 Barocius' text at this point includes the following sentence not contained in Friedlein: "Hence every rectilinear figure is either triangular, or quadrilateral, or multilateral in varying degrees." In the next line "or multilateral" is inserted after "quadrilateral."

trilateral or quadrilateral figure is rectilinear, for such a number of sides can occur in figures bounded by circumferences. So much for the classification of plane figures.

It has been said earlier that the straight line is a symbol of procession, motion, and infinity and resembles those generating and diversifying deities in the hierarchy that are çauses of change and motion. It follows that rectilinear figures are at home with those gods that preside over the generating activity of the forms as they go forth into all things. This is why the world of generation is primarily ordered in conformity with these figures; it derives its being from them, since its existence is dependent on motion and change.

XXIV-XXIX.[142] *Of trilateral figures an equilateral triangle is that which has its three sides equal, an isosceles triangle that which has two of its sides alone equal, and a scalene triangle that which has its three sides unequal. Further, of trilateral figures a right-angled triangle is that which has one of its angles a right angle, an obtuse-angled triangle that which has one of its angles obtuse, and an acute-angled triangle that which has its three angles acute.*

165

The classification of triangles here is based partly on their angles and partly on their sides. The classification based on sides, being familiar, comes first; then follows the classification based on angles, which is specifically characterizing; for the three angles—right, obtuse, and acute—belong only to rectilinear figures, whereas equality and inequality of sides are obviously found also in nonrectilinear figures. He says therefore that some triangles are equilateral, some isosceles, and some scalene; for a triangle either has all its sides equal, or all of them unequal, or only two of them equal. Then, starting afresh, he says that some triangles are right-angled, some obtuse-angled, and some acute-angled. The right-angled triangle is defined as having one of its angles a right angle, and the obtuse-angled similarly, for it is impossible that a triangle should have more than one right or one obtuse angle; and the acute-angled triangle is that which has all its

[142] 164.18 Deff. XX and XXI in Heiberg's text.

angles acute. It is not sufficient in this case to say that it has one acute angle, for thus all triangles would be acute-angled, since any triangle whatever has two acute angles, the acute-angled triangle alone having all three angles acute.

But it seems to me that, when the author of the *Elements* makes a separate classification based on the angles and another based on the sides, he is simply recognizing that not every triangle is also trilateral. For there are four-sided triangles, called "barb-like" by others,[143] but "hollow-angled" by Zenodorus.[144] Think of a three-sided figure having on one of its sides two lines projecting inwards. It will have an area bounded by the two outer [lines and the two inner ones, and it will have one angle contained by the two outer lines][145] and two others contained by the outer lines and the inner ones and situated at the extremities where these two pairs of lines converge. Such a figure is clearly a four-sided triangle.[146] It does not follow, then, whenever we find a figure with three angles—whether all acute, or one right, or one obtuse—that we have necessarily found a trilateral, that is, an isosceles triangle or some other three-sided figure; for it could even be four-sided. You could likewise obtain a quadrangle with more than four sides. So we cannot at once declare from the number of its angles how many sides a figure has. But enough of this.

The Pythagoreans assert that the triangle is the ultimate source of generation and of the production of species among things generated. Consequently the *Timaeus* says that the ideas of natural science, those used in the construction of the cosmic elements, are triangles.[147] They are divided into three kinds, bring into unity things that are in every way divided

166

[143] 165.23 I follow Friedlein's conjecture that παρ' αὐτοῖς is a corruption of παρ' ἄλλοις.

[144] 165.24 Zenodorus, a geometer of the second or first century B.C. who wrote a treatise on isometric figures, considerable parts of which have been preserved by Theon of Smyrna and by Pappus. See Heath II, 207-213; Gow, 271f.; Van der Waerden, 268f.

[145] 166.2f. There is another lacuna here. I have followed the conjectural restoration of Friedlein.

[146] 166.6 For a diagram and discussion of this figure see 328.21ff. and note at 329.7.

[147] 166.18 *Tim.* 53eff.

and changeable, are full of the indefiniteness of matter, and set up in advance the dissoluble bonds of material bodies, just as triangles themselves are contained by straight lines and have angles that bring together the plurality of these lines and provide them an imported fellowship and contact with one another. Rightly, then, Philolaus[148] dedicates the angle of the triangle to the four gods Kronos, Hades, Ares, and Dionysus, since he includes within their province the entire fourfold ordering of the cosmic elements derived from the heavens or from the four segments of the zodiacal circle. Kronos gives being to all the moist and cold essences, Ares engenders every fiery nature, Hades has control of all terrestrial life, and Dionysus supervises moist and warm generation, of which wine, being moist and warm, is a symbol. All these are distinct as far as their action on secondary things is concerned, but they are united with one another, and this is why Philolaus brings them to unity under one angle. If the differences between triangles contribute to the process of generation, it is reasonable to admit that the triangle is the chief agent in the production of sublunary things. The right angle furnishes them with their essence and bounds the measure of their being, and the idea of the right-angled triangle therefore is the essence-constituting factor for the generated cosmic elements. The obtuse angle furnishes them with extendedness in general, and the idea of the obtuse-angled triangle enlarges the enmattered forms and increases their variety and extent. The acute angle renders nature herself subject to division, and the idea of the acute-angled triangle prepares the way for the endless distinctions that come to be. In short, the idea of the triangle underlies the extended and completely divisible being of material bodies. So much we thought it necessary to observe regarding triangles.

From these classifications[149] you can understand that the

---

[148] 167.1   On Philolaus see note at 22.14; and for other references to him see 130.10, 173.11, 174.4.

[149] 168.4   This phrase evidently refers back to the early part of the commentary on this definition, 164.27-166.13, the material about the Pythagoreans being in a sense a digression.

species of triangle are seven in all, neither more nor less. The equilateral triangle is one only and is acute-angled; but each of the other two has three kinds. The isosceles is either right-angled, obtuse-angled, or acute-angled; and the scalene likewise has the same three forms. If, then, each of these exists in three species and equilateral triangles are of one kind only, let us say that there are seven kinds of triangle in all. You can also understand from the differences found in their sides the analogy they bear to the orders of being. The equilateral triangle, always controlled by equality and simplicity, is akin to the divine souls, for equality is the measure of unequal things, as the divine is the measure of all secondary things. The isosceles is akin to the higher powers that direct material nature, the greater part of which is regulated by measure, whereas the lowest members are neighbors to inequality and to the indeterminateness of matter; for two sides of the isosceles are equal, and only the base is unequal to the others. The scalene is akin to the divided forms of life that are lame in every limb[150] and come limping to birth filled with matter.

169   XXX-XXXIV.[151] *Of quadrilateral figures a square is that which is both equilateral and right-angled, an oblong that which is right-angled but not equilateral, a rhombus that which is equilateral but not right-angled, and a rhomboid that which has its opposite sides and angles equal to one another but is neither equilateral nor right-angled. Let · quadrilaterals other than these be called trapezia.*

Quadrilaterals ought first to be divided into two groups, one called parallelograms, the other nonparallelograms, and parallelograms into some that are both right-angled and equilateral, such as squares, others that are neither, such as rhomboids, and others either right-angled and not equilateral, such as oblongs, or equilateral and not right-angled, such as rhombi. For parallelograms necessarily have either both equality of sides and right-angledness, or neither of them, or one of them only; and the last is possible in two ways, so that

---

[150] 168.23   χωλεύουσι, a play on the meaning of σκαληνός, "lame."
[151] 169.1   Def. XXII in Heiberg's text.

170   parallelograms exist in four species. Of nonparallelograms some have only two sides parallel and the other sides not, and some have no sides at all parallel; the former are called trapezia, the latter trapezoids. Of trapezia some have the sides that join the parallels equal, others have them unequal; the former are called isosceles trapezia, the latter scalene. Hence there will be seven kinds of quadrilaterals: the square, the oblong, the rhombus, the rhomboid, the isosceles trapezium, the scalene trapezium, and the trapezoid.[152]

Posidonius makes a perfect division of rectilinear quadrilaterals by positing these seven species of them, as he does

171   also for the triangle.[153] Euclid, however, cannot make the division into parallelograms and nonparallelograms, since he has not spoken of parallel lines nor taught us what the parallelogram itself is. All trapezia and trapezoids he calls by the general name "trapezia," thus setting them off from the other four classes, to which he correctly assigns the property of parallelograms, that is, of having their opposite sides and angles equal. For the square, the oblong, and the rhombus have their opposite sides and angles equal; but he adds this only for the rhomboid, since to say that it is neither equilateral nor right-angled would be to set it forth by mere negations. When we are at a loss for specific characterizing definitions, we are compelled to use generic terms; and that he himself shows this character to be common to all parallelograms we shall learn later.

The rhombus resembles a square that has been shaken, and the rhomboid an oblong that has been set in motion; hence

[152] 170.12   The Munich codex has the following diagrams in the margin:

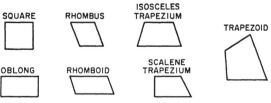

[153] 170.15   Cf. 168.5-12 where, however, Posidonius is not mentioned.

they do not differ from the square and the oblong with respect to their sides, but only in the obtuseness or acuteness of their angles, the others being right-angled. If you imagine the square or the oblong being pulled at opposite corners, you will find these angles contracting and becoming acute, the others spreading out and appearing obtuse. It seems that even its name is applied to the rhombus from this motion. If you imagine the square in revolution,[154] it will appear to be distorted at its corners, just as the circle when twirled like a sling appears to be an ellipse.

One might well ask why the square specifically has received the designation "quadrangle."[155] "Triangle" is the common term for all three-angled figures, including those not equiangular or equilateral; and "pentagon" is similarly used. Why, then, cannot "quadrangle" likewise be applied to other four-sided[156] figures? Our geometer indeed, when discussing these other figures, employs such phrases as "equilateral triangle"[157] and "a pentagon which is equilateral and equiangular,[158] which indicate that the triangle or the pentagon could be other than equilateral and equiangular. But the word "quadrangle" at once means the equilateral and right-angled figure. The reason is this. The square is the only area whose very nature exhibits an ideal with respect both to its sides and to its angles. [Its sides are equal, and] each of its angles, being a right angle, holds the measure of angles, a measure permitting neither increase nor decrease; because it is superior in both respects, then, it has properly received the generic designation. The triangle, even though it has equal angles, has them all acute, and in the pentagon they are all obtuse. Rightly, therefore, the square, being perfected by the equality

172

[154] 172.2 ρομβούμενον. Cf. ρέμβω, "twirl."
[155] 172.5 τετράγωνον. This term was used alike for "square" and for "quadrilateral" before and even after Euclid. His introduction of τετράπλευρον enabled ambiguity to be avoided (Heath, *Euclid* I, 188), yet the older use persisted. Even Proclus (166.10) has said that we could have τετράγωνα with more than four sides, and here the term can hardly mean "squares."
[156] 172.10 Proclus should have said "four-angled."
[157] 172.11 E.g. in Def. XXIV (Heiberg's XX).
[158] 172.12 E.g. in Bk. IV, Prop. 11.

173

of its sides and the rightness of its angles, has alone of all quadrilaterals received this designation; for to those species that excel we often apply eulogistically the designation of the genus.

The Pythagoreans thought that this more than any other four-sided figure carries the image of the divine nature. It is their favorite figure for indicating immaculate worth; for the rightness of the angles imitates integrity, and the equality of the sides abiding power. Change is the offspring of inequality, and steadfastness of equality; hence the causes of the firm foundation of all things and of pure and impartial power are naturally expressed by the square figure as an image of these properties. Philolaus, moreover, in another of his reflections calls the angle of the square the angle of Rhea, Demeter, and Hestia. For since the square is the substance of earth and the element nearest it, as we learn from the *Timaeus*,[159] and since the earth receives effluences and generative powers from all these goddesses, he rightly dedicates the angle of the square to these life-giving divine forces. For some call the earth Hestia or Demeter, and they say that it partakes of all that Rhea is, and in her are all the generating causes in earthly fashion. Hence he declares that the single bond of unity among these species of the divine is the angle of the square. They also liken the square to the whole of virtue, since it has four right angles, each of them perfect in the way in which

174

we say that each of the virtues is perfect and self-sufficient, namely, as a measure and a landmark for life, and all of them intermediates between the obtuse and the acute angles. We must not omit to observe that Philolaus dedicates the angle of the triangle to four gods and the angle of the square to three,[160] showing their penetration of one another and the communion of all in all, of odd numbers in the even and of even in the odd. Hence a tetradic triad and a triadic tetrad that partake

[159] 173.15 This passage implies that the construction of the primary bodies described in the *Timaeus* should be attributed to Philolaus. But see note at 22.14.

[160] 174.5 Cf. 130.10-12, 167.1-3.

of the generative and creative goods maintain the whole order of generated things. The number twelve, which is their product, ascends towards a single monad, the sovereignty of Zeus. Philolaus says that the angle of the dodecagon is the angle of Zeus, because Zeus holds together in a single unity the whole duodecimal number. In Plato likewise Zeus leads "the twelve" and has absolute dominion over all things.[161]

So much we thought it necessary to say about four-sided figures in order to bring out the thought of the author of the *Elements* and also to provide starting-points for speculative reflections to those who seek knowledge of the intelligible and invisible world.

175

XXXV.[162] *Parallel straight lines are straight lines which, being in the same plane and being produced indefinitely in both directions, do not meet one another in either direction.*

The basic propositions about parallels and the attributes by which they are recognized we shall learn later, but what parallel straight lines are is defined in the words above. They must lie in one plane, he says, and when produced in both directions do not meet but can be extended indefinitely. Lines that are not parallel may be produced to a certain distance without meeting, but what characterizes parallel lines is that they do not meet when extended indefinitely; and not simply this, they are capable of indefinite extension in both directions without meeting. Lines that are not parallel may be capable of indefinite extension on one side but not on the other; as they near each other on one side, they diverge more on the other. The reason is that two straight lines cannot enclose an area, and they would if they converged on both sides. Further, the definition rightly adds that the straight lines must be in the same plane; for if one of them should be in the given plane and the other above it, they would always be asymptotes to one another, whatever their position, but they would not for this reason be parallel.

[161] 174.16   *Phaedr.* 246ef.
[162] 175.1   Def. XXIII in Heiberg's text.

176 So the plane must be one, and the lines must be produced indefinitely in both directions and meet in neither. When these conditions obtain, they will be parallel straight lines.

This is the way Euclid defines parallel straight lines. But Posidonius says that parallel lines are lines in a single plane which neither converge nor diverge but have all the perpendiculars equal that are drawn to one of them from points on the other. Those lines between which the perpendiculars become progressively [longer or] shorter [intersect somewhere because they][163] converge upon one another; for a perpendicular can determine both the heights of figures and the distances between lines. Hence when the perpendiculars are equal, the distances between the straight lines are equal, but when they become greater or less, the distance increases or decreases and the lines converge on the side on which the perpendiculars are shorter.

But we must understand that absence of intersection does not always make lines parallel, for the circumferences of concentric circles do not intersect; the lines must also be extended indefinitely. This characteristic can be found not only in straight lines but in others as well. One can think of a helix inscribed around a straight line which can be prolonged with the straight line indefinitely and never meet it. Such cases Geminus rightly distinguishes from the former ones at the outset.[164] Some lines, he says, are finite and enclose a figure,

177 like the circle, the perimeter of an ellipse, the cissoid, and many others; others are unlimited and can be extended indefinitely, like the straight line, the section of a right-angled or an obtuse-angled cone,[165] and the conchoid. Again of those capable of being extended indefinitely some never enclose a figure, like the straight line and the above-mentioned conic sections, while others come together and, after making a

---

[163] 176.11 Barocius has a fuller text here. I have included in brackets the words taken from his translation.

[164] 176.26 The following account of Geminus' doctrine supplements an earlier one at 111.3. See note at that point.

[165] 177.5 I.e. the parabola and the hyperbola. See note at 111.8.

figure, then extend on indefinitely.[166] Of these some are asymptotic, namely, those which however far extended never meet, and others that do intersect are symptotic. Of asymptotic lines some are in the same plane with one another, others not; and of the asymptotes that are in the same plane, some are always equidistant from one another, others are constantly diminishing the distance between themselves and their straight lines, like the hyperbola and the conchoid. Although the distance between these lines constantly decreases, they remain asymptotes and, though converging upon one another, never converge completely. This is one of the most paradoxical theorems in geometry, proving as it does that some lines exhibit a nonconvergent convergence. Of the lines which are always equidistant from one another those straight lines which never make the interval between them less and which lie in the same plane are parallel. So much I have selected from Geminus' *Philokalia*[167] to elucidate the subject before us.

[166] 177.9   Heath (*Euclid* I, 160f.) suggests that the curve meant here is a variety of the conchoid. See also Tannery II, 23.

[167] 177.24   Is this the title of Geminus' comprehensive work or of one of its books, of which there must have been many? See Heath, *Euclid* I, 39.

POSTULATES AND
AXIOMS

THE FIRST principles of geometry are divided into three groups: hypotheses, postulates, and axioms. We have explained the difference between them in the earlier portions of this work,[1] but now let us examine particularly and more precisely the distinction between postulates and axioms, since they are the chief subjects of inquiry in the present section. Hypotheses, or what are called definitions, we have considered in the foregoing.

It is a common character of axioms and postulates alike that they do not require proof or geometrical evidence but are taken as known and used as starting-points for what follows. They differ from one another in the way in which theorems have been distinguished from problems.[2] Just as in a theorem we put forward something to be seen and known as a consequence of our hypotheses but in a problem are required to procure or construct something, so in the same way[3] axioms take for granted things that are immediately evident to our knowledge and easily grasped by our untaught understandings, whereas in a postulate we ask leave to assume something that can easily be brought about or devised, not requiring any labor of thought for its acceptance nor any complex construction. Hence clear knowledge without demonstration and assumption without construction distinguish axioms and

---

[1] 178.3 At 76.6ff. Heiberg's text has ὅροι instead of ὑποθέσεις and κοιναὶ ἔννοιαι instead of Proclus' ἀξιώματα. On these questions of terminology and the substantive issues underlying them respecting the foundations of Greek geometry see von Fritz, "Die APXAI in der griechischen Mathematik," in *Archiv für Begriffsgeschichte* I, 1955, 13-102, and Árpád Szabó, "Anfänge der Euklidischen Axiomensystems," in O. Becker (ed.), *Zur Geschichte der Griechischen Mathematik*, Darmstadt, 1965, 355-461.

[2] 178.13 See 77.7ff.

[3] 179.2 Reading with Barocius κατὰ ταὐτὰ for κατὰ ταῦτα in Friedlein.

postulates, just as knowing from demonstration and accepting conclusions by the aid of constructions differentiate theorems from problems.

Principles must always be superior to their consequences in being simple, indemonstrable, and evident of themselves. In general, says Speusippus, in the hunt for knowledge in which our understanding is engaged we put forward some things and prepare them for use in later inquiry without having made any elaborate excursion, and our mind has a clearer contact with them than sight has with visible objects; but others it is unable to grasp immediately and therefore advances on them step by step and endeavors to capture them by their consequences. For example, drawing a straight line from a point to a point is something our thought grasps as obvious and easy, for by following the uniform flowing of the point and by proceeding without deviation more to one side than to another, it reaches the other point. Again if one of the two ends of a straight line is stationary, the other end moving around it describes a circle without difficulty. But if we should wish to draw a one-turn spiral, we need a rather complicated device, for the spiral is generated by a complex of motions; and to construct an equilateral triangle will also require a special method for constructing a triangle. Geometrical intelligence will tell me that, if I think of a straight line one end of which is fixed and the other revolving about it, while a point is moving along it from the stationary end, I describe a monostrophic spiral; for when the end of the line which describes a circle has reached its starting-point at the same time as the point completes its movement along the line, they coincide and make me such a spiral. And again if I describe two equal circles and join their point of intersection with the centers of the circles and draw a straight line from one center to the other, I shall have an equilateral triangle. It is far from true, therefore, that these things can be done at first glance and by simple reflection; we should be content to follow the procedures by which the figures are constructed.

Whether such a construction is made easily or with difficulty, or whether a demonstration proceeds through more or

180

181

fewer middle terms, depends on the aptitudes of those who use these methods; but that a demonstration or a construction is needed at all results from the characteristic that conclusions lack the clarity of postulates and axioms. Both of these, postulate and axiom, must be simple and easy to grasp. But a postulate prescribes that we construct or provide some simple or easily grasped object for the exhibition of a character, while an axiom asserts[4] some inherent attribute that is known at once to one's auditors—such as that fire is hot, or some other quite evident truth about which we say that they who are in doubt lack sense organs or must be prodded to use them. So a postulate has the same general character as an axiom but differs from it in the manner described. For each of them is an undemonstrated starting-point, one in one way, the other in another, as we have explained.

Some persons, however, insist on calling them all postulates, just as they call all inquiries problems. Thus Archimedes at the beginning of his first book *On Equilibria*[5] says "we postulate that when equal weights are taken from equal lengths the remainders are equally balanced." Yet one might rather call this an axiom. Others designate all of them axioms, as they call a theorem everything that requires a demonstration. The same analogy, it seems, has led them to

182

transfer a term from a specific to a general use. Nevertheless, just as a problem differs from a theorem, so a postulate differs from an axiom, even though both of them are undemonstrated; the one is assumed because it is easy to construct, the other accepted because it is easy to know. This is the ground on which Geminus distinguishes postulate from axiom. Others would say that postulates are peculiar to geometry, while axioms are common to all sciences that deal with quantity and magnitude. For it is the geometer who knows that all right angles are equal and that a finite straight line may be produced in a straight line, whereas that things equal to the

---

[4] 181.9 Reading, apparently with Barocius, λέγει for λέγειν in Friedlein.
[5] 181.18 Adopting Hultsch's emendation τοῦ α ἰσορροπιῶν for τῶν ἀνισορροπιῶν (*Rheinisches Museum* N.F. xix, 1864, 450ff.).

same thing are equal to each other is a common notion used by the arithmetician and by all other scientists, each adapting the common truth to his particular subject-matter. But Aristotle, as we have said earlier,[6] maintains that a postulate is demonstrable and, even though not accepted by the learner, can still be taken as a starting-point, whereas the axiom is as such indemonstrable and everyone would be disposed to accept it, even though some might dispute it for the sake of argument.

These, then, are the three ways in which postulate and axiom are distinguished. According to the first—that which bases the distinction on the fact that the postulate produces and the axiom knows—clearly it is not a postulate that all right angles are equal. Nor is the fifth, that when two straight lines are intersected by a third making the two interior angles on one side of it less than two right angles, then the straight lines when extended will meet on that side on which the two angles are less than two right angles. For these statements are not assumed for the sake of any construction, nor do they demand that we produce anything; they only show a characteristic belonging respectively to right angles and to straight lines produced on the side on which the angles are less than two right angles. According to the second mode of distinguishing them, it will not be an axiom that two straight lines do not enclose an area, although some persons still list it as an axiom. For this is a character that belongs to the subject-matter of geometry, like the principle that all right angles are equal. According to the third, the Aristotelian method, everything that can be made convincing by proof will be a postulate, and whatever is indemonstrable an axiom. It was therefore in vain that Apollonius attempted to provide demonstrations for axioms. Geminus aptly comments that the one party have thought up demonstrations for indemonstrables and endeavored to establish what everybody knows by means of less well known middle terms, as Apollonius did when he tried to demonstrate the truth of the axiom that things equal to the

183

6 182.14 See 76.8 and note.

same thing are equal to each other,[7] whereas the other party include things that need demonstration among the undemonstrated matters, as did Euclid himself with his fifth and his fourth postulates. For some say that his fourth also is doubtful and needs to be demonstrated. Is it not ridiculous that theorems whose converses are demonstrable should be ranged among the indemonstrables? For that the interior angles made by intersecting straight lines are less than two right angles is demonstrated by Euclid himself in the theorem "in every triangle two angles taken together in any way are less than two right angles."[8] And it is also clearly demonstrable that the angle equal to a right angle is sometimes not a right angle.[9] We ought not to admit, then, says Geminus, that the converses of these propositions are indemonstrable. Thus it seems, according to his arrangement, that there are only three postulates, the other two and their converses requiring to be established by demonstration, and that it is superfluous to include among the axioms "two lines do not enclose an area" if it can be established by demonstration.

So much for the difference between postulates and axioms. Returning to axioms, we note that some of them are peculiar to arithmetic, some peculiar to geometry, and some common to both. That every number is measured by the number one is an arithmetical axiom; to geometry belong the principles that two equal straight lines will coincide with each other and that every magnitude can be divided indefinitely; but that two things equal to the same thing are equal to each other, and similar axioms, are common to both sciences. But each of them makes use of them only so far as its subject-matter requires, geometry for magnitudes, arithmetic for numbers. In the same way some postulates are peculiar to certain sciences, others are common. That a number can be divided into least

184

---

[7] 183.20　See 194.20ff.

[8] 184.2　"This seems to me a low point in the commentary. Euclid proves XVII because it is a weaker assertion than the parallel postulate. . . . His realization of the necessity for such a postulate despite the provability of its converse may have been his greatest contribution to geometry." (I.M.)

[9] 184.3　Proclus himself shows this at 189.12ff.

parts we should say is a postulate peculiar to arithmetic, that every finite straight line can be produced in a straight line peculiar to geometry, and that quantity is capable of indefinite increase common to both. For both number and magnitude are capable of such increase.

185 POSTULATES I-III. *Let it be postulated to draw a straight line from any point to any point, to produce a finite straight line continuously in a straight line, and to describe a circle with any center and distance.*

These three, because of their clarity and their demand that we produce something, are necessarily ranked among the postulates, at least according to Geminus. The drawing of a line from any point to any point follows from the conception of the line as the flowing of a point and of the straight line as its uniform and undeviating flowing. For if we think of the point as moving uniformly over the shortest path, we shall come to the other point and so shall have got the first postulate without any complicated process of thought. And if we take a straight line as limited by a point and similarly imagine its extremity as moving uniformly over the shortest route, the second postulate will have been established by a simple and facile reflection. And if we think of a finite line as having one extremity stationary and the other extremity moving about this stationary point, we shall have produced the third postulate; for the stationary point will be the center and the straight line the distance, and whatever length this line may have, such will be the distance that separates the center from all parts of the circumference.

If someone should inquire how we can introduce motions into immovable geometrical objects and move things that are 186 without parts—operations that are altogether impossible—we shall ask that he be not annoyed if we remind him of what was demonstrated in the Prologue about things in the imagination,[10] namely, that our ideas inscribe there the images of all things of which the understanding has ideas and that this unwritten tablet was the lowest form of "nous," the "passive."

[10] 186.3 See 51.13-54.14.

This statement, however, does not remove our difficulty, for the "nous" that receives these forms from elsewhere receives them through motion. But let us think of this motion not as bodily, but as imaginary, and admit not that things without parts move with bodily motions, but rather that they are subject to the ways of the imagination. For "nous," though partless, is moved, but not spatially; and imagination has its own kind of motion corresponding to its own partlessness. In attending to bodily motions, we lose sight of the motions that exist among things without extendedness. Partless things are free from material space and external movements, but another kind of motion and another space coordinate with their motion can be discerned in them. We say that the point has position in the imagination and do not ask how something can remain partless and still be moving somewhere and surrounded by space; for the space of extended things is extended, that of unextended things unextended. Consequently the forms peculiar to geometrical objects are quite other than the things whose existence comes from them. The motion of bodies is one thing, the motion of objects conceived in imagination is something else; and the space of extended objects is other than the space of partless beings. We must keep them separate and not confuse them, lest we disarrange the natures of things.

187

It appears that of these three postulates the first expresses in an image how existing things are contained among their more partless causes and bounded by them, and that they are comprehended by them on all sides even before they come to be. The straight line, for example, links already existing points one to another, is bounded by them and included between them. The second postulate shows how things can hold fast to their own origins and yet go out to all things, preserving continuity with their principles and not being separated from them, but ever driven by the all-powerful cause in them to move forth. And the third postulate shows that whatever goes forth turns back again to its own starting-point, for the revolution of the moving part of the line about its stationary end which generates the circle imitates the circular return.

But we must understand that the character of being produced indefinitely does not belong to all lines. It belongs
neither to the circular nor to the cissoid,[11] nor to any of the
figure-describing lines, nor even to all those that do not make
figures. For not even the monostrophic spiral can be produced
indefinitely, since it has its existence between two points; nor
can any other of the lines so generated. Nor is it possible to
join every point with every other by every line, for not every
line can exist between all points. So much for these matters;
let us pass on to what follows.

188   POSTULATE IV. *And that all right angles are equal to
one another.*

If we admit that this statement is self-evident and does not
require proof, it is not a postulate according to Geminus, but
an axiom; for it attributes an intrinsic property to right
angles and does not ask that something be produced by
simple reflection. Nor is it a postulate according to Aristotle's
classification, for in his opinion a postulate requires a proof.
But if we say it can be proved and seek to prove it, neither
then, in Geminus' opinion, will it be classed among the
postulates.[12]

Now the equality of right angles is manifest from our common notions; having the relation of a first term or bound[13]
with respect to the indefinite increase or decrease of the angles
on either side of it, the right angle is equal to every right
angle. For this is how we have produced the primary right
angle, by making equal the two angles on either side of the
upright straight line against which it stands. But if we must
provide a graphic proof of this postulate, let us assume two
right angles, ABC and DEF. I say they are equal. If they are
not, one of them will be greater than the other. Let this be the

[11] 187.21   Heath (*Euclid* I, 164f.) notes that Proclus' conception of
the cissoid and of the single-turn spiral is peculiar in that he thinks of
the former as a closed curve and similarly regards the latter as stopping
short at the point reached after one complete revolution of the straight
line (187.22-23). On the latter see also Tannery II, 39.
[12] 188.11   *Sc.* but rather among the theorems.
[13] 188.13   Reading with Barocius ὅρου instead of ὅρον in Friedlein.

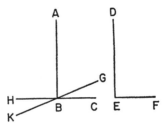

angle at B. Then if DE be made to coincide with AB, the line EF will fall within the angle, say, at BG.[14] Let BC be extended to H. Since ABC is a right angle, so also is ABH, and the two angles are equal to one another (for a right angle is by definition equal to its adjacent angle). Angle ABH, then, is greater than angle ABG.[15] Now let BG be extended to K. Since ABG is a right angle, its adjacent angle also will be a right angle, equal to ABG. The angle ABK, then, is equal to angle ABG, so that angle ABH is less than ABG. But it is also greater, which is impossible. Hence it is false that a right angle can be greater than a right angle.

This proof has been given by other commentators and required no great study. But Pappus[16] has rightly pointed out that the converse of this postulate is not true, namely, that the angle equal to a right angle is always a right angle. Only if the angle is rectilinear will it always be a right angle; it is possible to show that an angle with circular boundaries is equal to a right angle, and clearly we should not call such an angle a right angle. For in our classification of rectilinear angles we assumed that a right angle is produced by a straight line which stands upright with respect to the base line, so that an angle

[14] 189.1 Reading with Grynaeus and Barocius πιπτέτω instead of πιστεύω in Friedlein.

[15] 189.5 Reading $\overline{αβη}$ for the obviously erroneous $\overline{αβγ}$ in Friedlein.

[16] 189.12 Pappus of Alexandria lived at the end of the third and the beginning of the fourth century. He is the author of several commentaries, including one on Euclid, which have been lost, and of a *Collection* in eight books covering the whole of Greek geometry, which is extant and has been edited by Hultsch, 1876-1878. See Heath II, 355-439, and on Proclus' use of Pappus *Euclid* I, 24-27; also Gow, 304-311; and Van der Waerden, 286-290.

equal to a right angle will not always be a right angle unless it is also rectilinear. Let us imagine two equal lines AB and

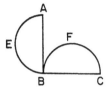

190   BC making a right angle at B, and let us describe with center and distance two equal semicircles upon them, AEB and BFC. Since the semicircles are equal, they will coincide with one another and the angle EBA will be equal to the angle FBC. Let ABF be added to each. Then the right angle as a whole will be equal to the lunular angle EBF, and yet this lunular angle is not a right angle. In the same way, if the angle at ABC is an obtuse or an acute angle, it can be shown that the lunular angle is equal to it, for this is the kind of circular angle that can always be coordinated with rectilinear

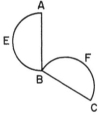 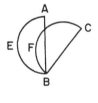

angles. But we should note this: in the case of the right angle and the obtuse angle we must add the angle included between the straight line AB and the circumference BF, whereas in the case of the acute angle it must be subtracted; for the straight line AB cuts the circumference BFC. Both these cases are set forth in the diagrams.

191   Let these proofs, then, be taken as showing both that all right angles are equal to one another and that not every angle equal to a right angle is a right angle. For if such an angle is not even rectilinear, how could it be called a right angle?

This postulate also shows that rightness of angles is akin to equality, as acuteness and obtuseness are akin to inequality. Rightness is in the same column with equality,[17] for both of them belong under the Limit, as does likeness also. But acuteness and obtuseness are akin to inequality, as is also un-likeness; for all of them are the offspring of the Unlimited. This is why those who look at the quantity of angles say that a right angle is equal to a right angle, while others, looking at the quality, say it is similar. For similarity has the same position among qualities that equality has among quantities.

POSTULATE V. *And that, if a straight line falling on two straight lines makes the interior angles on the same side less than two right angles, the straight lines, if produced indefinitely, will meet on that side on which are the angles less than the two right angles.*

This ought to be struck from the postulates altogether. For it is a theorem—one that invites many questions, which Ptolemy[18] proposed to resolve in one of his books—and requires for its demonstration a number of definitions as well as theorems.[19] And the converse of it is proved by Euclid himself as a theorem. But perhaps some persons might mistakenly think this proposition deserves to be ranked among the postulates on the ground that the angles' being less than two right angles makes us at once believe in the convergence and intersection of the straight lines. To them Geminus has given the proper answer when he said that we have learned from the very founders of this science not to pay attention to plausible imaginings in determining what propositions are to

192

[17] 191.8 See note at 7.3.
[18] 191.23 Claudius Ptolemaeus, an Alexandrian astronomer and geographer of the second century, author of the famous Σύνταξις (known to us through the Arabs as the *Almagest*), a comprehensive treatise on Greek astronomy in thirteen books. See Heath II, 273-297; Gow, 293-301; Van der Waerden, 271-274. For more about the book mentioned in this passage see Proclus 365.7-368.23, and Heath II, 295-297.
[19] 191.25 Cf. 365.7ff.

be accepted in geometry. Aristotle likewise says that to accept probable reasoning from a geometer is like demanding proofs from a rhetorician.[20] And Simmias is made by Plato to say, "I am aware that those who make proofs out of probabilities are impostors."[21] So here, although the statement that the straight lines converge when the right angles are diminished is true and necessary, yet the conclusion that because they converge more as they are extended farther they will meet at some time is plausible, but not necessary, in the absence of an argument proving that this is true of straight lines. That there are lines that approach each other indefinitely but never meet seems implausible and paradoxical, yet it is nevertheless true and has been ascertained for other species of lines. May not this, then, be possible for straight lines as for those other lines? Until we have firmly demonstrated that they meet, what is said about other lines strips our imagination of its plausibility. And although the arguments against the intersection of these lines may contain much that surprises us, should we not all the more refuse to admit into our tradition this unreasoned appeal to probability?

193

These considerations make it clear that we should seek a proof of the theorem that lies before us and that it lacks the special character of a postulate. But how it is to be proved, and with what arguments the objections to this proposition may be met, we can only say when the author of the *Elements* is at the point of mentioning it and using it as obvious.[22] At that time it will be necessary to show that its obvious character does not appear independently of demonstration but is turned by proof into a matter of knowledge.

AXIOMS I-V. *Things which are equal to the same thing are also equal to one another; and if equals be added to equals the wholes are equal; and if equals be subtracted from equals the remainders are equal; and the whole is greater than the*

[20] 192.11 *Nic. Eth.* 1094b26f.
[21] 192.13 *Phaedo* 92d.
[22] 193.7 See 364.13 and 371.10ff.

part; and things which coincide with one another are equal to one another.[23]

These are what are generally called indemonstrable axioms, inasmuch as they are deemed by everybody to be true and no one disputes them. Often indeed premises in general, of whatever sort they may be, whether they occur to us with genuine immediacy or require some supplementary teaching, are called "axioms." The members of the Stoa, at least, are accustomed to designate every simple affirmative proposition an "axiom,"[24] and whenever they write logical treatises for us they entitle them "About Axioms."[25] But some persons more accurately distinguish axioms from other premises and call "axiom" a premise that is immediate and self-evident because of its clarity, as Aristotle and the geometers say. According to them, axiom and "common notion" mean the same thing. We are therefore far from inclined to praise the geometer Apollonius for furnishing, as he thought, proofs of the axioms, doing the precise opposite[26] of Euclid; for Euclid enumerated what is demonstrable among the postulates, whereas Apollonius tried to discover demonstrations for indemonstrables. But demonstrables and indemonstrables differ in nature from one another, as we saw; and the sciences dealing with immediate premises that everywhere strike us because of their clarity belong to a different class from the sciences that employ demonstrations, for these get their starting-points from the former, taking and using them as they are needed for establishing their own conclusions.

194

[23] 193.14 Barocius' translation contains ten axioms; Heiberg's Greek text contains nine, of which four are bracketed as presumably later additions, leaving only the five listed by Proclus. But Axioms IV and V in Proclus are listed in the reverse order in Heiberg; and in Heiberg they carry the heading κοιναὶ ἔννοιαι ("common notions") instead of ἀξιώματα ("axioms"). On these variations in designation and in content see note at 178.3.

[24] 194.2 See note at 77.3.

[25] 194.3 I follow Barocius here instead of the unintelligible text of Friedlein. For this title see von Arnim II, 5.6f.

[26] 194.12 Reading with Barocius ἀπεναντίως for ἀπεναντίας in Friedlein.

The proof that Apollonius was persuaded he had discovered for the first of the axioms involves a middle premise that is not better known than the conclusion, if indeed it is not more doubtful. One can learn this from a passing glance at his proof. "Let A be equal to B," he says, "and the latter to C. I say that A is also equal to C. Since A, being equal to B, occupies the same space as it, and since B, being equal to C, occupies the same space as it, A also occupies the same space as C. Therefore they are equal." This argument involves two propositions that we must have previously accepted: one, that things which occupy the same space as each other are equal; the other, that things which occupy the same space as some identical third thing also occupy the same space as each other. Obviously these are far less clear than the proposed axiom.[27] For how do things fill the same space so as to be equal? Do they occupy it simultaneously as wholes, or successively in turn, or by some system of proportion? So it is altogether unacceptable to shift attention to space, which is far more unknown to us than the things in space. At any rate its nature is controversial and difficult to discover. Not to multiply words on this matter, we must present all axioms as immediate and self-evident, known from themselves alone, and trustworthy. He who adjoins a proof to things already abundantly evident does not confirm their truth but weakens the clarity that they have when we accept them without instruction.

This, then, we must accept in advance as the criterion of the peculiar character of axioms and understand that they all belong to the common genus of mathematics. For each of them is true not only of magnitudes, but also of numbers, of motions, and of times. This is necessarily so. For equal and unequal, whole and part, greater and less, are common characters of both discrete and continuous magnitudes. The investigation of time intervals also requires that all these propositions be accepted as obvious; and so does the study of motion, and number, and magnitude. In all these areas it is

195

A B C

196

---

[27] 195.11   Von Fritz (*Archiv für Begriffsgeschichte* I, 1955, 65, 100) suggests that Apollonius' purpose was rather to prove the transitivity of congruence for lines than to prove the first axiom.

true that things equal to the same thing are equal to each other, and so also is any other of the axioms that we may assume. These axioms are common, but each individual uses them with reference to his specific subject-matter and to the extent which his subject-matter demands. One man applies them to magnitudes, another to numbers, another to intervals of time. In this way, although the axioms are general, they lead to specific conclusions in each science.

Furthermore, there is no need to reduce them to the lowest possible number, as Heron[28] does when he proposes three only; for that the whole is greater than the part is also an axiom, which our geometer often invokes for aid in his proofs, and so too that things which coincide are equal, a principle that immediately hereafter contributes to the proof of the fourth theorem. Nor do we need to add others and then still others, some peculiar to geometry (in violation of what we have said about axioms as common principles), such as that two lines do not enclose an area, or some that are only corollaries of those mentioned, such as that doubles of the same thing are equal; for this last follows from the principle that if you add equals to equals the sums are equal. For when quantities equal[29] to half of a given quantity add this very half, they become double the same thing and are equal to one another by virtue of the equal additions. And by this principle not only doubles, but triples and any multiples of the same things, will evidently be equal.

With these axioms Pappus says we should include "If unequals be added to equals, the excess of one sum over the other is equal to the excess of one of the added quantities over the other," and its complement, "If equals be added to unequals, the excess of one sum over the other is equal to the excess of one of the original quantities over the other." Although these principles too are evident of themselves, yet they can be demonstrated in the following fashion. Let A and B be equals, and let unequals C and D be added to them, C being greater than D by E. Then since A is equal to B and

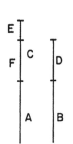

197

28 196.16 On Heron see note at 41.10.
29 196.27 Reading with Grynaeus ἴσα for ἴσον in Friedlein.

F equal to D, A and F together are equal to B and D. For if equals be added to equals the sums are equal. Hence C and A together exceed B and D by E only, the quantity by which alone C exceeds D. Again let C and D be unequals, let equals A and B be added, and let the excess of C over D be E. Then since A is equal to B and A and F together are equal to B and D, the sum of A and C will exceed the sum of B and D by E alone, the amount by which C exceeds D.

These results are consequences of the axioms laid down above and are rightly omitted in most copies.[30] And all the others that he[31] adds are anticipated in the definitions and follow from them: for example, that all parts of a plane, and all parts of a straight line, coincide with one another (for everything that is stretched to the utmost has this character); that a point divides a line, a line divides a plane, and a plane divides a solid (for all these figures are divided by the elements by which their adjacent parts are bounded); and that infinity in magnitude exists both by addition and by removal, though potentially in each case (for every continuous magnitude is capable of indefinite division and indefinite increase).

198

[30] 198.4   Standard texts? Or lists of axioms? 197.6 suggests such a standard list.
[31] 198.5   I.e. Pappus.

PROPOSITIONS:
PART ONE

NOW THAT we have summed up these matters, it remains[1] for us to examine the propositions that come
after the principles. Up to this point we have been dealing
with the principles, and it is against them that most critics of
geometry have raised objections, endeavoring to show that
these parts[2] are not firmly established. Of those in this group
whose arguments have become notorious some, such as the
Sceptics,[3] would do away with all knowledge, like enemy
troops destroying the crops of a foreign country, in this case a
country that has produced philosophy, whereas others, like
the Epicureans, propose only to discredit the principles of
geometry. Another group of critics, however, admit the principles but deny that the propositions coming after the principles can be demonstrated unless they grant something that is
not contained in the principles. This method of controversy
was followed by Zeno of Sidon,[4] who belonged to the school

of Epicurus and against whom Posidonius has written a whole
book and shown that his views are thoroughly unsound.

The disputes about the principles have been fairly well
disposed of in our preceding exposition, and Zeno's attack
will concern us a little later.[5] For the present let us briefly

[1] 199.2   I follow Barocius in reading λοιπόν for λοιπῶν. This section
of the text (to 200.18) in Barocius is continuous with the preceding
and constitutes the end of the *Principia*. Grynaeus also makes it continuous with the preceding but provides no separate heading for the
Propositions that follow.

[2] 199.5   To explain τὰ μέρη in Friedlein's text we must assume that
ταῦτα has dropped out just before these words.

[3] 199.9   Ἐφεκτικοί, the followers of Pyrrho of Elis, who advocated
withholding judgment in order to avoid falling into error. Diog. Laert.
I, 16; IX, 69-70.

[4] 199.15   Zeno of Sidon, an Epicurean of the late second and early
first century B.C., noted as a lucid and copious author. Diog. Laert.
VII, 35; X, 25; Cicero, *Academica* I, 46.

[5] 200.6   At 214.18ff.

review the definitions of theorem and problem, the distinction between them, the parts of each and the kinds into which they can be divided, and then turn to the exposition of the matters demonstrated by the author of the *Elements*. We shall select the more elegant of the comments made on them by the ancient writers, though we shall cut short their endless loquacity and present only what is most competent and relevant to scientific procedures, giving greater attention to the working out of fundamentals than to the variety of cases and lemmas which, we observe, usually attract the attention of the younger students of the subject.

I.[6] *On a given finite straight line to construct an equilateral triangle.*

Science as a whole has two parts: in one it occupies itself with immediate premises, while in the other it treats systematically the things that can be demonstrated or constructed from these first principles, or in general are consequences of them. Again this second part, in geometry, is divided into the working out of problems and the discovery of theorems. It calls "problems" those propositions whose aim is to produce, bring into view, or construct what in a sense does not exist, and "theorems" those whose purpose is to see, identify, and demonstrate the existence or nonexistence of an attribute. Problems require us to construct a figure, or set it at a' place, or apply it to another, or inscribe it in or circumscribe it about another, or fit it upon or bring it into contact with another, and the like; theorems endeavor to grasp firmly and bind fast by demonstration the attributes and inherent prop-

201

[6] 200.19 There is no point in reproducing Friedlein's separate numbering of problems and theorems, and I have merely assigned numbers to the propositions, as does Heiberg in his edition of the *Elements*. The distinction between theorem and problem is one to which Proclus attaches great methodological importance; but although he usually indicates at the beginning of his commentary on a proposition whether it is a problem or a theorem, yet in later references to it he usually calls it a theorem, or more simply designates it by a number, e.g. "the ninth," or "the fourth."

erties belonging to the objects that are the subject-matter of geometry.

Every kind of question that is a possible subject of inquiry is considered by geometry, some of them being referred to problems, others to theorems. Geometry asks the question "What is it?" and that in two senses: it wants either the definition and notion or the actual being of the thing. I mean, for example, when it asks "What is the homoeomeric line?" it wishes to find the definition of such a line, namely, "the homoeomeric line is a line all of whose parts fit upon each other," or to grasp the actual species of homoeomeric lines, that is, "it is either a straight line, a circular line, or a cylindrical helix." In addition, geometry asks "Does the object exist as defined?" This it does most of all in diorismi, examining whether the question proposed is or is not capable of solution, to what extent it is so capable, and in how many ways. And of course geometry asks "What sort of thing is it?" For when it investigates the properties that belong intrinsically to a triangle, or a circle, or to parallel lines, this is clearly an attempt to determine what sort of thing it is.

Many persons have thought that geometry does not investigate the cause, that is, does not ask the question "Why?" Amphinomus is of this opinion, though Aristotle originated it.[7] But you will find this question also included in geometry, says Geminus. For is it not the task of the geometer to inquire why it is that an indefinite number of equilateral polygonal figures can be inscribed in a circle, whereas in a sphere it is not possible to inscribe an indefinite number of polyhedra with equal sides and angles and composed of similar faces? For whose task would it be, if not the geometer's, to ask and find the answer to this question? It is true that, when the reasoning employs reduction to impossibility, geometers are content merely to discover an attribute; and again when they use a previous demonstration to prove a particular conclusion,

202

[7] 202.11 This reference to Aristotle is difficult to understand. The *Post. Anal.* insists that demonstration is reasoning that establishes the cause ($\alpha \iota \tau \iota \alpha$ or $\delta \iota \grave{\alpha} \ \tau \iota$): cf. esp. 85b23ff. And the $\alpha \iota \tau \iota \alpha$ that Aristotle demands appears to be identical with the conception of Geminus, as cited here.

the cause is not evident. But if the conclusion is universal and applies to all similar cases, the reason why is by that very fact made manifest.

203   So much for the questions that geometry considers. Every problem and every theorem that is furnished with all its parts should contain the following elements: an enunciation, an exposition, a specification, a construction, a proof, and a conclusion.[8] Of these the enunciation states what is given and what is being sought from it, for a perfect enunciation consists of both these parts. The exposition takes separately what is given and prepares it in advance for use in the investigation. The specification takes separately the thing that is sought and makes clear precisely what it is. The construction adds what is lacking in the given for finding what is sought. The proof draws the proposed inference by reasoning scientifically from the propositions that have been admitted. The conclusion reverts to the enunciation, confirming what has been proved.

So many are the parts of a problem or a theorem. The most essential ones, and those which are always present, are enunciation, proof, and conclusion; for it is alike necessary to know in advance what is being sought, to prove it by middle terms, and to collect what has been proved. It is impossible that any of these three should be lacking; the other parts are often brought in but are often left out when they serve no need. For example, both specification and exposition are

204   omitted in the problem "To construct an isosceles triangle having each of its base angles double the other angle."[9] And in most theorems there is no construction, because the exposition is sufficient, without the addition of anything else, to prove the proposed conclusion from the given. When, then, do we say the exposition is lacking? When the enunciation contains no statement of what is given. For although enunciation in general consists of what is given and what is sought, this is not always so. Sometimes it states only what is sought, that is, what must be known or constructed, as in the problem

8 203.4f.   The Greek terms here are respectively πρότασις, ἔκθεσις, διορισμός, κατασκευή, ἀπόδειξις, συμπέρασμα.
9 204.2   Euclid IV. 10.

just mentioned. For that problem does not announce what is the given from which we are to construct an isosceles triangle having each of its equal angles double the remaining angle, but simply that we are to construct such a triangle. At the same time even in this case we understand the proposal on the basis of preexisting knowledge, for as it happens we know the meaning of "isosceles," of "equality," and of "double"; and such preexisting knowledge, Aristotle says,[10] is the characteristic feature of all discursive learning. Nevertheless there is no specific hypothesis, as in other problems—for example, when we are required to divide a given finite straight line into two equal parts.[11] For here a straight line is given, and we are asked to divide it into two parts; so what is given is separate from what is sought. When, therefore, the enunciation contains both these elements, then we find both specification and exposition; but when the given is lacking, so are these others also. For the exposition is dependent on the given and the specification will be identical with the enunciation.[12] For what else could you say in defining the problem mentioned than that we are to construct an isosceles of such-and-such a sort? But this is what the enunciation said. Whenever, therefore, an enunciation does not contain a statement both of what is given and of what is sought, the exposition is silent because there is no given element to expound, and the specification is omitted in order not to repeat the enunciation. You could find many other such problems, particularly in the arithmetical books and in Book X, where we are asked, for example, to find two straight lines commensurate in square that have a mean proportional between them,[13] and many other cases of this sort.

Furthermore, everything that is given is given in one of the following ways: in position, in ratio, in magnitude, or in species. A point is given in position only; but the line and

205

[10] 204.17 *Post. Anal.* 71a1-2.
[11] 204.20 X below.
[12] 205.1 Omitting with Schönberger the period and the γάρ in the following sentence.
[13] 205.11 Euclid X. 28. The arithmetical books are VII, VIII, and IX.

the other figures may be given in all these ways. When we speak of a given angle [to be bisected] as rectilinear, we declare its kind, that is, show what sort of angle is given, namely, a rectilinear, so that we may not attempt to bisect a curvilinear angle by the same method. When we are required from two given unequal straight lines to cut off from the greater a length equal to the lesser,[14] our given is presented in magnitude; for greater and less, finite and unbounded, are predications peculiar to magnitude. When we say that if four magnitudes are in proportion they will also be in proportion alternately,[15] what is given is an identity of ratios among these four quantities. Whenever we are asked to place at a given point a straight line equal to a given line,[16] then the point is given in position; and since the position may vary, the construction admits of various possibilities. The given point may lie outside the straight line, or on it and at either one of its ends, or on the portion between its extremities. Since, therefore, the given may be understood in these four ways, clearly the exposition may be fourfold in kind. Sometimes two or three of the ways of being given are combined.

206

What is called "proof" we shall find sometimes has the properties of a demonstration in being able to establish what is sought by means of definitions as middle terms, and this is the perfect form of demonstration; but sometimes it attempts to prove by means of signs.[17] This point should not be overlooked. Although geometrical propositions always derive their necessity from the matter under investigation, they do not always reach their results through demonstrative methods. For example, when from the fact that the exterior angle of a triangle is equal to the two opposite interior angles it is shown that the sum of the interior angles of a triangle is equal to two right angles,[18] how can this be called a demonstration based on the cause? Is not the middle term used here only a sign? For even though there be no exterior

14 205.21   As in III.   15 206.1   As in V. 16.
16 206.4   As in II.
17 206.15   τεκμήρια. See Arist. Prior Anal. 70b1-3.
18 206.22   As in XXXII below.

angle, the interior angles are equal to two right angles; for it is a triangle even if its side is not extended. But when we demonstrate that the triangle constructed by the drawing of circles is equilateral, our approach is from the cause. For we can assert that it is the similarity and equality of the circles that causes the equality of the sides of the triangle.

Furthermore, mathematicians are accustomed to draw what is in a way a double conclusion. For when they have shown something to be true of the given figure, they infer that it is true in general, going from the particular to the universal conclusion. Because they do not make use of the particular qualities of the subjects but draw the angle or the straight line in order to place what is given before our eyes, they consider that what they infer about the given angle or straight line can be identically asserted for every similar case. They pass therefore to the universal conclusion in order that we may not suppose that the result is confined to the particular instance. This procedure is justified, since for the demonstration they use the objects set out in the diagram not as these particular figures, but as figures resembling others of the same sort. It is not as having such-and-such a size that the angle before me is bisected, but as being rectilinear and nothing more. Its particular size is a character of the given angle, but its having rectilinear sides is a common feature of all rectilinear angles. Suppose the given angle is a right angle. If I used its rightness for my demonstration, I should not be able to infer anything about the whole class of rectilinear angles; but if I make no use of its rightness and consider only its rectilinear character, the proposition will apply equally to all angles with rectilinear sides.

Let us view the things that have been said by applying them to this our first problem. Clearly it is a problem, for it bids us devise a way of constructing an equilateral triangle. In this case the enunciation consists of both what is given and what is sought. What is given is a finite straight line, and what is sought is how to construct an equilateral triangle on it. The statement of the given precedes and the statement of what is sought follows, so that we may weave them together as "If

207

208

there is a finite straight line, it is possible to construct an equilateral triangle on it." If there were no straight line, no triangle could be produced, for a triangle is bounded by straight lines; nor could it if the line were not finite, for an angle can be constructed only at a definite point, and an unbounded line has no end point.

Next after the enunciation is the exposition: "Let this be the given finite line."[19] You see that the exposition itself mentions only the given, without reference to what is sought. Upon this follows the specification: "It is required to construct an equilateral triangle on the designated finite straight line." In a sense the purpose of the specification is to fix our attention; it makes us more attentive to the proof by announcing what is to be proved, just as the exposition puts us in a better position for learning by producing the given element before our eyes. After the specification comes the construction: "Let a circle be described with center at one extremity of the line and the remainder of the line as distance; again let a circle be described with the other extremity as center and the same distance as before;[20] and then from the point of intersection of the circles let straight lines be joined to the two extremities of the given straight line." You observe that for the construction I make use of the two postulates that a straight line may be drawn from any point to any other and that a circle may be described with [any] center and distance. In general the postulates are contributory to constructions and the axioms to proofs. Next comes the proof: "Since one of the two points on the given straight line is the center of the circle enclosing it, the line drawn to the point of intersection is

209

---

[19] 208.17   Euclid's construction is as follows. Since Proclus follows Euclid's proof fairly closely in the commentary on this proposition, it is unnecessary to reproduce Euclid's reasoning here.

[20] 209.3   This and the following line in Friedlein have obviously been corrupted. Barocius had a better text, and I follow his translation.

equal to the given straight line. For the same reason, since the other point on the given straight line is itself the center of the circle enclosing it, the line drawn from it to the point of intersection is equal to the given straight line." These inferences are suggested to us by the definition of the circle, which says that all the lines drawn from its center are equal. "Each of these lines is therefore equal to the same line; and things equal to the same thing are equal to each other" by the first axiom. "The three lines therefore are equal, and an equilateral triangle [ABC][21] has been constructed on this given straight line." This is the first conclusion following upon the exposition. And then comes the general conclusion: "An equilateral triangle has therefore been constructed upon the given straight line." For even if you make the line double that set forth in the exposition, or triple, or of any other length greater or less than it, the same construction and proof would fit it.

210

To these propositions he adds: "This is what it was required to do," thus showing that this is the conclusion of a problem; for in the case of a theorem he adds: "This is what was to be demonstrated." For problems announce that something is to be done, theorems that some truth is to be discovered and demonstrated. In general, then, our geometer adds these words to his conclusions to show that what the enunciation stated has been accomplished, joining the end to the beginning in imitation of the Nous that unfolds itself and then returns to its starting-point. But he does not always add the same words: sometimes they are "This is what it was required to do" and sometimes "This is what was to be demonstrated," according to the difference between problems and theorems.

We have thus exercised ourselves and clarified all these distinctions by applying them to a single case, the first problem. The student should do this also for the remaining propositions, asking which of the principal elements are in-

21 209.23 Inserted from Euclid's text, which Proclus must have used in his exposition, in order to mark the contrast between the two conclusions drawn.

cluded and which are left out, in how many ways the given is formulated, and what are the principles from which we obtain the construction or the proof. For a comprehensive survey of these matters will provide no little exercise and practice in geometrical reasoning.

Now that we have made these distinctions, let us briefly run through certain things dependent on them, namely, lemma, case, porism, objection, and reduction.[22]

211   The term "lemma" is often used to designate any proposition invoked for the purpose of establishing another, as when we assert that a proof can be made from such-and-such a lemma.[23] But the specific meaning of "lemma" in geometry is "a proposition requiring confirmation." Whenever for a construction or a demonstration we assume something that has not been demonstrated but needs to be proved, in such a case, considering that the assumed proposition, though doubtful, is worthy of inquiry on its own account, we call it a lemma. It differs from a postulate and an axiom in being a matter for demonstration, whereas they are invoked in their own right without demonstration to establish other propositions. The best aid in the discovery of lemmas is a mental aptitude for it. For we can see many persons who are keen at finding solutions but do so without method. Thus Cratistus,[24] in our own day, was expert in arriving at the desired result from first principles, and with the fewest possible; but it was natural ability that led him to his discoveries. Nevertheless there are certain methods that have been handed down, the best being the method of analysis, which traces the desired result back to an acknowledged principle. Plato, it is said, taught this method to Leodamas,[25] who also is reported to

22 210.28f. The corresponding Greek terms are λῆμμα, πτῶσις, πόρισμα, ἔνστασις, ἀπαγωγή.

23 211.4   Proclus gives an example below (216.1ff.) of the use of a lemma and introduces and establishes a lemma at 319.5f. VII below is also a lemma, he says (264.15), preparatory to the proof of VIII.

24 211.16   Nothing more seems to be known of Cratistus.

25 211.22   Cf. Diog. Laert. III, 24. That Plato taught the method of analysis need not mean that he discovered it. See Heath I, 291f., and Euclid I, 134.

212

have made many discoveries in geometry by means of it. A second is the method of *diaeresis*, which divides into its natural parts the genus proposed for examination and which affords a starting-point for demonstration by eliminating the parts irrelevant for the establishment of what is proposed. This method also Plato praised as an aid in all the sciences.[26] A third is the reduction to impossibility, which does not directly show the thing itself that is wanted but by refuting its contradictory indirectly establishes its truth. Such is the scientific meaning of "lemma."

A "case" announces that there are different ways of making the construction, by changing the position of the points, lines, planes, or solids involved. Variations in case are generally made evident by changes in the diagram, wherefore it is called "case," because it is a transposition in the construction.[27]

"Porism" is a term applied to a certain kind of problem, such as those in the *Porisms* of Euclid.[28] But it is used in its special sense when as a result of what is demonstrated some other theorem comes to light without our propounding it. Such a theorem is therefore called a "porism,"[29] as being a kind of incidental gain resulting from the scientific demonstration.

An "objection" prevents an argument from proceeding on its way by opposing either the construction or the demonstration. Unlike the proposer of a case, who has to show that the proposition is true of it, he who makes an objection does not need to prove anything; rather it is necessary [for his opponent] to refute the objection and show that he who uses it is in error.

---

[26] 212.1 The method of division (διαίρεσις) is emphasized in almost all of Plato's later dialogues, particularly in the *Phaedrus, Sophist, Politicus,* and *Philebus.*

[27] 212.10 Perhaps because πτῶσις, the noun corresponding to the verb πίπτω, often means a "fall," e.g. of dice, as in Plato's *Rep.* 604c.

[28] 212.13 For further light on Euclid's lost *Porisms* see 301.21-302.13.

[29] 212.16 From πορίζω, "furnish," "provide." For further explanation see 303.5-17.

213

"Reduction" is a transition from a problem or a theorem to another which, if known or constructed, will make the original proposition evident. For example, to solve the problem of doubling the cube geometers shifted their inquiry to another on which this depends, namely, the finding of two mean proportionals; and thenceforth they devoted their efforts to discovering how to find two means in continuous proportion between two given straight lines. They say that the first to effect reduction of difficult constructions was Hippocrates of Chios,[30] who also squared the lune and made many other discoveries in geometry, being a man of genius when it came to constructions, if there ever was one.

So much for these matters. Now let us move on to the problem before us. It is evident to everyone that the equilateral is the most beautiful of triangles and most akin to the circle, which has all its lines from the center equal and a single simple line bounding it[31] from without. And the enclosing of the triangle by the two circles, by each of them indeed only in part—for it is inscribed in the whole of neither circle but only in the area consisting of segments of both[32]— seems to indicate in a likeness how the things that proceed from first principles receive perfection, identity, and equality from these principles. In this way too the things that move in a straight line are carried about in a circle through the eternal world-process, and souls, despite their movements from place

214

to place,[33] are likenesses of the unmovable activity of Nous because of their periodic return to their starting-points. It is said also that the life-giving source of souls is bounded by a twofold Nous. If, then, the circle is the likeness of intelligible being, and the triangle the likeness of the first soul because of the similarity and equality of its angles and its sides, it would seem reasonable to demonstrate it by means of circles as an

[30] 213.8   On Hippocrates see note at 66.4 above.
[31] 213.17   Reading with Barocius αὐτὸν for αὐτὸ in Friedlein.
[32] 213.20   Reading with Grynaeus and Barocius ἐκ τῶν for ἔκτον in Friedlein.
[33] 213.26   It is tempting to adopt Friedlein's emendation νοήσεις for Grynaeus' κινήσεις. But see Proclus, *Elements of Theology*, Prop. 198.

equilateral middle area cut off in them. And if, furthermore, every soul proceeds from Nous and returns to Nous and participates in Nous in a twofold fashion, for this reason also it would be proper that the triangle, which is a symbol of the three natures in the constitution of the soul, should take its origin from being comprehended by two circles.[34] Let these remarks, however, be taken only as reminders, through their likenesses, of the nature of things.

Since some persons have raised objections to the construction of the equilateral triangle with the thought that they were refuting the whole of geometry, we shall also briefly answer them. The Zeno whom we mentioned above[35] asserts that, even if we accept the principles of the geometers, the later consequences do not stand unless we allow that two straight lines cannot have a common segment. For if this is not

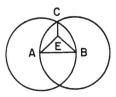

granted, the construction of the equilateral triangle is not demonstrated. Let AB be the straight line, he says, on which we are to construct the equilateral triangle. Let the circles be drawn, and from their point of intersection draw the lines CEA and CEB having CE as a common segment. It then follows that, although the lines from the point of intersection are equal to the given line AB, the sides of the triangle are not equal, two of them being shorter than AB. But if their equality is not established, neither are its consequences. Therefore, says Zeno, even if the principles be granted, the consequences do not follow unless we also presuppose that neither circumferences nor straight lines can have a common segment.

215

[34] 214.13   For understanding the "periodic return," the "life-giving source," the "twofold Nous," and the "three natures in the constitution of the soul" we must turn to Proclus' *Elements of Theology*, particularly Props. 184-211, though Plato's *Timaeus*, one of the chief sources of these doctrines, must always be kept in mind.
[35] 214.18   At 199.15.

To this we must reply first that in a sense it is presupposed in our first principles that two straight lines cannot have a common segment. For the definition of a straight line contains it, if a straight line is a line that lies evenly with all the points on itself. For the fact that the interval between two points is equal to the straight line between them makes the line which joins them one and the shortest; so if any line coincides with it in part, it also coincides with the remainder. For if each of the lines is stretched to the utmost, it must necessarily, because it is the shortest, coincide as a whole with the whole of the other. And, furthermore, this principle is also evidently assumed in the postulates. For the postulate that a finite straight line may be extended in a straight line shows clearly that the extended line is one and that its extension results from a single motion.

216 But if this be taken as a lemma and we demand that it be proved, let the line AB be, if possible, the common segment of AC and AD, and let a circle ACD be drawn with center at B and AB as distance. Then since ABC is a straight line through the center, AEC is a semicircle; and since ABD is a straight line through the center, AED is a semicircle. Hence AEC and AED are equal to one another, which is impossible.

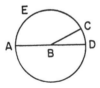

To this demonstration Zeno would reply that the proof we gave[36] that a diameter bisects its circle depends on our previous assumption that two circumferences cannot have a common segment. For we presupposed that one of the two circumferences would coincide with the other or else, not coinciding, fall either inside or outside it. But there is nothing, he says, to prevent its failing to coincide as a whole but coinciding in part. And as long as it has not been proved that the

[36] 216.12   At 157.10ff.

217

diameter bisects its circle, the proposition before us cannot be demonstrated. To this Posidonius gave the right answer when he made fun of the shrewd Epicurean[37] for not realizing that the proof is valid even though the circumferences coincide only in part. At the part where they do not coincide one circumference is inside, the other outside, and the same absurd consequences result when we draw a straight line from the center to the outer circumference. For the lines, because they are drawn from the center, will be equal, both that to the outer circumference, which is longer, and that to the inner circumference, which is shorter. Then either they completely coincide and are equal to one another, or one will coincide with the other in part and diverge in part, or no part of one will coincide with any part of the other; and if this last is the case, the one circumference will lie either outside or inside the other. All these alternatives are refuted in the same way. So much for this argument.

Zeno has also constructed another proof, as follows, which he tries to discredit. Let there be two straight lines, AC and AD, having a common segment AB, and let BE be drawn at right angles to AC. The angle EBC will then be a right angle. If, then, the angle EBD is a right angle, they will be equal,

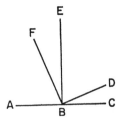

which is impossible; and if it is not a right angle, let FB be drawn at right angles to AD. Angle FBA is then a right angle; but angle EBA was also a right angle; therefore they are equal to each other, which is impossible. This is the proof; he attacks it as presupposing something that is established later,

[37] 216.21 Reading, apparently with ver Eecke, 'Επικούρειον for 'Επίκουρον.

that one can draw a line at right angles to a given straight line from a given point. Posidonius says that such a proof has never appeared in an elementary treatise and that Zeno is slandering the geometers of his time in accusing them of using a shabby proof. Nevertheless, he says, there is something to be said for this proof, since one of two straight lines can clearly be at right angles to the other; that is, any two straight lines can make a right angle. This indeed we presupposed in defining a right angle; for it is by virtue of this particular inclination alone that we construct the right angle. So let it be this one that we have by chance erected. Besides, he adds, Epicurus himself, and all other philosophers, admit that they have proposed many possible as well as many impossible hypotheses for the sake of examining their consequences.

So much for the equilateral triangle. We must also construct the others, and first the isosceles. Let the line AB be that on which an isosceles is to be constructed. Let circles be drawn as they were for the equilateral triangle, and let the line AB be prolonged in both directions to the points C and D. CB is then equal to AD. With B as center and distance CB

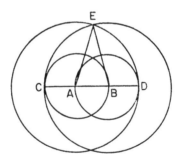

let the circle CE be drawn; and again with A as center and distance DA let the circle DE be drawn. From E, the point of intersection of the circles, let lines EA and EB be joined to the points A and B. Then since EA is equal to AD, and EB to BC,[38] and AD to BC, EA is also equal to EB. But they are

38 218.25   Inserting τῇ before β̄γ̄ in Friedlein.

also longer than AB. The triangle ABE is therefore isosceles; and this is what we were required to construct.

Now let it be further required to construct a scalene triangle upon the given straight line AB. Let circles be drawn with centers and distances as before. Let a point C be taken on the circle whose center is A, and let the connecting line AC be drawn; upon this let a point D be taken and the line DB

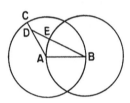

be drawn. Then since the center is A and AB is equal to AC, AB is longer than AD. B also is a center, and therefore EB is equal to AB. DB thus is longer than AB, and AB is longer than AD. The three sides DB, BA, and AD are therefore unequal. Hence the triangle is scalene, so that we have constructed the three kinds of triangle.

These matters are common knowledge. What is elegant in these constructions is that the equilateral triangle, which is equal on every side, can be constructed in only one way, whereas the isosceles, which has only two of its sides equal, can be constructed in two ways; for the given straight line is either shorter than either of the two equal sides, as in the triangle we constructed, or longer than both. And the scalene, having all its sides unequal, can be constructed in three ways; for the given straight line is either the longest or the shortest of the three or longer than one and shorter than the other. The reader can practise himself by examining at length or briefly each of these three hypotheses. For us what has been presented must suffice.

In general we shall see that some problems have a unique solution, others more than one, and some an indefinite number. We call "ordered," to use Amphinomus' term, those that have only one solution, "intermediate" those that have more than one but a finite number and "unordered" those

220

having an indefinite variety of solutions. How problems are handled that are capable of one or more than one solution is clear with regard to the triangles considered; the equilateral triangle is constructed in a single way, and of the others one has two solutions and the other three. Problems admitting of an indefinite number of solutions would be such as the following: "To divide a given straight line into three parts in continued proportion." If the line be divided in a ratio of two to one, and if then the square on the shorter length be applied[39] to the longer so as to fall short by a square figure, it will have been divided into three equal parts. But if the greater segment be more than double, say, triple the lesser and an area equal to the square on the lesser be applied to the greater in such a way as to fall short by a square, the line will have been divided into three unequal parts in continued proportion. Since there are an indefinite number of ways in which the line can be divided into two parts of which the greater is more than double or triple the lesser—for the series of multiples proceeds to infinity—there are consequently an indefinite number of ways in which the line may be divided into three parts in continued proportion.

221

We must also recognize that "problem" is used in several senses. Anything propounded may be called a problem, whether it be put forward for the purpose of instruction or of construction. But its special use in mathematics is to denote something proposed for theoretical construction, since the constructing carried out in mathematics is done for the purpose of theory. Frequently things incapable of solution are called problems; but more characteristically we use this designation for what is capable of solution and is neither excessive nor deficient. A problem such as the following is excessive: "To construct an equilateral triangle having its vertical angle two-thirds of a right angle." For this brings in an unneeded addition, since this property belongs to every equilateral triangle. Of excessive problems those that exceed by containing

---

[39] 220.19 On the "application of areas" used in this example see 419.15ff. and note at 420.23. The algebraic solution of the problem here discussed is neatly given by Heath, *Euclid* I, 128.

inconsistent or unreal conditions are called "impossibles," while those that contain realizable conditions are called "more than problems." A problem is deficient and is called "less than a problem" when it needs to have something added to make it definite and bring it into order and scientific determinateness, such as "To construct an isosceles triangle." This is insufficiently determinate and requires an addition specifying what sort of isosceles is wanted, whether one having its base greater or less than each of the equal sides, or one having its vertical angle double each of those at the base, like the half-square, or one having each of the base angles double the angle at the vertex, or one having these angles in some other ratio, triple or quadruple. One could vary the possibilities endlessly. These examples show that problems properly so-called aim at avoiding the indeterminateness that renders them capable of an indefinite number of solutions; nevertheless even those that are deficient are called problems, for the term is ambiguous. Clearly the very first problem in the *Elements* is in this respect superior in that it is neither excessive nor deficient nor indeterminate and thus having indefinitely many solutions; for such should be the character of what is to be an "element" of the others.

II. *At a given point to place a straight line equal to a given straight line.*

Some problems have no cases, while others have many; and the same is true of theorems. A proposition is said to have cases when it has the same force in a variety of diagrams, that is, can be demonstrated in the same way despite changes in position, whereas one that succeeds only with a single position and a single construction is without cases. For the presence of cases, whether in a theorem or a problem, generally shows itself in the constructions. Now our second problem has many cases. In it the point is given in position, and given only in this way; but the line is given both in species (for it is not simply a line, but this kind of line) and in position. We want to place a straight line equal to this straight line with its extremity at the given point, wherever the point may lie. It is evident in any case that the point is in the assumed plane of the straight line and not in a plane above it; for we must assume one plane for all the problems and theorems of plane geometry.

But someone may raise a difficulty: In what sense is it required that we draw a line equal to the given straight line? What if the given line is infinite? The statement of the given here applies equally to a finite and to an infinite line; and the given shows in its entirety what is set forth and proposed to us for inquiry. Euclid himself makes this clear by sometimes saying "On a given finite line to construct an equilateral triangle,"[40] and again "To a given infinite straight line to erect a perpendicular."[41] To anyone who raises this difficulty we must say: Has he not at once made clear, in asking that we place at a given point a line equal to a given line, that the given line is finite? In any case the line drawn from the given point will be bounded at that point itself, so that much more will that line be finite to which the drawn line is to be equal.[42] Consequently when he says "at a given point," he limits at the same time both straight lines, that which is given as well as that which is to be drawn equal to it.

It is clear that cases of this problem arise from differences in the position of the point. The given point lies either outside the given straight line or on it; and if it lies on it, it will be at one of the two extremities or between them; and if it lies off the line, it will either be at one side, so that the line joining it with the extremity of the straight line will make an angle with it, or lie in the direction of the given line, so that the line if prolonged will fall upon the point. Our geometer has taken a point lying off the line and at one side;[43] but for the

224

---

[40] 223.23 I.e. in I.   [41] 223.24 I.e. in XII.

[42] 224.2 "Proclus' argument is obviously unsatisfactory, since the line may be infinite in one direction. Euclid's use of the words 'finite' and 'infinite' is quite careless. Usually he omits both and assumes that he is dealing with finite lines. Nothing he proves requires the use of infinite lines." (I.M.) But see XII and Proclus' remarks in his commentary on it (284.4-17).

[43] 224.15 In II Euclid assumes the straight line BC and the point A, then draws AB, constructs the equilateral triangle DAB, and pro-

sake of practice we must consider all cases, and we shall choose the more difficult one to expound.

Let AB be the given straight line and C the given point, lying on the line between its two extremities; and let there be constructed, as in the proposition in the *Elements*, an equilateral triangle DCA on the line CA. Let DC and DA be produced. With A as center and distance AB let the circle BE be described, and again with D as center and distance DE the circle EF. Then since A is the center, AB is equal to AE, and for the same reason DE is equal to DG; and of these lines DC is equal to DA (for the triangle DAC is equilateral), and hence the remainder AE is equal to CG. And AE was equal to AB, as has been shown. CG is therefore equal to

225

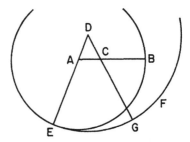

AB. Consequently to the given point C a line has been drawn equal to the line AB.

So many are the cases that arise from the position .of the given point; and there are still many more resulting from

---

duces DA to E and DB to F; then with center B and distance BC he describes the circle CGH, and again with center D and distance DG the circle GKL. Then since BC is equal to BG, and DL is equal to

DG, and in them DA is equal to DB, the remainder AL is equal to the remainder BG and therefore equal to BC. This is what was to be constructed.

variations in the construction of the equilateral, in the producing of its sides, and in the describing of the circles. Let
point A be taken as in the proposition in the *Elements*, along
with the straight line BC, and let AB be joined. Then let an
equilateral triangle be constructed with its vertical angle not
above (for there is not room)[44] but below, and let ADB be
this triangle. Then AD is either equal to BC or shorter or

FIG. I

longer. If it is equal (Fig. 1), the problem is solved. If it is
less, let a circle be drawn with center at B and distance BC,
and let the lines AD and BD be produced to F and G; and
with center at D and distance DG let a circle GE be described
(Fig. 2). Now since DG is equal to DE (for they are drawn
from a center) and AD is equal to BD (as sides of an equilateral triangle), the whole line AE is equal to the whole
line BG. But BG and BC are equal, as lines drawn from a

226

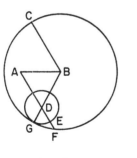

FIG. 2

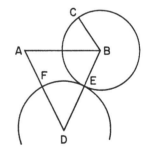

FIG. 3

44 225.16   A similar objection occurs at 275.7 and 289.21. Heath
(*Euclid* I, 23) notes that Heron in his commentary on the *Elements*
sometimes used constructions alternative to Euclid's to obviate objections of this sort.

center; therefore AE is equal to BC, which is what was to be constructed. But if AD is longer than BC (for this is the remaining alternative), let a circle CE be described with center B and distance BC (Fig. 3). Circle CE will therefore cut BD. Again with center D and distance DE let a circle be described. Its circumference FE will therefore cut line AD. Then since D is the center of circle FE, FD is equal to DE. But DA was also equal to DB; therefore the remainder AF is equal to BE. But BE is equal to BC, for they are drawn from a center. Hence AF is equal to BC and is drawn from A, which is what was to be constructed. There are many other cases, but these are enough to record for the present. With their help those who are diligent can exercise themselves on the others.

There are some, however, who would do away with the construction used in this proposition and its varieties, arguing as

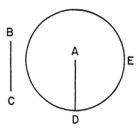

follows. Let A be the given point and BC the given straight line. With center A and distance equal to BC let a circle ED be described and a straight line be drawn from A to the circumference, namely, AD. This therefore is equal to BC, for its distance from the center was taken equal to BC, and the thing required has been done. Now anyone who reasons thus begs the question. For when he says that circle ED is described with center A and distance BC, he has already, in a way, taken a line equal to BC and placed its extremity at A; that is, he has begged the question[45] in using this extremity of the distance as center and the remainder as radius of the

<hr />

[45] 227.22 **Reading with Grynaeus and Barocius** φυλάττων for φυλάττον in Friedlein. On the impropriety of drawing the circle with a "compass-carried distance" (as De Morgan puts it) see Heath, *Euclid* I, 246.

228

circle described. But in the problem before us the center of the circle is not part of the distance, which lies elsewhere. We shall therefore in no wise adopt this method of proof.

III. *Given two unequal straight lines, to cut off from the greater a straight line equal to the less.*

This our third problem has two straight lines given unequal in length and requires that we take away from the greater a line equal to the less. This problem too has many cases. For either the given unequal straight lines are separate from one another, as in the proof presented by the author of the *Elements*,[46] or they meet at one of their extremities, or they cut each other, or the extremity of one cuts the other, and that in one of two ways, either the greater cutting the less or the less the greater. Now if they meet at one extremity, the proof is evident. For using the common extremity as center and the lesser line as distance,[47] you can describe a circle that will cut the greater line and take from it a line equal to the less, since whatever be the length of the greater line cut off by the circle that crosses it, this will be equal to the lesser line. But if one line at its extremity cuts the other, either the greater cuts the less, or vice versa; and if they cut one another, they are

229

cut into equal or into unequal parts by each other, or one is cut into equal and the other into unequal parts, and that in two ways. All these possibilities provide a marvellous variety for practice. Let us set forth a few of the many cases.

[46] 228.11   To solve III Euclid proceeds as follows. Taking as given AB and C, the two unequal lines, he uses the construction in the preceding problem to draw AD equal to C. Then with center A and distance AD he describes the circle DEF. Since AE is equal to AD

and AD to C, AE is equal to C and is the length required to be cut off from AB.

[47] 228.17   In this case the center is part of the distance used and hence is not subject to the disqualification that led to the rejection of the construction suggested at the end of the previous problem.

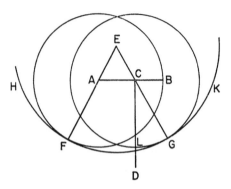

FIG. 1

Let AB and CD be unequal straight lines. Let CD be the greater, and let its extremity cut AB at C (Fig. 1). With A as center and distance AB let a circle BF be described and an equilateral triangle AEC be constructed on AC. Let lines EA and EC be produced. Again with E as center and distance EF let circle HGK be described. Again with C as center and distance CG let circle GL be drawn. Now since EF and EG are equal (for E is their center) and of these EA is equal to EC, the remainder AF is equal to CG. But AF is equal to AB (for A is their center), and CG to CL (for C is their center). Consequently a distance CL equal to AB has been cut off.

Now let CD be less than AB, and let its extremity cut AB at C. It cuts AB, then, either at the midpoint or not at the midpoint. First let it cut AB at the midpoint (Fig. 2). Then either CD is half of AB and AC is equal to CD; or it is less than half, and by drawing a circle with center C and distance CD you can cut off from AC a line equal to CD (Fig. 3); or it is greater than half, and by placing a line AF equal to CD

230

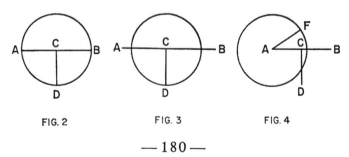

FIG. 2   FIG. 3   FIG. 4

with its extremity at point A and describing a circle with center A and distance AF you will cut off from AB a line equal to AF, that is, to CD (Fig. 4). But suppose CD cuts AB elsewhere than at the midpoint, and let AC be the part of it that is greater than half. If, then, CD is half or less than half of AB, using C as center and CD as distance you can cut off from AC a line equal to CD (Fig. 5). Or CD is greater than half of AB and is either equal to AC (in which case the problem is solved) or longer than it; and again by placing a

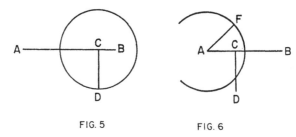

FIG. 5   FIG. 6

line equal to CD at A you can do the same thing; for with A as center and AF as distance you can describe a circle cutting off from AB a length equal to AF, that is, to CD (Fig. 6).

And if they cut one another, like CD and AB, let a circle AF be drawn with center B and distance BA (Fig. 7). Join BC, and let BC be produced to F. Then since the two straight

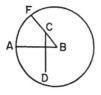

FIG. 7

lines BF and CD are unequal and CD at its extremity cuts BF, it is possible to take a length equal to BF from CD, or a length equal to CD from BF; for both cases have been demonstrated.[48] Consequently it is possible also to take away from

[48] 232.6 This passage has evidently occasioned considerable perplexity to the scribes, for the MSS variants are numerous. Its sense will be clear, however, if we recall that in 229.4-231.14 we considered

AB a length equal to CD, or from CD a length equal to AB. For AB and BF are equal to each other.

We have thus from a classification of cases tried to show their variety. The proof given by the author of the *Elements* is admirable, for it suits all the constructions mentioned; that is, it is possible, for any position, to place at the extremity of the greater line a length equal to the less and, by taking this extremity as center and the posited length as distance, to describe a circle that will cut off from the greater a length equal to the less, whether they cut one another, or only one cuts the other, or however otherwise they may be placed.

233

IV. *If two triangles have two sides equal to two sides respectively and have the angles contained by the equal straight lines equal, they will also have the base equal to the base, the triangle will be equal to the triangle, and the remaining angles will be equal to the remaining angles respectively, namely, those which the equal sides subtend.*

This is the first theorem we are given in the *Elements*. The propositions before it have all been problems, the first concerned with the construction of triangles, the second and third proposing to find a straight line equal to another straight line; and of these one constructs an equal line from the non-equal,[49] the other finds the equal by subtraction from the unequal. Now since equality, which is the first attribute in the category of quantity, has been provided us with respect to both triangles and straight lines,[50] our geometer follows up

---

cases that arise when CD at its extremity cuts AB, first under the supposition that CD is the longer of the two lines (229.5-18), and then (229.19-231.14) under the supposition that CD is the shorter. These are the two possibilities that recur in the present section (231.15-232.9) when CD at its extremity is taken as cutting BF (=AB), hence they do not need to be considered again. My translation does not follow Friedlein's text exactly, for I have adopted suggestions contained in his apparatus.

[49] 233.16   I.e. from the point. Cf. 234.23f.

[50] 233.20   What Proclus means to say here is, as the following lines show, that the *existence* of triangles and the *equality* of lines have been provided by the preceding problems.

these problems with this first theorem set forth above. For unless he had previously shown the existence of triangles and their mode of construction, how could he discourse about their essential properties and the equality of their angles and sides? And how could he have assumed sides equal to sides and straight lines equal to other straight lines unless he had worked these out in the preceding problems and devised a method by which equal lines can be discovered? Suppose someone, before these have been constructed, should say: "If two triangles have this attribute, they will necessarily also have that." Would it not be easy for anyone to meet this assertion with "Do we know whether a triangle can be constructed at all?" And suppose one went on to assert: "And if two triangles have the two sides equal to two sides, etc.," would not someone have questioned whether it is not possible that no straight lines should be equal to one another? And especially in geometrical forms that there should be inequality but no equality at all? We shall learn at least that the horned angle is always unequal, never equal, to an acute angle, that the same is true of the angle in a semicircle, and that the transition from the greater to the less does not always proceed through equality. It is to forestall such objections that the author of the *Elements* has given us the construction of triangles, a common method for the three kinds, and also the methods for producing equal lines, of which there are two, one that produces the equal line when a line does not previously exist at all, and the other that gets it by cutting it off from a longer line. These propositions are rightly preliminary to the theorem by which he proves that triangles having two sides equal to two sides respectively, and the angles contained by these equal sides equal, also have the base equal to the base, the area[51] equal to the area, and the other angles equal to the other angles.

There are three things proved and two things given about these triangles. One of the given elements is the equality of

---

[51] 235.7 ἐμβαδόν (cf. 236.22). "It is interesting to note that this word never occurs in Euclid's *Elements*. He just talks about figures being equal." (I.M.)

two sides (really two given sides, but obviously given in ratio to one another) and the equality of the angles contained by the equal sides. And the things to be proved are three: the equality of base to base, the equality of triangle to triangle, and the equality of the other angles to the other angles.[52] Since it would be possible for the triangles to have two sides equal to two sides and yet the theorem be false because the sides are not equal one to another but one pair to the other pair, he did not simply say, in his statement of the given, that the lines are equal, but that they are equal "respectively." For if it should happen that one of the triangles had one side of three units and the other of four, while the other triangle had one side of five units and another of two (the angle included between them being a right angle), the two sides of the one would be equal to the two sides of the other, since their sum is seven in each case. But this would not show the one triangle equal to the other; for the area of the former is six, of the latter five. The reason for this discrepancy is that the sides are not also equal respectively. We often fail to watch out for this in the distribution of plots of land; and many persons have taken the larger of two plots and got a reputation for justice as having chosen an equal portion, because the sum of the boundaries is the same in both cases. We must therefore take the sides as equal respectively, and whenever the author of the *Elements* adds this phrase, we should note that he does so for a reason. Even when speaking of the equality of the given equal angles, he adds "the angles contained by the equal sides" in order that we may not be misled by imprecise language into assuming that he means angles at the base. As to the "base" of a triangle, when no side has previously been named, we must suppose it to denote the side towards the observer, but when two sides have already been mentioned, it must mean the remaining side. So here the author of the *Elements*, having already taken two sides as equal to two sides, calls the other sides the bases of the triangles.

236

[52] 235.13f. Reading with Barocius ἤ for ἢ in each of its three occurrences in these lines.

Two triangles are said to be equal when their areas are equal. It can happen that two triangles with equal perimeters have unequal areas because of the inequality of their angles. "Area" I call the space itself which is cut off by the sides of the triangle, and "perimeter" the line composed of the three sides of the triangle. These are different things, and triangles with equal perimeters must also have the angles along one side equal if the areas are to be equal. It happens in some cases that, when the areas are equal, the perimeters are unequal and, when the perimeters are equal, the areas are unequal. Consider two isosceles triangles, each having its equal sides five units in length, but one having a base of eight, the other a base of six units. The person inexperienced in geometry would say that the triangle having the base of eight units is the greater, for its total perimeter is eighteen units. But the geometer would say that the area of both is twelve; and he can prove it by dropping a perpendicular from the vertex of each triangle and multiplying its length by half of the base. It is also possible, as I said, that triangles with equal perimeters have unequal areas, and some persons have wronged their associates in a distribution of lands by relying on the equality of perimeters and in fact getting a greater portion.

Base is said to be equal to base and generally a straight line to another straight line when the congruence of their extremities makes the whole of the one line coincide with the whole of the other. Every straight line coincides with every other, and in the case of equal lines their extremities also coincide. A rectilinear angle is said to be equal to a rectilinear angle when, if one of the sides containing it is placed upon one of the sides containing the other, the second side of the first coincides with the second side of the other. When the other sides fail to coincide, that angle is greater whose side falls outside, and that angle less whose side falls inside. For in the one case the one angle includes the other, in the other case it is included by the other. We shall infer the equality of two angles from the congruence of their sides in the case of rectilinear angles, as well as for others that are similar in form[53] . . . such as the lunular, the scraper-like, and the biconvex, since it is possible for angles to be equal without having their sides congruent. There is a certain lunular angle that is equal to a right angle, and it is impossible that circumferences should be congruent with straight lines.

This also must be understood in advance, that the side that lies opposite an angle is said to subtend it. Every angle in a triangle is contained by two sides of the triangle and subtended by the other. This is why our geometer has added to "the angles also are equal" the clause "which the equal sides subtend," so that we may not think it indifferent what angle we take and assert to be equal to any chance one of the other two angles of the triangle but should call equal the angles that equal sides subtend. And of the equal sides one subtends one angle and the other the other.[54]

So much, then, by way of preliminary explanations to clarify the theorem. For the proof we must also assume in advance that two straight lines cannot enclose a space. Our geometer takes this for granted.[55] For if the extremities of the

239

---

[53] 238.8 Something has dropped out between ὁμοειδῶν and οἷον. The three angles mentioned are not examples of ὁμοειδῆ (cf. 241.5-8), and they are obviously cited as examples of angles whose equality does not justify an inference of congruence. For these three varieties of angles see 127.7ff.; and for the lunular angle that is equal to a right angle see 189.21ff.

[54] 238.24 Omitting μία τῶν περιεχουσῶν, which makes sense only as a marginal notation explanatory of ἡ μὲν that has improperly crept into the text after ἡ δὲ.

[55] 239.2 Euclid's proof of IV runs as follows. Given are two triangles ABC and DEF having the two sides AB and AC equal respectively to the two sides DE and DF and the angle BAC equal to the angle EDF. If triangle ABC is applied to triangle DEF so that point A is placed on point D, and the straight line AB on DE, then B will coincide with E, because AB is equal to DE. Then if AB coincides with DE, the straight line AC will also coincide with DF, because the angle BAC is equal to the angle EDF. Hence C will also coincide with F, because AC is equal to DF. And since B also coincided with E, the base BC will coincide with base EF; for if it does not, two straight lines will enclose a space, which is impossible. Hence the whole triangle ABC will coincide with the whole triangle DEF and will be equal to it, and the remaining angles will also coincide with the remaining angles and be equal to them.

bases coincide, he says, the bases coincide; otherwise two straight lines will enclose a space. How do we know that this is impossible? Let ACB and ADB be two straight lines enclosing a space, and let them be prolonged indefinitely. On B as center and with distance AB let the circle AEF be de-

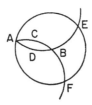

scribed. Since ACBF is a diameter, AEF will be half of the circumference. Again since ADBE is a diameter, AE will be half of the circumference. Hence AE and AEF are equal, which is impossible.[56] Two straight lines therefore do not enclose a space. This principle the author of the *Elements* recognizes in the first postulate when he says "To draw a straight line from any point to any point," which implies that it is always one straight line and not two that can join the two points. Several circular lines connecting the two points can be drawn on the same side, as well as on the opposite side. Thus it is that the extremities of a diameter are connected by two circumferences but by one straight line; and it is possible to draw an indefinite number of circular lines both inside and outside the semicircles uniting the given points. The reason is that the straight line is the least of all the lines that have the same extremities. Everywhere the least counts as a unit and measure of other things; so just as the right angle, being one, serves as measure of the infinitely many other angles (for through it we discover them), so also the straight line serves as measure of the lines that are not straight. So much for these matters.

240

The proof of this theorem, as anyone can see, depends

---

[56] 239.15 "Proclus' reasoning here is inconclusive, since E and F may coincide. It is perhaps worth noting that Heath (following Heiberg) thinks the reference to two lines enclosing a space is not genuine Euclid." (I.M.) See *Euclid* I, 249.

entirely on the common notions and grows naturally out of the very clarity of the hypotheses. Because two sides are equal respectively to two sides, they coincide with one another; and because the angles contained by these equal sides are equal, they also coincide. Since the angle coincides with the angle and the sides with the sides, the lower extremities of the sides also coincide; and if they coincide, the base coincides with the base; and if three sides coincide with three sides, so also does the triangle with the triangle and everything with everything. Visible equality, therefore, in things of the same form is manifestly the ground of the entire proof. For there are two axioms here that comprise the whole procedure of this theorem. One is that things which coincide are equal to one another. This is true without qualification and does not require a clarifying supplement. The author of the *Elements* uses it for establishing the equality of the bases, of the areas, and of the other angles; for these, he says, are equal because they coincide. The other is that things that are equal coincide with one another. This is not true in all cases, but only of things that are similar in form. Similar in form I call a straight line to a straight line, a circular segment to another segment of the same circle, and angles to other angles contained by similar lines similarly placed. Because the things of this sort that are given are equal, they coincide with one another. So the whole proof could be summarized as follows. Given these elements equal to those, that is, two sides to two sides and the angles contained by them, they coincide with each other; and if they coincide with one another, so does the base with the base and every part with every part; and if they coincide, they are equal. If, then, it is given that these elements are equal to those, it follows that every part is equal to every part. This shows us the primary method of identifying triangles that are equal in every respect. So much for the proof in general.

Carpus the engineer,[57] in his work on astronomy, has revived the discussion about problems and theorems—whether

[57] 241.19 On Carpus of Antioch see note at 125.25 above.

241

242

opportunely or not may be ignored for the present. In any case he falls upon this distinction and says that problems are prior in rank to theorems because problems discover the subjects whose attributes are under investigation. And the enunciation of a problem, he says, is simple, requiring no additional technical knowledge at all; it only demands that something clearly possible be done, such as constructing an isosceles triangle or, given two straight lines, cutting off from the greater a length equal to the less. What is unclear or difficult about these? But the enunciation of a theorem, he says, is a laborious matter and needs much precision and scientific acumen if it is not to appear redundant or lacking in some element of truth, as is illustrated by this the first of the theorems. And for problems one common procedure, the method of analysis, has been discovered, and by following it we can reach a solution; for thus it is that even the most obscure problems are pursued. But the handling of theorems is a difficult matter, and no one to this day, he declares, has been able to teach a uniform way of approaching them. Consequently the ease also with which a problem can be handled would make it the simpler form. Having made these distinctions, he proceeds:

243

> For these reasons, therefore, problems precede theorems even in the *Elements*. The *Elements* begins with them, and the first theorem is fourth in order, not because the fourth[58] is proved by the previous problems, but because, even if nothing from them is needed for its proof, they have to be given the precedence because they are problems and this is a theorem. In this theorem he relies entirely upon the common notions, and in a sense takes the same triangle as lying in different places; for congruence, as well as the equality which is inferred from it, is completely dependent on the clear judgment of sense-perception. Nevertheless, despite the fact that the proof used in the first theorem is of this sort, he rightly placed the problems before it, because they in general have the prior rank.

[58] 243.1   Reading with Barocius τέταρτον for πέμπτον in Friedlein.

Now problems rightly do come before theorems in order of presentation, especially for those who are coming to science from the arts concerned with sensible things; but in worth theorems are superior to problems. All of geometry, it appears, where it touches on the various arts, operates by way of problems; but where it borders on the highest science it rises by way of theorems from problems to theorems, from secondary to primary things, from the more practical arts to the more scientific insights. It is therefore vain to criticize Geminus for saying that a theorem is more perfect than a problem. Carpus himself gives problems the priority in order, but Geminus judges primacy in terms of worth and perfection. And as for the fourth proposition, we have explained in what way it requires the problems that precede it, by which we have learned how to construct the triangle and to discover equality. But we should add here[59] that, although this is the simplest and most fundamental of the theorems (for it is demonstrated without artifice from the primary notions alone), yet a theorem that demonstrates some property about triangles that have two sides equal to two sides and the contained angles equal is rightly placed after the problems by which the subjects of this property and the given elements in general have been constructed.

V. *In isosceles triangles the angles at the base are equal;*
and if the equal straight lines are produced further,
the angles under the base will be equal.

Some theorems are simple, others composite. By simple theorems I mean those whose hypotheses and conclusions are indivisible, having one thing given and one thing to be proved—as if, for example, the author of the *Elements* said "Every isosceles triangle has the angles at its base equal." A composite theorem is one consisting of a number of parts, having either its hypothesis composite, or its conclusion composite with its hypothesis simple, or both hypothesis and conclusion composite. Of composite theorems some are inter-

244

---

[59] 244.1 Reading with Barocius προσκείσθω for προκείσθω in Friedlein.

woven and others entire.[60] An entire theorem is one which, though composite, cannot be divided into simple theorems. The fourth, for example, has both a composite hypothesis and a composite consequent, but you cannot divide the given and make simple theorems. For if triangles have only the [two] sides equal or only the vertex angles, the conclusion does not follow. Interwoven theorems are such as can be divided into simple ones, like this: "Triangles and parallelograms with the same altitude are to one another as their bases"; for it is possible to divide it, saying "Triangles with the same altitude are to one another as their bases," and make the same statement for parallelograms. Of composite theorems some have a composite conclusion derived from a single hypothesis, others have composite hypotheses and draw a single conclusion from them all, and others have both conclusion and hypothesis composite. Thus in the fourth[61] the conclusion is composite, for three things are inferred in this theorem, namely, that the bases are equal, that the triangles are equal, and that the other angles, those subtended by the equal sides, are equal. But the hypothesis is composite in the common theorem about triangles and parallelograms with the same altitude. And both are composite in this: "The diameters of circles and of ellipses bisect both the areas and the lines that contain the areas." Of interwoven theorems some are universal, whereas others draw a general conclusion from particulars.[62] For instance, if we say "The diameter bisects the circle and the ellipse and the parallelogram," we do not take each of the interwoven subjects universally but make a general statement true of them all. But if we say "In a circle all the lines through the center bisect one another and make the angles of all the segments equal," we are speaking universally. In the case of the ellipse not all the angles of the segments are equal, but only those that are made by the axis. In general geometers have made such composite propositions both with a view to

[60] 244.23   ἀσύμπλεκτα.

[61] 245.15   Reading ἐν τῷ δ̄ for ἐνταῦθα, as Friedlein conjectures.

[62] 245.26   For this distinction between a universal conclusion and one that is merely general see Arist. *Post. Anal.* 73a25-74b4.

245

246

brevity and for the purpose of analysis; for often things left uncompounded do not lend themselves to analysis and only when put together provide an easy way of getting back to first principles.

With the foregoing observations in mind we must certainly call the fifth theorem composite, and composite in both members, in what is given and in what is sought. The author of the *Elements* indicates this by dividing the theorem, which itself is one, into two parts and setting out separately for each what is given and what is sought, saying "In isosceles triangles the angles at the base are equal," and then immediately afterwards, "and if the equal lines be produced further, the angles under the base are equal." We should not think of this as two theorems, but as one, though composite both in what is given and in what is sought. And each of the parts is true and complete, wherefore the analysis[63] also is true in each case. For if the angles at the base are equal, the triangle is isosceles; and likewise if the angles under the base are equal, the equal sides have been prolonged and the triangle is isosceles. But although the author of the *Elements* will establish the converse as regards equal angles at the base, he does not do so with respect to the equal angles under the base, although this also is true.

The reason for this omission we shall speak of later;[64] but let us first inquire why he even includes in this theorem the equality of the angles under the base. He is never going to use this result for the construction or the demonstration of any other problem or theorem. Since it will not be used later, why was it necessary to bring it into this theorem? To this question we must reply that, even if he was never intending to use "and the angles under the base of an isosceles triangle are equal" [in establishing later theorems], nevertheless it will be useful in meeting objections to them and refuting their adversaries. It is a mark of scientific and technical skill to arrange in advance for the undoing of those who attack what

[63] 246.23 I.e. the geometrical converse, or the inference of the premises from the conclusion.
[64] 247.6 I.e. at 248.11ff. and 258.14ff.

is going to be said and to prepare the positions from which one can reply, so that these previously demonstrated matters may later serve not only for establishing the truth, but also for refuting error. You can understand from this the usefulness of geometrical order for rhetoric. The man who is able to do this in his speeches, foreseeing the arguments that will be brought against the main points that he is going to make and, before they are used, preparing for their refutation by seemingly unnecessary material in his earlier statements, would be exhibiting the surest method of winning a debate. And this in fact is what the author of the *Elements* does; desiring to teach us, in advance of the theorems, the means by which we can refute objections to them by using the proposition demonstrated here, he also demonstrates "and the angles under the base of an isosceles triangle are equal" and so prepares the way for the refutation of unfounded objections. As we proceed it will be clear that we can meet objections both to the seventh and to the ninth theorems by this principle. This also explains why the sixth does not contain also the converse of this part of the fifth, since this part has no usefulness as a leading theorem[65] but only incidentally contributes to our understanding of the science as a whole.

If anyone should demand that we demonstrate the equality of the base angles of an isosceles without prolonging the equal sides[66]—for it is not necessary to demonstrate their equality

[65] 248.13 On the distinction between a "leading theorem" and its converse see 254.6ff.

[66] 248.18 Euclid's proof of V depends on producing the equal sides AB and AC of the given isosceles triangle ABC. Taking a point F at random on BD, cutting off from AE a length AG equal to AF, and joining FC and GB, he proves by IV the equality of the two triangles AFC and AGB, then the equality of triangle BFC to triangle

CGB. Thus angle FBC is equal to angle GCB and angle BCF to angle CBG. Then since the whole angle ABG was proved equal to

through the equality of the angles under the base—we can show the proposition to be true by altering the construction slightly and putting the outer angles inside the isosceles. Let ABC be an isosceles triangle, let any chance point, say D, be taken on AB and a length AE equal to AD be taken from AC, and let the lines BE, DC, and DE be drawn. Then since

249

AB is equal to AC and AD is equal to AE and angle A is common, BE will be equal to DC, and the remaining angles equal to the remaining angles, so that angle ABE is equal to angle ACD. Again since DB is equal to EC and BE is equal to DC and angle DBE is equal to angle ECD and the base DE

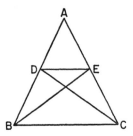

is common, all corresponding parts are equal, so that angle EDB is equal to angle DEC and angle DEB is equal to angle EDC. Then since angle EDB is equal to angle DEC, when equal angles DEB and EDC have been subtracted, their remainders, angles BDC and CEB, are equal. But DB and DC are sides equal respectively to EC and BE, and BC is the common base, and all corresponding parts are equal;[67] so the remaining angles, those subtended by the equal sides, are equal. Angle DBC therefore is equal to angle ECB; for one of them, DBC, is subtended by DC and the other, ECB, by EB. Hence the angles at the base of an isosceles triangle are equal even when the equal sides are not prolonged.

Pappus has given a still shorter demonstration that needs no supplementary construction, as follows. Let ABC be isosceles

---

angle ACF, and in these angles CBG is equal to BCF, the remaining angle ABC is equal to the remaining angle ACB, the angles at the base of triangle ABC. And angle FBC was above proved equal to angle GCB, and they are under the base.

[67] 249.13   For ἄρα in Friedlein read ἴσα; cf. lines 7-9.

250 with side AB equal to side AC. Let us think of this triangle as two triangles and reason thus: Since AB is equal to AC and AC is equal to AB, the two sides AB and AC are equal to the two sides AC and AB, and the angle BAC is equal to

the angle CAB (for they are the same); therefore all the corresponding parts are equal, BC to CB, the triangle ABC to the triangle ACB,[68] the angle ABC to the angle ACB, and angle ACB to angle ABC. For these are angles subtended by the equal sides AB and AC. Hence the angles at the base of an isosceles are equal. It looks as if he discovered this method of proof when he noted that in the fourth theorem it was by uniting the two triangles so that they coincide with each other, thus making them one instead of two, that the author of the *Elements* perceived their equality in all respects. In the same way, then, it is possible for us, by assumption, to see two triangles in this single one and so prove the equality of the angles at the base.

We are indebted to old Thales for the discovery of this and many other theorems. For he, it is said, was the first to notice and assert that in every isosceles the angles at the base
251 are equal, though in somewhat archaic fashion he called the equal angles similar. But even more should we admire the men of more recent times, of whom Geminus is one, who have demonstrated something even more universal, namely, that equal straight lines from any point falling upon a homoeomeric line make equal angles, so that whether the triangle have a straight line or a circumference or a cylindrical helix as

---

[68] 250.8 Reading $\overline{\gamma\beta}$ for $\overline{\beta\gamma}$; and in the next line $\overline{\alpha\gamma\beta}$ instead of $\overline{\alpha\beta\gamma}$ (first occurrence in the line).

its base the angles at its base are equal. Geminus uses this
theorem in showing that there are three and only three lines
that are homoeomeric: the straight line, the arc of a circle,
and the cylindrical helix. And this is the genuine universal[69]
to which this character primarily belongs, just as to have two
of its angles greater than the third is an attribute that belongs
essentially to every triangle, as will be shown later. Consequently, although every isosceles triangle has its base angles
equal, this attribute does not belong universally to the isosceles, but to the straight lines falling upon a homoeomeric line,
for it is they that primarily have the property of subtending
equal angles.

VI. *If in a triangle two angles are equal, the sides which
subtend the equal angles will also be equal.*

This is the first of the theorems to exhibit the two procedures of conversion and reduction to impossibility. For it
is the converse of the theorem preceding it, and its proof
employs the reduction to impossibility. We must explain
both of them, so far as they are relevant to our present
undertaking.

Conversion among geometers has two meanings. In the
strict and primary sense it occurs when two theorems interchange their conclusions and their hypotheses with each other,
that is, when the conclusion of the first becomes the hypothesis
of the second and the hypothesis of the first is adduced as
conclusion of the second. For example: "In an isosceles triangle the angles at the base are equal" (here the hypothesis
is "isosceles triangle," and the equality of the angles at the
base is the conclusion); and "Triangles having equal angles at
the base are isosceles." The latter is precisely what the sixth
theorem asserts, taking as hypothesis the equal angles at the
base and as conclusion the equality of the sides that subtend
those equal angles. The other form of conversion involves only
a certain interchange among the component parts. If, for
example, a theorem is composite and arrives at a conclusion
from several hypotheses, we take the conclusion and one

252

[69] 251.12   See note at 245.26.

253

hypothesis and reach a conclusion consisting of one or more of the other hypotheses. It is in this sense that the eighth theorem will be the converse of the fourth. The fourth states "When the sides and angles are equal, the bases that subtend them are equal"; the[70] other "On equal bases equal sides contain equal angles." Of these components "on equal bases" was the conclusion of the former, while the positing of equal sides was one of its hypotheses. Of these two forms of conversion the primary type is uniform and determinate, but the other is varied and can run to a great number of theorems; there is not a single converse, but many, because of the plurality of hypotheses in the composite theorem. Often, however, we make a single converse of a theorem whose hypothesis consists of two or more members, when they are not all determinate, but some indefinite.

But we must also note in this connection that many conversions are made fallaciously and are not true converses. For example, every hexagonal number is triangular,[71] but it is not true that every triangular number is hexagonal. The reason is that the former character is of more general, the latter of more

254

particular application; and one can be asserted of the other only as true in every instance. But propositions about attributes that a subject has primarily and essentially can be converted. These matters also have engaged the attention of the mathematicians in the circle of Menaechmus and Amphinomus.

Among converse theorems themselves we are accustomed to call some "leading theorems" and others "converses." When, for example, we posit a genus and demonstrate its property, this we call a leading theorem; but whenever, contrariwise, we make the property our hypothesis and our conclusion the genus to which this property belongs, such a theorem we call a converse. "Every isosceles triangle has its

[70] 253.3   Reading τὸ for τὰ in Friedlein.
[71] 253.18   Triangular numbers are obtained by adding successive members of the series of integers (1, 2, 3, 4, . . .), i.e. 3, 6, 10, 15, . . . ; hexagonal numbers are obtained by adding successive members of the arithmetical progression with a difference of 4 (1, 5, 9, 13, 17, . . .), i.e. 6, 15, 28, 45, ... See Heath I, 76-79.

base angles equal" is a leading theorem, for its hypothesis is what is naturally primary, namely, the genus itself, the isosceles triangle; but "every triangle having two angles equal also has the subtending sides equal and is isosceles" is a converse, for it exchanges the subject for its attribute, making the latter its hypothesis and proving the former from it. So much we had to say regarding geometrical conversions.

Although reductions to impossibility lead us always to something clearly impossible, that is, to something whose contradictory is generally admitted, sometimes they lead us to principles inconsistent with the common notions or postulates or definitions, and sometimes to results that contradict something previously proved. This sixth theorem, for example, shows its consequences to be impossible because it contravenes the common notion that says the whole is greater than the part.[72] By contrast the impossibility at which the eighth arrives is one that would overthrow not a common notion, but something that has been demonstrated in the seventh theorem; for what the seventh denies the eighth shows to be affirmatively asserted by those who do not accept its conclusion.

Every reduction to impossibility takes the contradictory of what it intends to prove and from this as a hypothesis proceeds until it encounters something admitted to be absurd and, by thus destroying its hypothesis, confirms the proposition it set out to establish. In general we must understand that all mathematical arguments proceed either from or to the starting-points, as Porphyry[73] somewhere says. Those that proceed from the starting-points are themselves of two kinds, as it happens, for they proceed either from common notions, that is, from self-evident clarity alone, or from things previously demonstrated. Those that proceed to the starting-points are either affirmative of them or destructive. But those that affirm first principles are called "analyses," and their reverse procedures "syntheses" (for it is possible from those principles to proceed in orderly fashion to the thing sought, and

255

[72] 255.2 See Euclid's demonstration in note to 256.15.
[73] 255.14 On Porphyry see note at 56.24.

this is called "synthesis"); when they are destructive, they are called "reductions to impossibility," for it is the function of this procedure to show that something generally accepted and self-evident is overthrown.[74] There is a kind of syllogism in it, though not the same as in analysis; for the structure of a reduction to impossibility accords with the second type of hypothetical argument. For example, if in triangles that have equal angles the sides subtending the equal angles are not equal, the whole is equal to the part. But this is impossible; therefore in triangles that have two angles equal the sides that subtend these equal angles are themselves equal. So much regarding reductions to impossibility.

As we said, the author of the *Elements* employs conversion in the enunciation, which takes the conclusion of the fifth theorem as the given and its hypothesis as what is to be proved, and reduction to impossibility in the construction and the proof.[75] Should it be objected that in cutting off from AC a length equal to AB we should not cut it off from C but from A, we can adopt this hypothesis and arrive at the same impossibility. Let AD be equal to AB. Then let BA be pro-

[74] 255.26 I.e., if the proposed premise is accepted. Thus what is overthrown is this initial premise (the ἀρχή in this sense), not one of the generally accepted and self-evident ἀρχαί. Analysis confirms the initial premise, if it can be confirmed, by tracing it back to first principles, whereas the reduction to impossibility destroys it by showing that its consequences contradict first principles or their consequences.

[75] 256.15 For the proof of VI Euclid assumes triangle ABC with angle ABC equal to angle ACB, with unequal sides AB and AC, of which AC is the longer. By cutting off from AC a length DC equal to AB and joining DB, he proves that triangle DBC is equal to triangle

ABC, the less to the greater, which is impossible. Hence it is incorrect to assume that AB is not equal to AC, and it is therefore equal to it. In Heiberg's text AB is taken as the greater side, but I have modified his diagram to make it accord with the text that Proclus seems to have had before him. The validity of the proof is of course not affected.

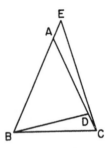

longed, and let AE be equal to DC. The whole of BE is then
equal to AC. Let EC be drawn. Now since AC is equal to BE
and BC is a common side, two sides are equal to two sides,
and the angle at B is equal to angle ACB by hypothesis. All
corresponding parts are therefore equal, by the fourth, so that
triangle EBC is equal to triangle ABC, the whole to the part,
which is impossible.

Now that the answer to this objection is clear, the next
thing is to demonstrate the other part of the converse, for
the author of the *Elements* has in the sixth theorem as a whole
converted only a part of the fifth, and we must add the converse that remains. This will be that which takes as hypothesis that the angles under the base of a triangle are equal and
proves that the triangle is isosceles. Let ABC be the triangle
and the sides AB and AC be extended and the angles under
the base be equal. I say that ABC is isosceles. For let the
point E be taken on AE, let CF be made equal to BE, and
let lines EC, BF, and EF be drawn. Then since BE is equal to
CF and the line BC is common, two sides are equal to two
sides, and angle EBC is equal to angle FCB, for they are

257

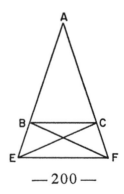

258

the angles under the base. All parts are therefore equal to all the parts, by the fourth. Their bases EC and BF are thus equal, the angle BEC is equal to angle CFB, and angle CBF to angle BCE, for they are subtended by equal sides. Now the whole of angle EBC was assumed equal to the whole angle FCB, and of these the part FBC is equal to ECB; therefore the remainder EBF is equal to the remainder FCE. And BE is equal to CF, and BF is equal to EC, and these two sides contain equal angles, and all corresponding parts are equal, so that angle BEF is equal to angle CFE. Consequently the sides AE and AF are equal, by this very sixth theorem which is here demonstrated.[76] And from these two sides equal segments, BE and CF, have been cut off, so that the remainders AB and AC are equal. Therefore triangle ABC is isosceles. Thus if a triangle has its two angles equal, it is isosceles; and if with its sides prolonged it has equal angles under the base, again the given triangle is isosceles.

What, then, is the reason why the author of the *Elements* did not also convert this second part? Is it not that it was irrelevant even in the fifth theorem to show that the angles under the base are equal, this being introduced for the purpose of meeting other difficulties, and that to prove a triangle is isosceles when the angles under its base are equal neither constitutes a leading demonstration[77] nor helps him in the solution of other questions? Besides, this converse is evident from the following theorems, and they furnish him the points of departure for proving that, when the angles under the base are equal, the triangle is isosceles. For if every straight line meeting another and making two angles with it makes these angles equal to two right angles,[78] then when the angles under the base are given as equal, the angles on the base will of course also be equal; and since they are equal, the sides that subtend them will also be equal. Using as

259

---

[76] 258.7 Perhaps this remark is intended to forestall the criticism (see Heath, *Euclid* i, 257) that he assumes the result of VI in proving the converse of the second part of V.

[77] 258.19 For the meaning of "leading" see 254.7ff.

[78] 259.3 XIII below.

he does this theorem throughout his treatise, he was able to infer that, when angles under the base are equal, the triangle is isosceles, if ever it was needed for the proof of any theorem. And very soon hereafter he will have clearly proved that, if a straight line stands upon a straight line and makes angles with it, it will make either two right angles or angles whose sum is equal to two right angles. The theorems that come before that one is established do not need this converse, and those that come after can be proved by means of it, if the need should arise.

VII. *Upon an identical straight line, if two straight lines have been constructed upon it, there cannot be constructed two other straight lines equal respectively to the former two and having the same extremities but meeting at a different point on the same side.*

This theorem has a character that is rare and not often found in scientific premises; for to be framed negatively and not affirmatively hardly suits their nature. At any rate the enunciations of geometrical and arithmetical theorems are usually affirmative. The reason is, as Aristotle says,[79] that the universal affirmative proposition is best fitted for science, since it is more self-sufficient, needing no negative premise to supplement it, whereas the universal negative needs an affirmative if it is to be proved. For without an affirmative premise there is no proof nor syllogism; and this is why the sciences that demonstrate do so affirmatively for the most part and rarely make use of negative conclusions.

The enunciation of this theorem is remarkably full and precise, and the many added phrases that make it irrefutable and unambiguous safeguard it against the attacks of pettifogging critics. First of all, our geometer stipulates "on the same straight line" to prevent our misleading the users of this premise by showing on a second straight line two straight lines constructed equal respectively to the first two. In the second place, given a single straight line, he does not say that

260

[79] 260.1   *Post. Anal.* 79a17-32.

the two straight lines constructed on it are simply equal to the two straight lines (there would be no impossibility in that), but that they are equal respectively. Would it be extraordinary if we should construct a second pair of lines equal together to the first pair, with one of its members longer and the other shorter? "Respectively" expresses the impossibility intended. Thirdly, he adds "meeting at another point." What if we took two lines equal respectively to the two lines already constructed and congruent with them and thus constructed both them and the two given straight lines meeting at the same vertical point? For if the straight lines are equal, of course their extremities coincide. Fourthly, there is the phrase "on the same side." Could we not on the same given straight line make the first two straight lines extend on one side and the second two on the other, so as to make the straight line the common base of two triangles with opposite vertices? In order that we may not impute our own error to the author of the *Elements* if we are thus misled, he adds "on the same side." Fifthly, he subjoins "having the same extremities as the given straight lines." For it would be possible on the same straight line to construct two lines equal respectively to the two given lines and meeting at a different point on the same side, using the whole of the straight line and constructing the two lines upon it, when the two lines constructed do not have the same extremities respectively as the two given lines, but different ones. If we imagine two diagonals of a square constructed on one of its sides, there will be two lines equal to two lines, a side and a diagonal equal to the parallel side and the other diagonal, yet these equal lines will not have the same extremities; for neither the parallel sides nor the diagonals, though equal, will have the same extremities. If we observe all these qualifications, the enunciation is correct and the reasoning in its proof unassailable.[80]

261

262

80 262.3 Euclid's proof of VII runs thus: Given two straight lines AC and CB constructed on the straight line AB and meeting at the point C, if possible let two other straight lines AD and DB be con-

But perhaps some persons, notwithstanding all these scientific restrictions, will be bold enough to object and say that what our geometer calls impossible is possible, even under the conditions laid down. Let AB be the straight line, and on it let lines AD and DB be constructed equal to lines AC and CB, and let AD and DB be inside the others, so that they meet at different points, C and D, and have the same extremities, A and B, as the given straight lines. Suppose, further, AC equal to AD and CB to DB. To those who make this

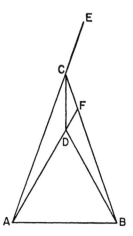

objection we shall reply by drawing the line DC and extending AC and AD. From this construction it is clear that triangle ACD is isosceles, since AD is equal to AC according to their thesis and the angles under its base are equal, namely, ECD

---

structed on the same straight line AB, meeting at another point D and equal to the former two respectively. Let CD be drawn. Then since

AC is equal to AD, angle ACD is equal to angle ADC; therefore angle ADC is greater than angle DCB, and hence angle CDB is still greater than angle DCB. Again since CB is equal to DB, angle CDB is equal to angle DCB. But it was also proved much greater than it, which is impossible.

263

and FDC.[81] Angle FDC is therefore greater than angle BCD; much more, then, is angle BDC greater than BCD. But again since DB is equal to CB and the angles BDC and BCD at the base are also equal, then the same angle is both much greater than and equal to another, which is impossible. This, we see, is precisely what we said when commenting on the fifth theorem, that the equality of the angles under the base would be useful, if not for the demonstration of later theorems, at least for the solution of objections to them. For now we have refuted the objection by inferring that, if AC and AD are equal, the angles ECD and FDC will also be equal. It will be evident in the case of other theorems also, that this principle contributes in the same way to the solution of difficulties raised.

But suppose someone should say: "Let straight lines BD and BC be constructed on line AB equal to lines AC and AD, BC equal to AC and BD to AD, at different points, A and B, and having the same extremities, C and D, with lines AC and AD." What can we say to this argument? Obviously that both the given straight lines constructed on the straight line AB and the lines equal to them must be constructed on the same straight line AB, for this is what the author of the *Elements* says in his enunciation. But the straight lines AC

264

and AD are not constructed on the straight line AB; they are constructed at a point on the straight line AB, but not on the line. So the lines constructed on the straight line AB—that is, AC, BC and AD, BD—are different from the lines posited in

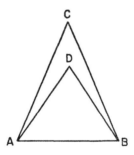

[81] 262.22 Friedlein's conjectural addition to the text is not really necessary, and I have left it untranslated.

this argument and from the lines equal to them, though the lines constructed were supposed to be equal to the lines constructed on the straight line AB.[82] So much by way of answer to this question.

It is clear that the author of the *Elements* demonstrates this theorem by means of the reduction to impossibility and that the impossibility contravenes the common notions that the whole is greater than the part and that the same thing cannot be both greater than and equal to another. It appears also that this theorem is a lemma preparatory to the eighth theorem, for it contributes to the proof of it and is neither an element in the strict sense nor elementary.[83] For its use does not extend widely. At least we shall find that our geometer makes very sparing use of it.

265

VIII. *If two triangles have two sides equal to two sides respectively and the base equal to the base, they will also have the angles equal which are contained by the equal straight lines.*

The eighth theorem is the converse of the fourth, but not a converse of the primary type, for it does not take as its conclusion the whole of the hypothesis of the fourth and as its hypothesis the whole of the conclusion. Instead it unites a part of the hypothesis of the fourth with a part of the conclusion and demonstrates one part of what was given. "Having two sides equal to two sides" is a hypothesis in both theorems; but "having the base equal to the base" was a part of the conclusion of the former, though it is given in this one.

---

[82] 264.7 To understand Proclus' answer it is necessary to realize that συνίστασθαι ἐπί is the conventional phrase in Greek geometry for constructing a triangle on a line, using the line itself as the base of the triangle. Hence a Greek geometer should understand that the problem of constructing two lines upon a given line and meeting at a point means drawing them from the extremities of the given line. Heath, *Euclid* I, 259. Thus in Proclus' diagram the lines constructed on AB, in this sense of the terms, are the two pairs AC, BC and AD, BD, whereas the lines posited in this argument are AC and AD, to which BC and BD are respectively equal.

[83] 264.16 For the meaning of "lemma," "element," and "elementary" see 211.1ff. and 72.3ff.

And "having the [contained] angles equal" was given in the fourth, but the eighth seeks to prove it. So the conversion is effected merely by an interchange between the given elements and elements in the conclusion.

If someone should want to know why it is eighth in order and does not come immediately after the fourth as its converse, as the fifth is followed by the sixth which is its converse (indeed most converse theorems follow their leading theorems and are demonstrated immediately after them), we must say that the eighth needs the seventh. For it is proved by use of the reduction to impossibility, and the impossibility involved is such as becomes known to us from the seventh. And this in turn needed the fifth for its proof. Necessarily, therefore, the seventh and the fifth were established before the theorem now being proved. And since the converse of the fifth could easily be proved from first principles, it was properly placed immediately after the fifth, both because of its kinship to it and because the impossibility shown by the reduction depends on the common notions and not, as in the eighth, upon another theorem. Propositions that contravene the common notions are clearer means of refutation than those that contradict theorems; for the latter are grasped by demonstration, whereas our knowledge of the former is superior to demonstration.

The author of the *Elements*, then, demonstrates the present theorem by means of the seventh which he has just proved.[84]

84 266.16   Euclid demonstrates VIII as follows: Given two triangles ABC and DEF having sides AB and AC equal respectively to sides DE and DF and the base BC equal to the base EF, he applies triangle ABC to triangle DEF, placing B on E and BC on EF. Point C will

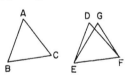

coincide with F, because BC is equal to EF. Then AB and AC will also coincide with DE and DF; for if not, they will fall beside them, like EG and GF, which is the construction that is shown by the preceding theorem to be impossible. So angle BAC will also coincide with angle EDF and will be equal to it.

But the school of Philon say that they can prove the eighth without the use of the seventh.[85] Let us suppose, they say, two triangles ABC and DEF having two sides equal to two sides and the base BC equal to the base EF. Place the triangle

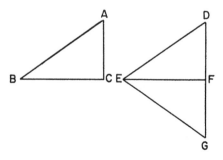

267

ABC in the same plane with DEF, making their bases coincide, but on the other side of the line EF, so that their vertices are opposite. Instead of ABC let the triangle thus placed be EFG, with EG equal to DE and FG equal to DF. FG will then lie either on a straight line with DF or not on a straight line, and if not on a straight line, making with it either an angle opening inwards or an angle opening outwards. First let us suppose it makes a straight line. Then since DE is equal to EG and DFG is a single straight line, the triangle DEG

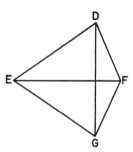

is isosceles, and the angle at D is equal to the angle at G. If FG is not on a straight line with DF, let it make an angle

[85] 266.18 Philon of Byzantium, probably of the second century B.C. See Heath II, 300-302. Philon's exposition and construction are somewhat confused in Friedlein. I have drawn upon Barocius for my translation of the first few lines.

opening inwards, and let line DG be drawn. Then since DE and EG are equal and they have a common base DG, angle EDG is equal to angle EGD. Again since DF is equal to FG and they have a common base DG, angle FDG is equal to angle FGD. But angle EDG was equal to angle EGD, and hence the whole of angle EDF is equal to the whole of angle EGF, which is what it was required to demonstrate. Thirdly, let FG make an angle opening outwards with DF, and let line DG be joined outside. Then since DE and EG are equal and they have a common base DG, angles EDG and EGD are equal. Again since DF and FG are equal and they have a common base DG, angle FDG is equal to angle

268

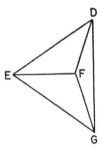

FGD. But the whole angles EDG and EGD were equal to each other; therefore their remainders, angles EDF and EGF, are equal to each other, and we have found what was enunciated, having demonstrated the theorem for every position of the line FG without making any use of the seventh. Was it not, then, superfluous, they say, for the author of the *Elements* to introduce it? For if it was only for the sake of the eighth that we brought it in, and if the eighth can be demonstrated without it, is not the seventh manifestly useless?

To this we must reply, as others have done before us, that the demonstration of the seventh is of the greatest utility to astronomers in the area of eclipses. For by the use of this theorem they say they can show that three successive eclipses cannot occur at equal intervals from one another, that is, the third separated from the second by the same interval of time as the second from the first. For example, if the second

269

has occurred six months and twenty days after the first, the interval before the third occurs cannot be this length of time but must be either longer or shorter. They say that this is demonstrated to be the case by the seventh theorem, and that this is not the only theorem the author of the *Elements* has demonstrated because it contributes to astronomy as a secondary aim, but many other theorems and problems as well. Take the last theorem in Book IV, which shows how to inscribe the side of a fifteen-angled figure in a circle—what reason can anyone suggest for his proposing it other than the bearing of this problem on astronomy? For by inscribing this fifteen-angled figure in the circle through the poles we get the interval between the poles of the celestial equator and those of the zodiacal circle, which are separate from each other by the length of the side of a fifteen-angled figure. It seems, then, that the author of the *Elements*, looking to astronomy, has given us proofs of many matters that prepare us for that science. And seeing that this seventh is proved from the fifth and that it provides an uncomplicated proof of the eighth, he gave it this position; for although Philon's procedure is elegant, its use of a variety of cases makes it unsuited for an elementary treatise. So much for our answer to this question.

If anyone should wonder why he did not add to the eighth the other details included in the fourth, namely, the equality of the triangles and of the remaining angles, our answer is that, when the angles at the vertex were proved to be equal, the equality of all parts to one another followed through the fourth. This, then, was the only thing it was necessary to prove independently; the rest could be inferred as consequences of it.

It appears that what makes the angles at the vertex equal is both the equality of their containing sides and the equality of the bases. For when the bases are not equal, the angles do not remain the same, even though the containing sides are supposed [equal],[86] but the shorter or longer the base, the smaller or larger the angle. Nor when the bases are the same but the

270

[86] 270.9 Reading with Barocius ἴσων just before ὑποκειμένων in Friedlein.

sides become unequal does the angle remain the same; rather it becomes greater as the sides become less and less as they become greater, for the angles undergo a change of character the reverse of that of their containing sides. Imagine yourself dropping sides to a base of fixed length. If you decrease the length of the sides, you will increase the size of the angle which they contain by making the interval between them greater; but if you raise them up and add to their length, you make smaller the angle which they contain, for they meet at a greater distance, since their vertex is further from the base. It is safe to say, then, that both the identical length of the bases and the equality of the sides determine the equality of the angles.

271   IX. *To bisect a given rectilinear angle.*

He mingles theorems with problems and interweaves problems with theorems and, by using both, achieves a full treatment of the elements, now providing the subjects, now investigating their attributes. So having shown in the previous theorems that in a single triangle the equality of the sides implies the equality of the angles, and conversely, and having done likewise for two triangles (except that for two triangles the method of conversion was different from that for one), he now turns back to problems with the demand to bisect a given rectilinear angle.

Clearly the angle here is given in kind, for it is a "rectilinear" that is mentioned, not any chance angle. The bisection of angles in general is not a matter for an elementary treatise, since it is even questioned whether bisection of an angle is always possible. One could doubt, for instance, whether we can bisect the horned angle. Determinate also is the ratio of the cut required, and this again with good reason. To divide in any ratio that might be chosen—as into three, or four, or five equal parts—goes beyond the present means of construction. We can divide a right angle into three parts by using some of the theorems that follow, but we cannot thus divide an acute angle without resorting to other lines that are mixed in kind.

272   This is shown by those who have applied themselves to the

problem of trisecting a given rectilinear angle. Nicomedes[87] made use of conchoids—a form of line whose construction, kinds, and properties he has taught us, being himself the discoverer of their peculiarities—and thus succeeded in trisecting the rectilinear angle generally. Others have done the same thing by means of the quadratrix of Hippias and that of Nicomedes, they too using mixed lines, namely, the quadratrices. Still others have started from the spirals of Archimedes and divided a given rectilinear angle in a given ratio. The thoughts of these men are difficult for a beginner to follow, and so we pass them by here. We can perhaps examine them more appropriately in the third book of the *Elements*, where the author bisects a given circumference.[88] There one finds the same method of inquiry employed not only for bisecting, but also for trisecting; and his procedures for dividing a circumference into three equal parts use the same lines as those used by the ancients.[89] Rightly, then, our geometer,[90] who has mentioned only the straight line and the circumference, bisects only rectilinear angles and circumferences. Since the species that arise from them by mixture are difficult to enumerate and explain without a meticulous examination, he passes them by, omitting all such questions as require the use of mixed lines and restricting his inquiry to the primary and simplest forms and to the matters that can be constructed or studied by their means alone. Of this sort is the problem before us, to bisect a given rectilinear angle. For his construction here he uses one postulate and the first and third theorems, and for the demonstration only the eighth theorem. For

273

---

[87] 272.3   Nicomedes, of the third century B.C. For the little that we know of him see Heath I, 225f., II, 199; and for the trisection of the angle and the constructions referred to just below, *Euclid* I, 265-267.

[88] 272.16   Cf. Euclid III. 30.

[89] 272.20   The reference to trisecting is puzzling. "The text here is misleading, because the methods used throughout the *Elements* are insufficient for trisecting a circumference, i.e. an arc of a circle." (I.M.)

[90] 272.20f.   The text of these lines has been corrupted. I assume with Friedlein that γεομέτρης, or some similar word, has dropped out after ὁ in line 20; and for the rest I have made use of Barocius, who found, or constructed, a more readable text.

problems always require a demonstration, as we said earlier, and get their scientific character from it.

Some may perhaps object to our geometer that the equilateral triangle constructed by him does not have its vertex between the two straight lines, but may have it on either of them or outside both, and that this becomes clear from the *Elements*.[91] Let the angle BAC be the angle which it is required to bisect. Take a point B on AB, cut off a length CA equal to BA, let BC be joined, and on this line construct an equilateral triangle BCD. Clearly this point D will lie either between lines AB and AC, or on AB, or on AC, or outside of both. The author of the *Elements* takes it as lying between them. To this they object and make difficulties with his proof, saying that it may lie on one of the straight lines or outside both of them. Let us suppose, then, that D lies on AB, in such a fashion as to make BCD equilateral. DB is then equal to DC, the angles at the base, CBD and BCD, are equal, and thus the whole angle BCE is greater than angle CBD. But

274

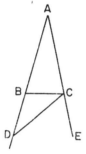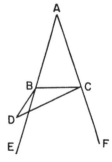

since BA is equal to CA, triangle ABC is isosceles and has the angles under its base BC equal. Hence angle BCE is equal to angle CBD; but it was greater, which is impossible. It is therefore not possible that the vertex of the equilateral triangle should lie on the line ABD. In the same way we can prove that it does not lie on line ACE. Let it then lie outside both, if possible. Then since BD is equal to CD, the angles at the base, BCD and CBD, are equal. Therefore angle BCD

91 273.15 Euclid's proof of IX can be so readily reconstructed from Proclus' description that there is no need to reproduce it in a footnote.

is greater than angle CBE, and hence angle BCF is greater still than angle CBE. But it was also equal (for it lies under the base BC of the isosceles triangle ABC), and this is impossible. Consequently point D does not lie outside the two straight lines on this side. Similarly it can be shown that it does not lie outside them on the other side. You see again that we have refuted the objections by using the theorem that an isosceles triangle has the angles under its base equal. This is that proposition of which we have said earlier that many assertions contrary to science can be shown to be unsound and easily refutable by its means; and this useful function it discharges for our geometer.

275

If someone should say that there is no room under the base BC,[92] we shall have to construct the equilateral triangle on the same side as BA and AC. In that case it is necessary that its sides either coincide with BA and AC, if these are supposed equal to BC; or lie outside them, if BA and AC are shorter than BC; or lie inside, if they are longer than BC. First let

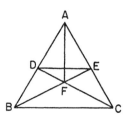

them coincide and BAC be itself an equilateral triangle. Let D be a point on BA; from AC cut off a length AE equal to AD; and let the lines DE, BE, CD, and AF be joined. Then since BA is equal to AC and AD to AE, the two sides BA and AE are equal to the two sides AC and AD, and they enclose the same angle, so that all corresponding parts are equal and angle DBE is equal to angle ECD. And since line DB is equal to line EC, and line BE to line CD, all corresponding parts are equal, so that angle DEB is equal to angle EDC, for they are subtended by equal sides. Line DF is then equal to line

276

[92] 275.7 Reading εἶναι with Barocius and Grynaeus instead of εἰδέναι in Friedlein. See note at 225.16 above.

EF, by the sixth. Then since AE is equal to AD and AF is common and DF is equal to EF, the angle DAE has been divided into equal parts, which is what it was required to do. Now suppose the sides of the equilateral triangle lie outside

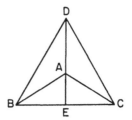

lines BA and AC. Let these sides be BD and DC, and let DA be drawn and produced to E. Then since BD and DC are equal and DA is a common side, and BA and AC are equal, angle BDA is equal to angle CDA, by the eighth. Again since BD and DC are equal and enclose equal angles with the common side DE, as has been demonstrated, base BE is equal to base EC, by the fourth. Then since BA is equal to AC and AE is common, angle BAE is equal to angle CAE, which is what it was required to prove. But if the sides of the equi-

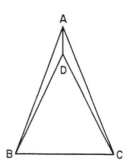

lateral triangle, BD and DC, fall within the lines BA and AC, again let AD be joined. Then since BA is equal to AC and AD is common and base BD is equal to base DC, angle BAD is therefore equal to angle CAD, by the eighth. So the angle at A has been bisected, however the equilateral triangle be placed.

277

— 215 —

Now that we have completed these demonstrations, let us take up the theorems that follow, adding only that an angle in a hypothesis may be given in one of four ways: either in position, as when we say that it lies on this straight line and at this point and is so given in the hypothesis; or in kind, as when we say it is a right angle, or acute, or obtuse, or in general rectilinear or mixed; or in ratio, as when we say it is double or triple another, or simply larger or smaller; or in magnitude, as when we say it is the third of a right angle. The angle here is given in kind alone.

X. *To bisect a given finite straight line.*

This too is a problem. It posits a finite straight line, since a line unlimited in both directions can in no wise be made determinate, and if it is without limit in one direction only, any division will cut it into unequal segments, wherever the point of section be taken, for the part that extends to infinity will necessarily be greater than the remainder, which is limited. The remaining alternative, then, is that a line which is to be bisected must be taken as finite in both directions.

278

This problem may move some persons to suppose that geometers assume in advance as a hypothesis that a line does not consist of indivisible parts. For if it did, a finite line would consist of either an odd or an even number of parts. But if it has an odd number of parts, it seems that when a line is bisected the indivisible is bisected, since otherwise one segment would consist of a larger number of indivisible parts and be greater than the other. Consequently it will not be possible to bisect a given line if its magnitude consists of indivisible parts. But if it is not composed of indivisible parts, it will be divisible to infinity. This, then, they say, appears to be an agreed principle in geometry, that a magnitude consists of parts infinitely divisible. To this we shall give the reply of Geminus, that geometers do assume, in accordance with a common notion, that what is continuous is divisible. The continuous, we say, is what consists of parts that are in contact, and this can always be divided. But they do not assume that what is continuous is also divisible to infinity; rather they demonstrate it from appropriate principles. For when geometers demonstrate that there is incommensurability among magnitudes and that not all magnitudes are commensurable with one another, what else could we say they are demonstrating than that every magnitude is divisible indefinitely and that we can never reach an indivisible part which is the least common measure of magnitudes? This, then, is demonstrable, but it is an axiom that every continuum is divisible; hence a finite line, being continuous, is divisible. This is the

279 notion that the author of the *Elements* uses in bisecting the finite straight line, not the assumption that it is divisible to infinity. That something is divisible and that it is divisible to infinity are not the same. One could use this problem also to refute the doctrine of Xenocrates[93] that asserts indivisible lines. For in general if there exists a line, it is either a straight line and can therefore be bisected, or circular and greater than some straight line—for every circular line has some straight line shorter than itself—or mixed and hence even more subject to division, since its simple components are divisible. But these matters must be reserved for study elsewhere.

Our geometer bisects a given finite straight line by using for his construction the first and the ninth, and for his proof the fourth only, since he shows the bases to be equal by means of the angles.[94] Apollonius of Perga bisects a given finite straight line in the following way. Let AB, he says, be the

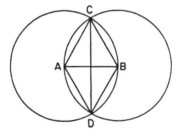

[93] 279.5 Xenocrates of Chalcedon, a disciple of Plato and head of the Academy after the death of Speusippus.
[94] 279.16 Euclid constructs on the given line AB an equilateral triangle ABC, bisects the angle ACB, and then by IV proves that the line CD which bisects the angle also bisects the line AB.

finite straight line which is to be bisected. With A as center
and distance AB let a circle be described, and again another
circle with B as center and distance BA; and let the points of
intersection of the circles be joined by line CD. This line bisects
280   the line AB. For let lines CA and CB be drawn; then each
of them is equal to AB, DA and DB are equal for the same
reason, and CD is a common base; therefore angle ACD is
equal to angle BCD, so that AB is bisected in accordance with
the fourth. Such is the kind of proof of the present problem
given by Apollonius. It too starts from an assumed equilateral
triangle, but instead of proceeding from the bisected angle at
C it proves that the line is bisected because of the equality of
the bases. The proof given by the author of the *Elements* is
therefore much better, since it is simpler and proceeds from
the principles.[95]

XI. *To draw a straight line at right angles to a given
straight line from a given point on it.*

Whether we take the straight line as limited in both directions, or unlimited in both, or unlimited in one and limited in
the other, with the point lying on it, the construction with
which our geometer solves this problem succeeds.[96] For even
if the given point lies on the extremity of the line, we can
produce the same construction by extending the straight line.
Clearly the point here is given in position, and in position only
as lying on the straight line; but the straight line is given only
281   in kind, for its length, ratio, and position are not determined.
The author of the *Elements* proves this proposition by using
the first and third theorems and one of the postulates (namely,
the first), and also the eighth theorem and the definition of a

[95] 280.11 ἀπὸ τῶν ἀρχῶν must mean here "in proper order from
principles." The criticism of Apollonius' procedure is not that he does
not have first principles, but that he does over again what has already
been done in IX. For a similar use of this phrase see 326.13 and 336.8.

[96] 280.19 To prove XI Euclid assumes a straight line AB and a
point C on it, takes another point D at random on AC, cuts off on
CB a length CE equal to DC, and on DE constructs an equilateral
triangle FDE; he then proves by VIII and Def. X that FC is at right
angles to AB.

line at right angles. If anyone should demand that we take the point at the extremity of the line without extending the line beyond it and draw the right-angled line from this point, we can show that this also is possible. Let the line be AB and the given point A, and let any point C on AB be taken and from it a line CE at right angles to AB be constructed in the

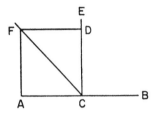

manner taught in this theorem. And let a distance CD be cut off on CE equal to AC and the angle at C be bisected by the line CF, and from D let a line be drawn at right angles to CE meeting CF at F and from F to A the line FA be joined. I say that the angle at A is a right angle. For since CD is equal to AC, the side CF is common, and they enclose equal angles—for the angle at C has been bisected—then DF is equal to FA, and all corresponding parts are equal, by the fourth, so that the angle at A is equal to the angle at D and is therefore a right angle. Thus our problem is solved. But the author of the *Elements* has no need of this device, for he stipulates that the line be drawn "at right angles," not "at a right angle." We should not, then, take the point at the extremity of the line if the straight line is to make angles, not an angle only, with the given line.

Apollonius draws the line at right angles in the following way:[97]

[Let AB be the given line and C a point on it.] Take any point D on AC, cut off from CB a length CE equal to CD, then with center at D and distance DE let a circle be described, and again another circle be described with center E and distance DE; and let a line be drawn from F to C.

[97] 282.9   The words in brackets are added from Barocius.

I say that this is the line at right angles. For if we draw lines FD and FE, they will be equal; and equal also are CD and CE, and FC is common, so that the angles at C are equal, by the eighth. They are therefore right angles.

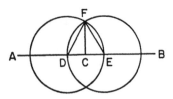

Once more you see that this proof, which requires the drawing of circles, is more complex than that given by the author of the *Elements*, since it was possible at once to erect an equilateral triangle on DE and establish the theorem.[98] All other features in the two proofs are identical. The proof by means of the semicircle is not even worth mentioning, for it assumes many of the later propositions and completely departs from the order of the *Elements*.[99]

XII. *To a given infinite straight line, from a given point which is not on it, to draw a perpendicular straight line.*

This problem was first investigated by Oenopides,[100] who thought it useful in astronomy. In archaic fashion, however,

[98] 282.23 I.e. the method of constructing an equilateral triangle has already been found in I, and to repeat it here is unnecessary. See note to 280.11.

[99] 283.3 From what Proclus says about this proof we can imagine that it ran something like this. Given a line AB and a point C on it from which it is required to construct a line at right angles, take a point D, not on AB, nor on the extension of AB, and with D as

center and DC as distance draw a circle intersecting AB at a point E other than C. From E draw through D a diameter of the circle to F. The line drawn from C to F will be at right angles to AB, by III. 31.

[100] 283.7 On Oenopides see note at 66.2.

he calls the perpendicular a line drawn "gnomonwise," because the gnomon also is at right angles to the horizon. This "perpendicular" differs from the line drawn "at right angles" only in relation to its point of origin, as the word for perpendicular (κάθετος) indicates, and not in substance.[101] Again there are two kinds of perpendicular, plane and solid. Whenever the line and the point from which a perpendicular is dropped are in the same plane, it is called a plane perpendicular, but when the point is above and outside the assumed plane, a solid perpendicular. The plane perpendicular is drawn to a line, the solid to a plane. Hence the latter necessarily makes right angles not with one line, but with all the lines[102] in the plane, for the perpendicular was drawn to the plane. In this problem, then, the author of the *Elements* is proposing that a plane perpendicular be drawn; for it is proposed that it be drawn to a straight line, and the argument proceeds on the assumption that all the elements involved are in a single plane.

284

Now when considering the construction of a line at right angles, we had no need of the infinite, since the point was taken as lying on the line itself; but in the case of the perpendicular the given line is assumed to be infinite, since the point from which the perpendicular is to be dropped lies somewhere outside the line. If the line were not infinite, it would be possible so to take the point that it would lie outside the given line, but on a straight line with it, so that the line when prolonged would fall on it; and thus the problem could not be solved. For this reason he posits the straight line as infinite, so that if the point is taken only on one or the other side of the line, there will be no place left in which it can lie in a straight line with the given straight line and thus will lie outside and not on it.

[101] 283.12f. I.e. the line "at right angles" in the preceding proposition was to be *erected on* the line, whereas the "perpendicular" is to be *dropped upon it*. Reading φησὶ with Grynaeus and Barocius instead of φασὶ in Friedlein, and κάθετος ("plumb line") instead of κάθοδος ("descent"). There is a play on these words at 290.17ff.

[102] 283.21 *Sc.* that meet it, as Proclus has more correctly said at 135.22. Cf. Euclid XI, Def. 3.

This is the reason, then, why the line to which the perpendicular is to be dropped is given as infinite. But it is worth inquiring in what sense in general the infinite has existence. It is clear that, if there is an infinite line, there will also be an infinite plane, and infinite in actuality if the problem is to be a real one. That in sensible things there is no magnitude indefinitely extended in any direction has been sufficiently shown by the inspired Aristotle[103] and by those who derive their philosophy from him. For it is not possible for the body moving in a circle to be infinite, nor any other of the simple bodies, for the place of each is determinate. But neither is it possible that there should be an infinite of this sort among separate and indivisible ideas; for if there is no extension nor magnitude in them, there can hardly[104] be infinite magnitude.

It remains, then, that the infinite exists in the imagination, only without the imagination's knowing the infinite. For when the imagination knows, it simultaneously assigns to the object of its knowledge a form and limit, and in knowing brings to an end its movement through the imagined object; it has gone through it and comprehends it. The infinite therefore, is not the object of knowing imagination, but of imagination that is uncertain about its object, suspends further thinking, and calls infinite all that it abandons, as immeasurable and incomprehensible to thought. Just as sight recognizes darkness by the experience of not seeing, so imagination recognizes the infinite by not understanding it. It produces it indeed, because it has an indivisible power of proceeding without end, and it knows that the infinite exists because it does not know it. For whatever it dismisses as something that cannot be gone through,[105] this it calls infinite. So if we supposed the infinite line to be given in imagination, exactly like triangles, circles, angles, lines, and all the other geometrical figures, should we not ask in wonder how a line can ac-

[103] 284.24   *Phys.* 204a8-206a8; *De Caelo* 271b1-276a17.

[104] 285.5   Reading with Grynaeus and Barocius σχολῇ γ' ἄν instead of σχολή, εἰ in Friedlein.

[105] 285.18   ἀδιεξίτητον, like διέξεισι in 285.9, is an echo of Aristotle's *Phys.* 204a2-7, 207b29.

tually be infinite and how, being indeterminate, it associates with determinate notions? But the understanding from which our ideas and demonstrations proceed does not use the infinite for the purpose of knowing it, for the infinite is altogether incomprehensible to knowledge; rather it takes it hypothetically and uses only the finite for demonstration; that is, it assumes the infinite not for the sake of the infinite, but for the sake of the finite. If our imagination could see that the given point does not lie on the extension of the finite line and is so separated from it that no part of the line could underlie the point, the demonstration would no longer need the infinite. It is therefore that it may use the finite line without risk of refutation or doubt that it posits the infinite, relying on the boundlessness of imagination as the source which generates it.

This is enough for the present concerning the hypothesis of the infinite. Let us now move on to the objections that have been brought against the construction used in this problem.[106] Let the straight line be taken as infinite, they say, and the point C from which the perpendicular is to be drawn and the point D on the other side of the line from C be given, as our geometer says; but let us have the circle cutting the line AB at A and B and also at F, as in the position diagrammed.[107] To this argument we reply that what it says is impossible. For let AB be bisected at H, join CH and extend the line to the circumference [at D], and let CA and CB [and CF][108] be joined. Then since these are lines from a center, and AH is equal to HB, and CH is a common side, all corresponding parts are equal. CH therefore makes right angles at H. Again

---

[106] 286.15   Euclid's proof of XII goes as follows: Given an infinite straight line AB and a point C not on it, he takes a point D at random on the other side of the line and with center C and distance CD describes a circle cutting the given line at G and E. He then bisects GE, draws a line from C to the midpoint H, and proves by VIII that CH is perpendicular to AB.

[107] 286.22   This is a legitimate objection to Euclid's unexpressed assumption that the circle cuts AB in only two points. Proclus tries to answer the objection, but his refutation is inconclusive; see note at 289.6.

[108] 287.2   The words in brackets are supplied from Barocius.

since CA and CB are equal, they make equal angles at points A and B. But CA is also equal to CF, so that angle CAF is equal to angle CFA; and CB is equal to CF, so that angle

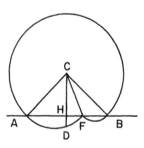

CFB is equal to angle CBF. Then since the angles at A and B are equal, angle CFA is equal to angle CFB, and they are adjacent and consequently right angles. Both of the angles at H are right angles, and hence CH is equal to CF; but CF is also equal to CD, for they are lines from a center; therefore CH is equal to CD, which is impossible. Consequently the circle does not cut the straight line AB at another point.

But if anyone should say that the circle described bisects AB at F, again we can show the same impossibility. [Let all the lines be drawn as before, and][109] let FB be bisected at H. Then since AF and FB are equal, CF is common, and base CA is equal to CB, all corresponding parts are equal, so that the angles at F are right angles. Again since FH is equal to HB and CH is common and the base CF (let it be drawn) is equal to CB (for they are lines from a center), the angles at

288

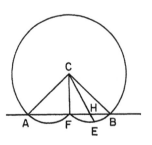

109 287.17 From Barocius.

H are right angles, for they are equal and adjacent. Then since each of the angles CFH and CHF is a right angle, CF is equal to CH. But CF is equal to CE, for they are lines from a center. Therefore CH is equal to CE, which is impossible.

There remains the third objection to be dealt with. Let the described circle, they say, cut the straight line both at A and B and at F and H. Then if we bisect AB at K and join CA, CF, CK, and CB, we can show that this is impossible. For since AK and KB are equal, CK common, and the bases CA and CB equal, the angles at A and B are equal, and the angles at K are right angles. But each of the sides CA and CB is equal to CF; therefore the angles at F are right angles, for they are equal and adjacent; and hence CF is equal to CK, for they subtend right angles. But CF is equal to CD, for they are lines from a center; therefore CD is equal to CK, which is impossible. It is thus not possible for the described circle to cut the straight line AB either at one or at two points other than points A and B.[110]

289

These, then, are the objections. There are also cases in the construction involved in this problem, and these we must treat separately from the objections. A case and an objection are not identical; the case proves the same thing in another way, but an objection is adduced to show absurdity in the proof objected to. By not discriminating between these, commentators have introduced them all together and have not made it clear whether they are asking us to diagram cases or objections. We therefore distinguish them and adduce the cases after the objections. Let AB be the infinite straight line and C the

---

[110] 289.6 On the insufficiency of Proclus' refutation see Heath, *Euclid* I, 272f. "His method of proof only enables us to show that, if the circle meets AB in one more point besides G, E, it must meet it in more points still. We can always find a new point of intersection by bisecting the distance separating any two points of intersection." One consequence would be that there are an infinite number of perpendiculars from C to AB. "This in fact is possible under the Riemann hypothesis; but for a proof that it is not possible in Euclidean space, we have to wait until XVI. This involves no difficulty, since XII is not used before." But Heath also shows how, if it had been necessary, Euclid could have demonstrated this assumption by means of the "invaluable" VII.

290

given point. Someone may say there is no room on the other side of the line, but only on the side on which C lies.[111] Taking, then, point D on the straight line, and with C as center and CD as distance, we can describe a circumference DEF; then bisecting DF at H, we can join CD, CH, and CF. Then since DH is equal to HF, CH common, and CD equal

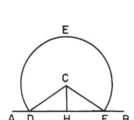 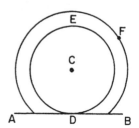

to CF (for they are lines from a center), the angles at H are equal adjacent angles and are therefore right angles. CH, then, is the perpendicular to DF. Finally, if anyone should say that the described circle does not cut the straight line AB but is tangent to it, like the circle DE, we can take a point F[112] outside and, with C as center and distance CF, can reach the desired result as in the case expounded above.

Enough about the cases of the problem, which have been presented to provide exercise for our readers. If I may add some reflections on these two problems, it seems that the line erected at right angles is an imitation of life lifting itself to the upper world from the hollows here below, rising undefiled and remaining uninclined towards worse things, whereas the perpendicular (κάθετος) is a likeness of life following the path downwards (κάθοδος)[113] and holding itself free of the indeterminateness in the world of generation. For the right angle is a symbol of undeviating energy, held in control by equality, definiteness, and boundaries. This is clearly why the *Timaeus*, in its account of the divine soul, calls the "circle

291

[111] 289.21   See note at 225.16.

[112] 290.10   τὸ ē in Friedlein and Barocius is obviously an error. What is needed is a point outside the circle DE. In line 12 read τῷ γϛ̄ for τῷ γε̄.

[113] 290.20f.   See note at 283.12f.

— 226 —

of the other," that which contains the ideas of sensible things, a right circle.[114] In our souls it is fractured in all sorts of ways and undergoes complex distortions because of its connection with generation, but in the universe it is firmly set, unwavering and undefiled, above sensible things. And if the infinite straight line is a symbol of the whole world of becoming in infinite and indeterminate change and of matter itself which possesses no boundary nor shape, and if the point lying outside carries the likeness of partless being devoid of anything material, most certainly, then, the perpendicular dropped from above would be an imitation of life proceeding immaculately from the One and Indivisible into the world of generation. And if, furthermore, the perpendicular cannot be otherwise shown than through the use of circles, this too would furnish an indication of the stability imparted through Nous to living things. Although life itself, being essentially change, is indeterminate, yet it acquires definiteness and is filled with pure power when it partakes of Nous and goes forward with it.

XIII. *If a straight line set up on a straight line makes angles, it will make either two right angles or angles equal to two right angles.*

Our author has returned to theorems, in consequence of what he has demonstrated in the problems. He has drawn a perpendicular to a straight line at right angles to it, and the next step was to inquire what angles will be made and how they will be related to the straight line if the line standing on it is not perpendicular.[115] This theorem shows generally

292

[114] 291.3   Cf. *Tim.* 37b: ὁ τοῦ θατέρου κύκλος ὀρθὸς ἰών and Proclus' *Commentary on the Timaeus* II, 209.18, Diehl: τὸν ὀρθὸν κύκλον.

[115] 292.2   Euclid proves XIII in the following manner: Given a straight line AB set up on line CD and making angles CBA and ABD. If these angles are equal, they are two right angles, by Def. X. If not, let BE be drawn from B at right angles to CD, by XI. Therefore CBE and EBD are two right angles. Now since CBE is equal to the two angles CBA and ABE, let angle EBD be added to each. Therefore angles CBE and EBD are equal to the three angles CBA, ABE, EBD, by Axiom II. Again since angle DBA is equal to the two angles DBE and EBA, let angle ABC be added to each; therefore angles DBA

that every straight line standing on a straight line and making angles with it makes either two right angles, if it stands upright without inclining towards either end, or angles equal to two right angles, if it inclines towards one and away from the other end of the given straight line. For whatever amount it takes away from one right angle by inclining to one side it adds to the other by diverging from that side.

We should notice how much concern for precision our geometer shows in this proposition also. For he does not simply say that every straight line standing on another straight line makes either two right angles or angles equal to two right angles but adds "if it makes angles." For suppose it stands at the extremity of the straight line and makes one angle with it. Would it be possible for this to be equal to two right angles? Obviously not, for every rectilinear angle is less than two right angles, just as every solid angle is less than four right angles. Even if you take the angle that seems the most obtuse, you can only give it such a magnitude as will still fall short of the measure of two right angles. We must then so erect the straight line that it makes angles.

This, as I said, is a mark of his scientific precision. But what does he intend when he adds that it makes "either two right angles or angles equal to two right angles?" For when it makes two right angles, it makes angles equal to two right angles, since all right angles are equal to one another. Is it not that the one expression denotes an attribute common to both equal and unequal[116] angles, the other a property of equal angles only? Whenever both a general and a special attribute can be affirmed truly of something,[117] we are accustomed to indicate its character by the special attribute; but whenever we cannot hit upon this, we are satisfied with

293

---

and ABC are equal to the three angles DBE, EBA, and ABC. But the angles CBE and EBD were also proved equal to the same three angles; therefore (by Axiom I) angles DBA and ABC are equal to angles CBE and EBD, i.e. to two right angles.

[116] 293.2 The sense required supports Friedlein's suggestion for the insertion of καὶ τῶν ἀνίσων.

[117] 293.4 Putting Friedlein's comma after κοινόν, not after ἀληθεύῃ.

the general character for the clarification of the things under consideration. That adjacent angles are equal to [two][118] right angles is a general statement that applies indeed to right angles, but not to right angles alone; but that they are right angles is a statement distinctive of their equality. Consequently to say only that the angles are equal to two right angles is to imply unequal angles, for to them alone is it truly applicable, not to equal angles. And this the author of the *Elements* has logically distinguished by "[equal] to two right angles"; for this phrase in and by itself denotes a pair of unequal angles.

From this we can see how equality is a measure and a boundary of inequality as well. For even though the diminution and increase of the obtuse and acute angles is indefinite and undetermined, yet this increase and diminution are said to be limited and bounded by the right angle. And though each of them departs in a different direction from likeness to the right angle, yet both of them by a certain unity of nature refer back to the standard of the right angle; and since they are unable to equal the simplicity of the right angle, they attain equality when it is doubled. The dyad, which is in itself indefinite, is a paradigm of their indeterminateness. Here it seems we have a manifest image of the forthgoing of the primary causes which stand as a single boundary line ever the same about the indefiniteness of generation. For how otherwise could the world of generation, which partakes of the more-and-less and undergoes limitless change, be brought into harmony with the intelligible world, and in a sense made like to it, than by participation in those causes[119] which with their productive powers are always going forth and duplicating themselves? For in their simplicity and partlessness they completely transcend the world of generated things. So much we can derive from this theorem for understanding the whole of things.

294

[118] 293.7   Inserting δυσὶν before ὀρθαῖς, as in lines 11 and 14 below.
[119] 294.9   Reading with Barocius ἢ before instead of after διὰ τῆς μεθέξεως.

XIV. *If with any straight line, and at a point on it, two
straight lines adjacent[120] to one another and not lying
on the same side make the adjacent angles equal to two
right angles, the two straight lines will be in a
straight line with one another.*

This is the converse of the theorem just demonstrated.
Converses always follow their leading theorems. The previous theorem constructed a straight line on another and
showed that it makes the adjacent angles either two right
angles or equal to two right angles. This theorem assumes
angles making two right angles at a straight line and shows
that it is one straight line that makes them at the straight
line mentioned. What was given in the former is the conclusion in this, and it is proved by the reduction to impossibility. This is the method ordinarily used for proving the
converse of a theorem, although in problems, at least, our
geometer admits also leading constructions.

In this theorem too we can observe an unexcelled level of
precision in scientific expression. In the first place, after saying "if with any straight line," he adds "and at a point on it."
Suppose, since the straight line has two extremities, a line
drawn at one end and another at the other made the angles
on the straight line equal to two right angles. Could they for
this reason be on a straight line with each other? How could
they be, being drawn from different points of the straight
line? For this reason he adds "and at a point on it," intending
that the two angles should lie at one point. Secondly, since
it would be possible for straight lines drawn from the same
point on a straight line not to be adjacent to one another (for
one could assume countless straight lines at a single point),
he adds "two straight lines adjacent to one another." Thirdly,
since lines adjacent to one another can be considered as
either on the same side or on both sides, but those adjacent on
the same side cannot be in a straight line with one another, he
has ruled this out and bidden us take the adjacent lines as

[120] 294.16 The ἐξῆς in Proclus is not found in our text of Euclid.
Most mathematicians would regard it as unnecessary. Schönberger
remarks that Proclus is "more popish than the Pope."

296

situated on opposite sides; for those are what can be proved to lie on a straight line. Let lines BC and BD be two lines standing on the straight line AB and on the same side of it. These, moreover, are adjacent to one another, for there is no other straight line between them. Things are adjacent to one another when there is nothing of the same sort between

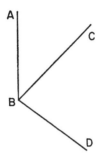

them. For example, we call columns adjacent to one another when there is no other column between them. Of course there is air between, but nothing of their kind. Because these lines lie on the same side, they cannot have the property of being in a straight line, even though the angles they make with AB may be equal to two right angles—for there is nothing to prevent the angle at ABD from being one and one-third of a right angle and angle ABC being the remaining two-thirds.

So much for the enunciation. In the construction he uses one postulate, the second (that a finite straight line can be extended in a straight line),[121] just as in the proof he uses the preceding theorem and two axioms (things equal to the same thing are equal to one another, and if equals be subtracted from equals the remainders are equal); and for the reduction to impossibility he uses the axiom that the whole is greater than the part, for when the one common angle had been subtracted, the whole was equal to the part, which is impossible.[122]

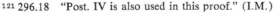

[121] 296.18 "Post. IV is also used in this proof." (I.M.)

[122] 296.24 Euclid's proof of XIV is as follows: With straight line AB and at the point B on it, let two straight lines BC and BD not lying on the same side make the adjacent angles ABC and ABD equal to two right angles. Then BD is in a straight line with CB. For if it is

297   That it is possible for two adjacent lines drawn at the same point on a straight line and lying on the same side of it to make angles on the straight line equal to two right angles we can demonstrate thus, after Porphyry. Let AB be a straight line. Take any chance point on it, say C, and let CD be drawn at right angles to AB, and let angle DCB be bisected by CE.[123] Let a perpendicular EB be dropped from E, let it be extended, and let BF be equal to EB and CF be joined. Then since EB is equal to BF, BC is common, and these sides

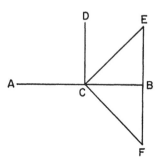

contain equal angles (for they are right angles), base EC is equal to base CF, and all corresponding parts are equal. Angle ECB is therefore equal to angle FCB. But angle ECB is half of a right angle, for a right angle was bisected by EC; hence FCB is half of a right angle. Angle DCF is therefore one and one-half of a right angle. But DCE is half of a right angle; therefore on line CD and at point C on it there are two adjacent straight lines CE and CF lying on the same side of it and making with it angles equal to two right angles, CE making an angle equal to half of a right angle, and CF an

298

---

not, let BE be in a straight line with BC. Then since AB stands on the straight line CBE, angles ABC and ABE are equal to two right angles. But angles ABC and ABD are also equal to two right angles; therefore angles CBA and ABE are equal to angles CBA and ABD. Let angle CBA be subtracted from each; then the remaining angle ABE is equal to the remaining angle ABD, the less to the greater, which is impossible. Therefore BE is not in a straight line with CB. Similarly we can prove that neither is any other line except BD. Therefore BD is in a straight line with CB.

[123] 297.10   "This argument is strangely stated. AB is taken as given, and CD drawn perpendicular to it. But CD should be given, and BC drawn perpendicular to it." (I.M.)

angle equal to one and one-half of a right angle. Thus to prevent our drawing the impossible conclusion that CE and CF, which make angles with DC equal to two right angles, lie on a straight line with one another, our geometer has added the phrase "not lying on the same side." Hence the lines that make with a line angles equal to two right angles must lie on opposite sides of the line, though starting at the same point, one extending to this and the other to that side of the straight line.

XV. *If two straight lines cut one another, they make the vertical angles equal to one another.*

Vertical angles are different from adjacent angles, we say, in that they arise from the intersection of two straight lines, whereas adjacent angles are produced when one only of the two straight lines is divided by the other. That is, if a straight line, itself undivided, cuts the other with its extremity and makes two angles, we call these angles adjacent; but if two straight lines cut each other, they make vertical angles. We call them so because their vertices come together at the same point; and their vertices are the points at which the lines[124] converging make the angles.

299

This theorem, then, proves that, when two straight lines cut one another, their vertical angles are equal. It was first discovered by Thales, Eudemus says, but was thought worthy of a scientific demonstration only with the author of the *Elements*. Not all the principal parts are present in this demonstration, for the construction is lacking. But the proof, which is indispensable, depends on the thirteenth theorem and uses also two axioms: one, that things equal to the same thing are equal to each other; the other, that if equals be subtracted from equals the remainders are equal.[125]

[124] 298.23 ἐπίπεδα here must be an error; εὐθεῖαι or γραμμαί is obviously required, with a corresponding change in the preceding participle. It is strange that neither Barocius, nor Taylor, nor Schönberger, nor ver Eecke seems to feel any scruples against reading "planes."

[125] 299.11 "Proclus fails to notice that Post. IV is also used in this proof." (I.M.)

Euclid's theorem is clear;[126] and its converse is another equally clear: if upon a straight line we assume two straight lines not on the same side and making the vertical angles equal, these lines lie on a straight line with each other. Let

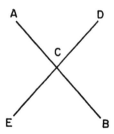

AB be a straight line and C be any point on it, and at C let two straight lines CD and CE be taken not on the same side and making angles ACD and BCE equal. I say that CD and CE lie on a straight line. For since CD stands on AB, it makes angles DCA and DCB equal to two right angles. But angle ACD is equal to angle BCE. Hence DCB and BCE are equal to two right angles. Then since on a straight line BC two adjacent straight lines CD and CE, not on the same side, make with it adjacent angles equal to two right angles, CD and CE lie on a straight line with one another. The converse of the present theorem is therefore demonstrated. Our geometer, it seems, omitted this converse because it is easily proved by the same method of reduction to impossibility which we used to prove the preceding theorem. Taking the same hypotheses as above, I say that CD lies on a straight line with CE. For if it does not, let us take CF as lying on a straight line with CD. Then since two straight lines, AB and DF, cut one another, they make the vertical angles equal; hence angles ACD and BCF are equal. But angles ACD and BCE were

300

---

[126] 299.12   Euclid's proof of XV (adapted to the lettering of Proclus' diagram) points out that AC standing on ED makes angles ECA and ACD equal to two right angles and that EC standing on AB makes angles ECA and ECB equal to two right angles; hence angles ECA and ACD are equal to angles ECA and ECB. Subtracting ECA from each, ACD is equal to ECB. Similarly ACE can be proved equal to DCB.

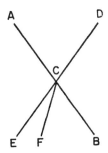

equal. Therefore angle BCE is equal to angle BCF, the greater to the less, which is impossible. Consequently there is no other straight line that is on a straight line with CD, and CD and CE therefore are on a straight line, assuming the vertical angles to be equal. Since this is the same proof as that adduced in the fourteenth theorem, would it not have been superfluous to bring in this converse? For the sake of practice, however, we have established it both by the reduction to impossibility and by direct proof.

301

It seems that the force of this fifteenth theorem comes from the homoeomery of straight lines and from their being stretched to the utmost, since lines so characterized and crossing one another necessarily have the same inclinations to one another on both sides. Circular lines, and in general lines that are not straight, do not necessarily make their vertical angles equal when they cut one another, but sometimes equal and sometimes unequal. For instance, if two equal circles cut one another through their centers, or even at some point other than their centers, they make the lunular angles at the vertex equal, but of the other angles—that is, the biconvex and the biconcave—one is the greater.[127] But in the case of straight lines their being stretched to the utmost makes equal [on both sides] the divergence of the segments of the one from those of the other.

[127] 301.15 See diagram at 127.14. ACD and BCE in this diagram are equal, each of them, according to 333.15, being equal to two-thirds of a right angle. Of the other two angles one, presumably ACB, is greater than the other, DCE.

PORISM. *From this it is clear that, if two straight lines*
cut one another, they make the four angles
equal to four right angles.

"Porism" is a geometrical term and has two meanings.[128]
We call "porism" a theorem whose establishment is an incidental result of the proof of another theorem, a lucky find[129]
as it were, or a bonus for the inquirer. Also called "porisms"
are problems whose solution requires discovery, not merely
construction or simple theory. We must see that the angles
at the base of an isosceles triangle are equal, and our knowledge in such cases is about already existing things. Bisecting
an angle, constructing a triangle, taking away or adding a
length—all these require us to make something. But to find
the center of a given circle, or the greatest common measure
of two given commensurable magnitudes, and the like—these
lie in a sense between problems and theorems. For in these
inquiries there is no construction of the things sought, but a
finding of them. Nor is the procedure purely theoretical; for it
is necessary to bring what is sought into view and exhibit it
before the eyes. Such are the porisms that Euclid composed
and arranged in three books.[130]

But of such porisms we shall not speak here; the porisms
in the *Elements* are theorems that come to light along with
the demonstrations of other theorems, without being themselves the object of the preceding inquiry, like the one stated
here. For the question under investigation was whether the
vertical angles are equal if two straight lines cut one another;

302

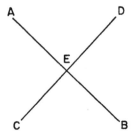

[128] 301.22   Cf. 212.12ff.
[129] 301.24   ἕρμαιον, "gift of Hermes," "windfall."
[130] 302.13   On the attempts to reconstruct from this and other
evidence the contents of Euclid's lost *Porisms* see Heath I, 431-438.

303

and the proof of this also carries with it the proof that the four angles are equal to four right angles. For when we said "Let AB and CD be two straight lines cutting one another at point E; then since AE stands on CD, it makes the adjacent angles equal to two right angles; and again since BE stands on CD, it makes the adjacent angles equal to two right angles," then together with the conclusion drawn there we also proved that the angles about E are equal to four right angles.

A porism, then, is a theorem whose truth becomes evident without effort through the proof of another problem or theorem. For we appear to hit upon porisms as it were by accident, not as answers to problems or inquiries; hence we likened them to lucky finds. And it may be that the masters in mathematics gave them this designation in order to show ordinary people, who get excited over some apparent gain, that these, and not the sort of things they suppose, are the true windfalls and gifts of the gods. For they are produced by the resources we have within us; our prolific capacity for knowledge adds them to the results of the preceding inquiries, thus revealing the inexhaustible richness of the world of theorems.

Such, then, is the way in which the peculiar character of porisms is to be described. They can be classified, first, by the sciences in which they appear: some porisms belong to geometry, others to arithmetic. The one before us is geo-
·metrical, that which occurs at the end of the second theorem in the seventh book arithmetical. Secondly, by the propositions

304

that precede them: some follow on problems, others on theorems. The present one results from a theorem, but that in the second [proposition of the seventh] book[131] comes from a problem. Thirdly, according to their methods of proof: some are established by direct proof, others by reduction to impossibility. The one before us is made evident by direct proof, whereas that which is implied in the proof of the first theorem

---

[131] 304.2   Barocius translates as if he read ἐν τῷ δευτέρῳ τοῦ ἐβδόμου βιβλίου, which probably indicates a better text at his disposal and certainly is what Proclus intended. The only porism in the second book (that following II. 4) belongs to a theorem, not a problem. The porism Proclus intends to refer to here is that mentioned four lines earlier, at 303.23.

of the third book comes to light by a reduction to impossibility. There are many other ways of classifying porisms, but these are enough for us at present.

The porism that we are now discussing, in teaching us that the space about a point can be divided into angles equal to four right angles, forms the basis of that paradoxical theorem which proves that only the following three polygons can fill up the space about a point: the equilateral triangle, the square, and the equilateral equiangular hexagon. The equilateral triangle, however, must be taken six times, for six angles, each two-thirds of a right angle, will make four right angles; the hexagon three times, for each angle of a hexagon is equal to one and one-third of a right angle; and the square four times, for each angle of a square is a right angle. Hence six equilateral triangles, meeting at their angles, complete the four right angles, and similarly three hexagons, and four squares. All other polygons, however they may be put together at their angles, either fall short of or exceed four right angles; only these, in the numbers mentioned, can equal four right angles. This theorem is Pythagorean.

This porism also enables us to prove that, if more than two lines—three, or four, or any number you like—cut one another at a single point, the angles that result will be equal in sum to four right angles, for they divide up the space of four right angles. It is clear also that the angles will always be double the number of the straight lines. Thus if two straight lines cut one another, there will be four angles equal to four right angles; if three lines intersect, six angles; and if four, eight; and so on indefinitely, for the number of the straight lines is always doubled. But the angles, though increasing in number, decrease in size, because the magnitude that they divide remains the same, namely, four right angles.

XVI. *In any triangle, if one of the sides is produced, the exterior angle is greater than either of the interior and opposite angles.*

Some persons have cited this enunciation elliptically, without "if one of the sides is produced," and thereby have

306

given occasion—perhaps to others and certainly to Philippus,[132] as Heron the engineer tells us—to criticize it. For the triangle as such never has an exterior angle. But all who desire to prevent this criticism state the proposition with the addition of the omitted clause, since this accords with our geometer's custom. For example, in the fifth theorem, wishing to demonstrate that the angles under the base of an isosceles triangle are equal, he adds: "when the equal sides are produced," the angles under the base are equal. Though this theorem may appear elliptically in other texts, it certainly was written in its full form by the author of the *Elements*.

What, then, does the enunciation say?[133] That in any triangle, if you produce one of its sides, you will find the exterior angle constructed on it to be greater than either of the opposite interior angles.[134] A little later[135] it will be demonstrated that it is equal to both of them, and this proves that it is greater than either. Of necessity he compares it with the opposite angles, not with the adjacent one; for the exterior angle can be either equal to or less than the angle adjacent to it, whereas it is always greater than either of the others. For example, if the triangle is a right-angled one and you think of one of the sides around the right angle as produced, the exterior angle will be equal to its adjacent angle. And if it is an obtuse-angled triangle, it will be possible for the adjacent inner angle to be greater than the exterior angle.

---

[132] 305.24   This Philippus is probably the Philippus of Mende mentioned at 67.23.

[133] 306.9   Putting the question mark after πρότασις, as does Barocius, and removing it in line 12.

[134] 306.12   Euclid proves XVI as follows: Let ABC be a triangle, and let one side BC be produced to D. Let side AC be bisected at E, let B and E be joined and BE produced in a straight line to F, making EF equal to BE. Let F and C be joined and AC be drawn through to G. Then in triangles ABE and CFE sides AE and EB are equal to sides CE and EF respectively, and angle AEB is equal to angle FEC (by XV); therefore the triangles are equal, and angle BAE is equal to angle ECF. But angle ECD is greater than angle ECF, and hence angle ACD is greater than angle BAC. Similarly if BC be bisected, angle BCG (i.e. angle ACD, its vertical angle) can be proved greater than angle ABC.

[135] 306.13   In XXXII.

But with respect to the opposite angles—for the angle adjacent to the exterior one is but one of the angles within the triangle, whereas there are two opposite angles—the exterior angle is greater than either of them, though not greater than the angle adjacent to it.

307 Some, however, have united two theorems, this and the one next to be demonstrated, and expressed the enunciation as follows: "In any triangle, if one side is produced, the exterior angle of the triangle is greater than either of the interior and opposite angles, and any two of the interior angles taken together are less than two right angles." There is some excuse for their uniting the theorems, in that our geometer himself does so later in the case of equal angles: "In any triangle the exterior angle is equal to the two interior and opposite angles, and the three angles of a triangle are equal to two right angles."[136] So here they think it appropriate in a similar case to unite the conclusions and make the enunciation composite. It is clear that what is proposed for demonstration will be composite, and the hypothesis, at least if it is presented with the above-mentioned addition, will also be composite, for we must suppose two things, the given triangle and one side produced. And if it is stated without this addition, it is potentially composite, though actually simple; for even if this addition is not posited, it must always be understood as part of the given, since the very fact of supposing that there is an external angle assumes that the side has been produced. So much for this.

308 The present theorem enables us to infer that it is impossible to have three equal straight lines falling from the same point upon the same straight line. For let three straight lines, AB, AC, and AD, drawn from one and the same point to the straight line BD, be equal. Then since AB is equal to AC, the angles at the base are equal, and angle ABC is therefore equal to angle ACB. Again since AB is equal to AD, angle ABD is equal to angle ADB. But angle ACB was equal to angle ABC; therefore ACB is equal to ADB, the

[136] 307.12   In XXXII.

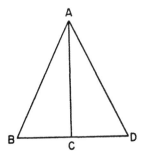

exterior angle to the interior and opposite angle, which is impossible. Thus three equal straight lines cannot be drawn from the same point to the same straight line.

By means of this theorem we can also demonstrate that which says that, if a straight line falling on two straight lines makes the exterior angle equal to the interior and opposite angle, these straight lines cannot form a triangle, nor can they intersect, since this angle will be both greater than the other and equal to it, which is impossible. Let AB and CD be straight lines, and let BE, falling on them, make equal angles ABD and CDE. Then AB and CD will not intersect. 309 For if they intersect, the angles remaining equal, angle CDE will be equal to angle ABD, though it is exterior and greater than the interior and opposite angle. Necessarily, then, if they intersect, the angles no longer remain equal, but the angle at D in every case increases.[137] For if, while AB remains fixed,

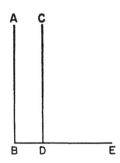

[137] 309.4   More accurately, "becomes relatively larger." In only two of the three cases considered below does the angle at D actually increase.

you think of CD as moving towards it in order that they may intersect, you will make the divergence at angle CDE greater, for the more CD moves towards AB, the more it diverges from DE. And if you think of CD as remaining fixed and AB as moving towards it, you will make angle ABD smaller, for AB simultaneously moves towards CD and BD. And if you make both of them move towards one another, you will find that AB, in moving towards BD, also contracts its angle, while CD, in moving towards AB, diverges from DE and thus increases the angle CDE. Of necessity, then, if a triangle is produced and AB and CD intersect, the exterior angle will be greater than the opposite interior one; for if the interior angle remains the same, the exterior is increased, and if the exterior remains the same, the interior is decreased, or both change,[138] the interior contracting and the exterior expanding. The cause of these changes is the motion of the straight lines, the one moving towards the side where it makes the interior angle, the other moving away from the side where it makes the exterior angle. From this you can infer how constructing things brings before our eyes the true causes of the conclusions.

310

XVII. *In any triangle two angles taken together in any manner are less than two right angles.*

The present theorem demonstrates generally that any two angles of a triangle are less than two right angles; and the sequel determines by how much they are less, namely, by the third angle of the triangle. For the three angles are equal to two right angles,[139] so that the two of them will be less than two right angles by the third angle. The proof given by the author of the *Elements* follows an obvious path, since it uses the previous theorem.[140] But here, as in the previous theorem,

[138] 310.1 Delete *et* in Friedlein.
[139] 310.16 This anticipates the conclusion of XXXII.
[140] 310.19 Euclid proves XVII as follows: Given the triangle ABC with side BC produced to D, the exterior angle ACD is greater than the interior and opposite angle ABC, by the preceding theorem. Adding ACB to each, we see that angles ACD and ACB are greater than angles ABC and BCA. But angles ACD and ACB are equal to two

311

we must look at the construction of the triangles in order to discover the cause of this present character. So again let AB and CD be lines at right angles to BD. If there is to be a triangle, AB and CD must incline towards one another. But

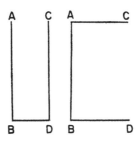

their inclination decreases the interior angles, so that they become less than two right angles, for they were right angles before the inclination. Similarly if we think of lines AC and BD as standing on AB at right angles to it, the same consequences will follow from the lines' being inclined to one another, and the angles on AB will become less than two right angles. And similarly for the remaining side.

This, then, is the cause, not that the exterior angle is greater than either of the interior and opposite angles. For it is not necessary that a side be produced, nor that there be any exterior angle constructed; but it is necessary that any two of the interior angles be less than two right angles. And how can what is not necessary be a cause of what is necessary?[141] The cause is, as I said, the factor stated, namely, the inclination of the straight lines towards the base, which decreases

312

the angles. Since the author of the *Elements* demonstrates the conclusion by means of the exterior angles, now let us establish the same result without producing one of the sides. Let ABC be a triangle, let any chance point D be taken on BC,

---

right angles; therefore angles ABC and BCA are less than two right angles. Similarly we can prove that angles BAC and ACB are less than two right angles, and so also angles CAB and ABC.

[141] 311.21   In Aristotle's theory of demonstration a necessary conclusion can only be derived from necessary premises; see *Post. Anal.* 73a24 and *passim*. Necessity in the strict sense is not identical with formal necessity.

and let AD be joined. Then since one side, BD, of the triangle ABD has been produced, the exterior angle ADC is greater than the interior angle ABD. Again since one side, DC, of triangle ADC has been produced, the exterior angle ADB is greater than the interior angle ACD. But the angles about AD

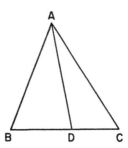

are equal to two right angles, by the thirteenth. Therefore the angles ABC and ACB are less than two right angles. Similarly we can prove that angles BAC and BCA are less than two right angles by taking a point on AC and joining a straight line from B to the point taken. And once more we can show that angles CAB and ABC are less than two right angles by taking a point on AB and joining a straight line to this point from C. The conclusion, then, has been demonstrated by the same theorem without producing any of the sides of the triangle.

313   With the help of this theorem we can also prove that it is impossible to draw two perpendiculars from the same point to an identical straight line. For let AB and AC be two perpendiculars dropped from A upon BC. Angles ABC and ACB are therefore right angles. But since ABC is a triangle, any two of its angles are less than two right angles. Hence ABC and ACB are less than two right angles. But they are also

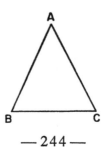

equal to two right angles because the lines are perpendicular, which is impossible. Consequently it is not possible to draw two perpendiculars from the same point to the same straight line.

XVIII. *In any triangle the greater side subtends the greater angle.*

We have learned through the fifth and sixth theorems that the equality of the sides of a triangle makes equal the angles subtended by them and that the equality of the angles likewise shows the subtending sides to be equal. But that the inequality of the sides implies the inequality of the subtended angles, and conversely, we learn from this and the following theorem, that is, the eighteenth and the nineteenth. The former proves that the greater side subtends the greater angle, the latter that the greater angle is subtended by the greater side. They are converses of each other, considering in contrary subjects the same attributes as do the fifth and sixth theorems. But obviously with scalene triangles we shall take the terms "greater and lesser sides" as relative and distinguish between the greatest, the intermediate, and the least sides; and likewise for the angles. In the case of isosceles triangles it will be enough to distinguish merely the greater and the less, for it is one side that is unequal to the two others, either greater or less, just as in the treatment of equilateral triangles these propositions have no place at all.[142]

You see how the propositions that demonstrate equality of angles or sides suit both equilateral and isosceles triangles, and those that demonstrate inequality suit both scalene and isosceles. The reason is that some triangles are the product of equality alone, some of inequality alone, and some of both, having one character by virtue of equality and another because of inequality. And there are some beings that are akin to the Limit, others to the Unlimited, and others that are gen-

314

[142] 314.12 I adopt Schönberger's supposition that 315.4-10 has been misplaced in the text and was intended to follow at this point. If this supposition is correct, it should be read before the following paragraph.

erated from both by the principle of the Mixed. Thus this triad of principles permeates everything: lines, angles, figures, and among figures the three-sided, the four-sided, and all their successors. But the Limit in geometrical forms is sometimes manifested through likeness, sometimes through equality; the Unlimited sometimes through unlikeness, sometimes through inequality; and the Mixed sometimes arises out of likenesses and unlikenesses, and sometimes out of equalities and inequalities.[143] The reason is that geometrical figures belong to the categories of quantity and quality.

These being the two attributes[144] indicated, it is clear that, when the author of the *Elements* says "in any triangle," he does not mean the equilateral triangle, but "any triangle that has a greater and a lesser side." For we must consider what is given as the leading element, and the conclusion must be thought of as conforming to it. Thus "Whatever triangle has a greater and a lesser side, this will have its greater side subtending its greater angle."

In his construction our geometer takes the triangle ABC and the side AC as greater than the side AB; and in order to prove that the angle at B is greater than the angle at C, he cuts off from AC a length AD equal to AB.[145] One could maintain that the length cut off should be at C; so let us

315

---

[143] 315.2 Reading instead of Friedlein's text the fuller one implied in Barocius: ἐξ ὁμοιοτήτων καὶ ἀνομοιοτήτων, ἐξ ἰσοτήτων καὶ ἀνισοτήτων.

[144] 315.4 The two attributes are "greater" and "less," the terms discussed in 314.5-12, which this comment seems intended to follow (see note at 314.12). The text of the present passage has been corrupted as well as displaced, but the general meaning is clear.

[145] 315.14 In the proof of XVIII Euclid assumes a triangle ABC having the side AC greater than side AB, takes a length AD on AC equal to AB, and draws BD. Then, by XVI, angle ADB is greater than angle DCB. But angle ADB is equal to angle ABD; therefore ABD is

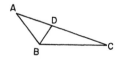

also greater than ACB, and hence ABC is greater still than ACB, which is what was to be proved.

316

prove the proposition before us on this hypothesis, after Porphyry. Let DC be the length equal to AB, let AB be produced to E, and let BE be equal to AD. The whole of AE is then equal to AC. Let EC be joined. Then since AE is equal to AC, angle AEC is equal to angle ACE, by the fifth.

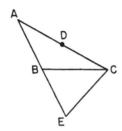

Therefore angle AEC is greater than angle ACB. But angle ABC is greater than angle AEC, for the side EB of triangle CBE has been produced, and angle ABC, being exterior, is greater than the opposite interior angle. Hence angle ABC is even greater than angle ACB, which is what it was required to prove.

Such are the geometrical proofs. But clearly the cause of this attribute is the increase or decrease in length of the side itself that subtends the angle. For when it is greater it spreads out the angle, and when smaller it diminishes and contracts the angle also. This is because of the full tension of the straight line; being stretched to the utmost, it changes the size of the angle according to its own increase or decrease. I make these statements about a single triangle, since it is possible for the same angle to be subtended by a longer or a shorter line and for the same straight line to subtend larger or smaller angles. For let ABC be any isosceles triangle, and let a point D be taken on side AB, a length AE equal to AD be taken on side AC, and DE be joined. The angle at A is then subtended by both DE and BC, one of which is longer and the other shorter; and by the same reasoning it is possible to take countless other straight lines of varying lengths as subtending angle A. Again let ABC be an isosceles triangle, having BC shorter than BA and AC. On BC let an equilateral

317

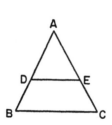 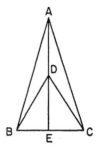

triangle BDC be constructed, and let AD be drawn and produced to E. Then since angle BDE is an exterior angle of
triangle ABD, it is greater than angle BAD; and in the same
way angle CDE is greater than angle CAD. The whole angle
BDC is then greater than angle BAC, and the same straight
line subtends both the larger and the smaller angle. And it has
been shown that the same angle can be subtended by longer
or shorter lines. In one and the same triangle, however, one
line subtends one angle, the greater line always the greater
angle and the shorter line the smaller angle. The cause of this
we have seen.

XIX. *In any triangle the greater angle is subtended by
the greater side.*

This theorem is the converse of the previous one. In each
case both what is given and what is sought are simple. The
conclusion of that is the hypothesis of this, and the hypothesis of that is the conclusion of this. That theorem precedes
because it takes as given the inequality of the sides, and this
one follows because its hypothesis is unequal angles; for the
sides of rectilinear angles are thought of as containing,[146] the
angles as contained. And in that theorem the method of proof
is direct, but in this it proceeds by a reduction to impossibility.
Our geometer proves the impossibility by division.[147]

318

---

[146] 318.13 "As containing," or "as superior to," the verb περιέχειν
having this double meaning, on which Proclus appears to play here.
[147] 318.16 I.e. by distinguishing the possible alternatives and disproving each of them in turn, except the conclusion to be established.
The material given in quotation marks is not an exact quotation, but
rather a paraphrase, of Euclid's proof as we have it in our text.

"When the angles are unequal," he says, "I say that the subtending sides also are unequal, and the greater side subtends the angle given as greater. For if the side that subtends the greater angle is not greater than, it is equal to or less than the other side. But if it is equal to it, the angles they subtend are also equal, by the fifth; and if it is less, the angle it subtends is the lesser angle, by the preceding theorem, for it has been demonstrated that the greater side subtends the greater angle and the lesser side the lesser angle. But the relation of these angles is the reverse. Consequently the side that subtends the greater angle is greater than the other side."

319

But it is also possible to demonstrate the present proposition without this use of division if we first prove a little lemma, as follows:[148]

If an angle of a triangle is bisected and the straight line bisecting it meets the base and divides it into unequal parts, the sides that contain the angle will be unequal, and the greater will be that which meets the greater segment of the base, and the less that which meets the lesser.

Let ABC be a triangle. Let the angle at A be bisected [by the line AD],[149] let AD divide BC into unequal segments, and let CD be greater than BD. I say that AC is greater than AB.

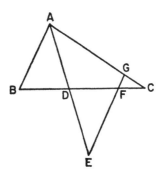

―――――――――――――

[148] 319.4 This alternate proof and the lemma on which it depends seem to have been contained in Heron's commentary. See Heath, *Euclid* I, 285. Cf. 346.13.

[149] 319.13 The phrase in brackets comes from Barocius, as does the bracketed phrase in line 18.

Let AD be produced [to E] and DE be made equal to AD. And since CD is greater than DB, let DF be laid off equal to BD, and let EF be drawn and produced to G. Then since AD is equal to DE and BD to DF, two sides are equal to two sides, and they contain equal vertical angles. Therefore the base AB is equal to base EF, and all corresponding parts are equal, so that angle DEF is equal to angle DAB. But this is equal to DAG, so that side AG is also equal to EG, by the sixth. Therefore AC is greater than EF. And EF is equal to AB. Hence AC is greater than AB, which is what it was required to prove.

320

Taking this lemma as established, we can prove that the greater side subtends the greater angle. Let ABC be a triangle having the angle at B greater than the angle at C. I say that AC is greater than AB. Let BC be bisected at D, let AD be drawn and produced so that DE is equal to AD, and let BE

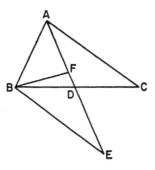

be joined. Then since BD is equal to DC and AD to DE, two sides are equal to two sides, and they contain equal vertical angles. The base BE is therefore equal to AC, and all corresponding parts are equal, so that angle DBE is equal to the angle at C. But the angle at C is less than angle ABD.[150] Hence angle DBE is less than angle ABD. Now let angle ABE be bisected by the line BF. EF will then be greater than AF. Then since in triangle ABE the angle at B has been bisected by BF and EF is greater than AF, BE is greater than AB, by the previously proved lemma. But BE has been shown to be

321

[150] 321.1   I.e. by hypothesis.

equal to AC. Therefore AC is greater than AB, and the conclusion wanted has been proved.

It was obviously from a desire to avoid complexity in the order of demonstration that the author of the *Elements* avoided this method of proof, preferring to proceed by division and reduction to impossibility because he wished to establish the converse of the preceding theorem without anything intervening. The eighth theorem, which is the converse of the fourth, introduced considerable confusion in making the conversion difficult to recognize. It is preferable to prove a converse by the reduction to impossibility while preserving continuity than to break the continuity with the preceding demonstration. This is why he almost always proves a converse by the reduction to impossibility.

322   XX. *In any triangle two sides taken together in any manner are greater than the remaining side.*

The Epicureans are wont to ridicule this theorem, saying it is evident even to an ass and needs no proof; it is as much the mark of an ignorant man, they say, to require persuasion of evident truths as to believe what is obscure without question. Now whoever lumps these things together is clearly unaware of the difference between what is and what is not demonstrated. That the present theorem is known to an ass they make out from the observation that, if straw is placed at one extremity of the sides, an ass in quest of provender will make his way along the one side and not by way of the two others. To this it should be replied that, granting the theorem is evident to sense-perception, it is still not clear for scientific thought. Many things have this character; for example, that fire warms. This is clear to perception, but it is the task of science to find out how it warms, whether by a bodiless power or by physical parts, such as spherical or pyramidal particles. Again it is clear to our senses that we move, but how we move is difficult for reason to explain, whether through a partless medium or from interval to interval, and in this case how we can traverse an infinite number of intervals, for every magnitude is divisible without end. So with respect to a

323

triangle let it be evident to perception that two sides are greater than the third; but how this comes about it is the function of knowledge to say.

This is enough by way of answer to the Epicureans. We must give a brief account of the other proofs of this proposition that the followers of Heron and Porphyry have constructed without producing the straight line, as the author of the *Elements* does.[151] Let ABC be a triangle and let it be required to prove that AB and AC are greater than BC. Let the angle at A be bisected [by the line AE].[152] Then since angle AEC is an exterior angle of triangle ABE, it is greater than angle BAE. But angle BAE is equal to angle EAC. Therefore angle AEC is greater than angle EAC, so that side AC is greater than side CE. By the same reasoning AB is

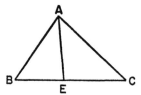

greater than BE, for angle AEB is an exterior angle of triangle AEC and greater than angle EAC, that is, greater than BAE, so that AB is also greater than BE. Consequently AB and AC are greater than the whole of BC. We can construct a similar proof for the other sides.

Again let ABC be a triangle. If it is an equilateral triangle, two of its sides will of course be greater than the third; for of three equal quantities two of them, however chosen, will always be double the third. If it is isosceles, it has a base either less or greater than either of the equal sides. If the base is less,

324

[151] 323.9   In proving XX Euclid assumes the triangle ABC, draws BA through to D, making DA equal to AC, and joins DC. Then since DA is equal to AC, angle ADC is also equal to angle ACD, by V; therefore angle BCD is greater than angle ADC. And in triangle DCB angle BCD is greater than angle BDC; therefore by XIX side DB is greater than side BC. But DA is equal to AC; therefore BA and AC are greater than BC. Similarly we can prove that AB and BC are also greater than CA, and BC and CA greater than AB.

[152] 323.12   From Barocius.

again two sides will be greater than the third. But if the base is greater, let BC be greater, and let a length BE be cut off equal to each of the other sides, and let AE be joined. Then since angle AEC is an exterior angle of triangle AEB, it is

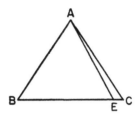

greater than angle BAE. By the same reasoning angle AEB is greater than angle CAE. The two angles about AE are therefore greater than the whole angle at A, and one of them, BEA, is equal to angle BAE, since AB is equal to BE. Therefore the other, angle AEC, is greater than angle CAE, so that AC is greater than CE. But AB was equal to BE, and hence AB and AC are greater than BC. If ABC is a scalene triangle, let AB be the greatest side, AC the side of middle length, and BC the least. Now the greatest taken together with either of the others is obviously greater than the third, for by itself it is greater than either. And if we want to prove that AC and BC are greater than AB the greatest, we shall proceed as in the case of the isosceles by cutting off from the greatest side a length equal to one of the others, joining its extremity with C, and making use of the properties of exterior angles.

325   Again let ABC be any triangle. I say that AB and AC are greater than BC. For if they are not, they will be either equal to BC or less than it. Suppose them to be equal, and take away a length BE equal to AB. The remaining length EC is thus equal to AC. Then since AB is equal to BE, they subtend equal angles. Likewise since AC equals EC, they also subtend equal angles. Hence the angles at E are equal to those at A, which is impossible.[153] Now suppose AB and AC to be

[153] 325.11   Because the angles at E are equal to two right angles? "I suspect the impossibility is that angle CEA is greater than BAE and AEB is greater than CAE, by XVI. Proclus uses this kind of reasoning in his next argument (325.16ff.)." (I.M.)

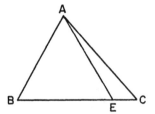
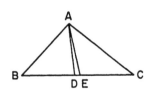

less than BC, and let BD be laid off equal to AB and CE equal to AC. Then since AB is equal to BD, the angle BDA is equal to angle BAD; and since AC is equal to CE, angle CEA is equal to angle EAC. Hence the two angles BDA and CEA are equal to the two angles BAD and EAC. But since angle BDA is an exterior angle of triangle ADC, it is greater than angle EAC, for it is greater than DAC. By the same reasoning, since angle CEA is an exterior angle of triangle ABE, it is likewise greater than angle BAD, for it is greater than angle BAE. Angles BDA and CEA are [therefore][154] greater than the two angles BAD and EAC. But they were equal to them above,[155] which is impossible. Therefore AB and AC are neither equal to BC nor less than it, but greater. Similarly for the other sides.

326

XXI. *If on one of the sides of a triangle, from its extremities, there are constructed two straight lines within the triangle, the straight lines so constructed will be less than the remaining two sides of the triangle but will contain a greater angle.*

The meaning of the enunciation is evident, the proof given by our geometer is clear, and the theorem follows from first principles.[156] It depends on two theorems, the one proved

[154] 326.1 ἄρα is missing here. Barocius' translation suggests that he found it in his text.

[155] 326.2 Taking ἐκεῖ as having been displaced from its proper position after αὐταῖς.

[156] 326.13 Euclid's proof of XXI is: Given BD and DC two straight lines constructed on side BC of triangle ABC and within the triangle, let BD be drawn through to E. Then the two sides AB and AE of triangle ABE are greater than BE, by XX. Let EC be added to each; therefore AB and AC are greater than BE and EC. Again in triangle CED the two sides CE and ED are greater than CD. Let

before this and the sixteenth. For to prove that the lines constructed within the triangle are shorter than the lines of the triangle outside them, he requires the theorem that in any triangle two sides are greater than the third; and for showing that the angle they contain is greater than that contained by the outer lines, he uses the proposition that in any triangle the exterior angle is greater than the interior and opposite angle.

327

It will furnish evidence of his geometrical precision and at the same time be a reminder of the paradoxes in mathematics if we show that it is possible to construct two lines within a triangle on one of its sides—not on the whole of it, but on a part—which will be greater than the outer lines and on the other hand will contain an angle that is less than that between the outer lines. The demonstration of this will make clear why it was necessary that the author of the *Elements* add that the lines constructed within the triangle must start "from the extremities" of the common base, that is, must be constructed on one of the sides as a whole, not upon a part of the whole. At the same time, as I said, it will reveal one of the paradoxes in geometry. For is it not a paradox that the lines constructed on the whole of the side are less than the outer lines, whereas those constructed on a part of it are greater?

Let us suppose, then, a right-angled triangle ABC, having the angle at B right; and let any point D be taken on BC, and let AD be joined. AD is then greater than AB.[157] Let a length DE be cut off on AD equal to AB, and let EA be bisected at F and FC be joined. Then since ACF is a triangle, AF and FC are greater than AC. But AF is equal to FE, and

---

DB be added to each; therefore CE and EB are greater than CD and DB. But AB and AC were proved greater than EB and CE; thus AB and AC are still greater than CD and DB. Again in triangle CDE the exterior angle BDC is greater than angle CED, by XVI. Likewise in triangle ABE the exterior angle CEB is greater than angle BAC. But angle BDC was proved greater than angle CEB; therefore it is still greater than angle BAC.

[157] 327.17 "Angles ABD and ADB are less than two rights, by XVII. Therefore ADB is less than a right, so less than ABD. Therefore AD is greater than AB, by XIX." (I.M.)

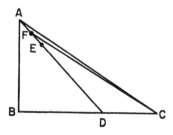

hence FE and FC are greater than AC. But DE is equal to
AB; therefore FC and FD are greater than AB and AC, and
they are within the triangle. Again let ABC be an isosceles
triangle having its base BC greater than either of the equal
sides, and let a length BD be cut off on BC equal to AB. Let
AD be joined, and a point E on AD be taken at random and
joined to C by EC. Then since AB is equal to BD, angle
BAD will be equal to angle BDA; and since angle BDA is an

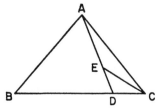

exterior angle to triangle EDC, it will be greater than the
interior and opposite angle DEC, so that angle BAD is greater
than angle DEC. Hence angle BAC is even greater than angle
DEC. And angle BAC is contained by the outer lines, and
angle DEC by the inner lines. Consequently DE and EC have
been constructed within the triangle containing an angle less
than that contained by the outer lines, and the proposition has
been demonstrated without the use of the parallel lines of the
commentators.[158] It is therefore necessary that the lines con-

---

[158] 328.16   It would be inappropriate to prove this proposition by
using the theory of parallel lines, which has not yet been established,
if some other procedure is available. Proclus' language suggests that
some of the commentators had been guilty of this error. According to
ver Eecke (see his note *ad loc.*), Pappus was one of them. The point
is well illustrated later at 340.5ff., where Proclus, in defending Euclid

structed begin at the extremities of the base. For the lines constructed on a part of the base have been shown to be sometimes longer than the outer lines and to contain a smaller angle. When they are thus constructed from the extremities, they reveal the shape of the so-called barb-like triangle. This also is one of the paradoxical problems in geometry, to find a four-sided triangle, such as BAC. Though bounded by four sides BA, AC, CE, and EB, it has three angles, one at B, another at A, and a third at C. Consequently the figure here presented is a four-sided triangle.[159]

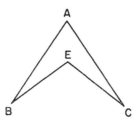

XXII. *Out of three straight lines which are equal to three given straight lines to construct a triangle: thus it is necessary*[160] *that two of the straight lines taken together in any manner should be greater than the remaining one.*

We have gone over to problems again. He asks us to construct, given three straight lines two of which are greater than the third, a triangle with sides equal to the given straight lines. He sees, first, that it is impossible to construct a triangle from lines that already have a prescribed position, but possible

---

against certain objections, is compelled to assume what is to be proved later, viz. this very theory of parallels, in order to show that the theorems whose omission is complained of could not have been established by Euclid at this stage of his exposition.

[159] 329.7   Cf. 165.22ff. It was the practice of Greek geometers not to recognize as an "angle" any angle not less than two right angles; consequently the reentrant angle is ignored, and angle BEC is regarded as outside the figure. Cf. 268.2, where the angle made by DF and FG in the diagram is said to be "outside." See Heath, *Euclid* I, 263f.

[160] 329.10   For δεῖ δὲ in Friedlein read δεῖ δή, since this is the reading in our Euclid text and is required by the sense. See Heath, *Euclid* I, 293.

only from lines equal to them, and, secondly, that of the lines which are to make the triangle two "taken together in any manner" must be greater than the other, for in any triangle two sides are greater than the remaining one, as has been demonstrated. This is why he adds that two of the lines from which we start must in every case be greater than the third, or a triangle will not result from lines equal to them; and, besides, this added clause alone serves to refute the objections brought against his construction.

330

This problem, then, belongs among the determinate, not the indeterminate ones; for, like theorems, some problems are indeterminate and some determinate. If we simply say "Out of three straight lines equal to three given straight lines to construct a triangle," this is indeterminate and insoluble; but if we add "two of them taken together in any manner are greater than the third," it is determinate and soluble. This fact also comes out: just as theorems are distinguished according to truth or falsehood, so problems are differentiated according to their manifest possibility or impossibility.

A little attention to the construction will teach us that the objections brought against it can be answered by the restriction stated. We shall follow the words of our geometer. "Let A, B, and C be straight lines, of which two, taken together in any manner, are greater than the other." And now we must make the required construction. "Let a straight line DE be laid out terminating at D in one direction but without

331

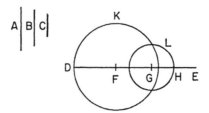

limit in the other. Let there be laid off a length DF equal to A, a length FG equal to B, and a length GH equal to C. With F as center and distance FD let a circle K be described; again with G as center and distance GH let a circle L be described; and let the two circles cut one another." This is what the author of the *Elements* assumes.[161] Someone may ask: "From where does he get this? For perhaps the circles are only tangent to each other, or not even in contact. For of three positions, they must have one: they either cut each other, or are tangent, or are separate from each other." Now I say that they necessarily cut one another. Let us first suppose them to be tangent to one another. Then since F is the center of circle K, DF is equal to FN; and since G is the

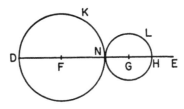

center of circle L, GH is equal to GN; then the two lines DF and GH are equal to the one line FG. But they were assumed to be greater than it, since A together with C is greater than B and these lines are equal to those. Again, if it be possible, let the circles be separate from one another, like K and L. Then since F is the center of circle K, DF is equal to FN; and since G is the center of circle L, GH is equal to GM. Consequently FG as a whole is greater than DF and GH. For FG exceeds DF and GH by NM. But DF and GH were posited as

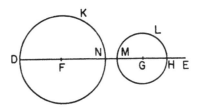

greater than FG, as lines A and C are greater than B; for DF is equal to A, FG to B, and GH to C. It is necessary, then, that circles K and L cut one another. So the author of the

---

[161] 331.9   From this point Euclid proceeds to solve XXII by drawing lines from F and G to the point of intersection of the circles and then proving that the triangle which they make is the triangle required.

Elements was right in assuming that the circles intersect, since he had also posited that of the three straight lines two of them taken together in any way are greater than, neither equal to nor less than, the other one. If the circles are tangent, the two lines must be equal to, and if separate must be less than the third.

333

XXIII. *On a given straight line and at a point on it to construct a rectilinear angle equal to a given rectilinear angle.*

This too is a problem, and the credit for its discovery belongs rather to Oenopides, as Eudemus tells us. It requires us to construct on a given straight line and at a given point on it an angle equal to a given rectilinear angle. For necessary reasons our geometer adds that the given angle is rectilinear, since it is not possible to construct on a straight line an angle equal to any angle whatever. It has been proved that only two of the circular angles are equal to rectilinear angles, namely, the "axe," which has been shown to be equal to any rectilinear angle, and the lunular angle which is equal to two-thirds of a right angle.[162] This species of lunule is produced when two

---

[162] 333.15ff. This passage has confused all previous translators. The first of these two species of circular angles that are equal to rectilinear ones has been discussed at 189.23ff, where its equality to any rectilinear angle is demonstrated. But there it is called μηνοειδές, "moon-like" (190.8). Here, however, it is called the "axe" (πέλεκυς), whereas the second species mentioned here is called μηνοειδές. This has led Barocius, Taylor, and ver Eecke in their translations to imply that the terms have been inadvertently interchanged, and Schönberger explicitly to assert it. The truth is, however, that Proclus is referring to two species of lunular angles, one that is equal to any rectilinear angle (cf. 190.12-14), and the other equal to two-thirds of a right angle. The latter is a special case which results from two circles intersecting each other through their centers, a point which Proclus mentions here to differentiate it from the others. Both types were presumably also called πέλεκυς, a fact which leads to the apparent confusion in Proclus' exposition. Taylor (II, 126) and ver Eecke (284) give a proof which may be that which Proclus intended, showing that the second species of lunule mentioned is equal to two-thirds of a right angle. Let the circles AC and BD be drawn passing through their respective centers A and B; and from the center C, with a radius CB equal to AB, describe the arc ABD and draw lines CB, CD, and CA. Then since

334

circles cut one another through their centers. The requirement that the construction be on a straight line makes determinate the kind of angle to be constructed, not indeterminate, but either rectilinear or mixed; and since no mixed angle can be equal to a rectilinear, it is evident that this angle is rectilinear.

The author of the *Elements* has done what is required by using without qualification the preceding problem to construct a triangle from three straight lines equal to the three given lines;[163] but you could get the construction of the triangle in a more instructive manner, as follows. Let AB be a

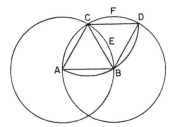

ACB is an equilateral triangle, as CBD is also, each of the angles ACB and BCD will be equal to two-thirds of a right angle; and since the mixed angle BCE is equal to the mixed angle DCF, the angle ECF will be equal to angle BCD, i.e. to two-thirds of a right angle.

163 334.6 Euclid solves XXIII as follows: Given the straight line AB and the rectilinear angle DCE, to construct on AB at A an angle equal to DCE, he takes points D and E at random on CD and CE respectively, joins DE, then constructs a triangle FAG on AB whose sides AF, AG, and FG are equal respectively to CD, CE, and DE. Then, by VIII, angle DCE is equal to angle FAG. Heath (*Euclid* I, 295) remarks that the construction of the triangle assumed in this

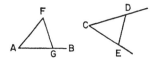

proposition is not exactly the construction used in XXII. "We have here to construct a triangle on a certain finite straight line AG as base; in XXII we have only to construct a triangle with sides of given length without any restriction as to how it is to be placed." Hence the construction of XXII cannot be used here without qualification. Proclus modifies the method of XXII and thus constructs the triangle in what he considers a "more instructive fashion." He appears later (335.15f.) to plume himself modestly on this contribution he has made to Euclid's construction.

straight line, A the given point on it, and CDE the given rectilinear angle. Now let us do what is required. Let CE be joined and AB be produced in both directions, to F and G, and let FA be equal to CD, AB to DE, and BG to CE. With A as center and distance FA let the circle K be described; and again, as in the preceding problem, with center

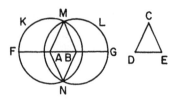

B and distance BG let circle L be described. These circles therefore intersect one another, as has been proved. Let them then intersect at points M and N, and from M draw lines to the centers, and likewise from N. Then since FA is equal to AM and to AN, and FA is equal to CD, AM and AN are each equal to CD. Again since BG is equal to BM and to BN, and BG is equal to CE, therefore BM and BN are each equal to CE. But AB also is equal to DE. The two lines AB and AM are therefore equal to DE and DC, and the base BM is equal to CE; hence angle MAB is equal to the angle at D. Again the two lines AN and AB are equal to the two lines CD and DE, and the base BN is equal to the base CE; hence angle NAB is equal to angle CDE. The task required has been done twice: we have constructed not one only, but two angles equal to the given angle, one on each side of the line AB, so that in the sequel, on whichever side we choose to make the construction, our result will be sure and unimpeachable.

This is our contribution to the construction given by the author of the *Elements*. But we do not commend the proof given by Apollonius, since it requires theorems proved in Book III. He takes CDE as any angle and AB as the straight line. With D as center and distance CD he draws an arc CE, and likewise with A as center and distance AB the arc BF. He then cuts off a length BF equal to CE, joins AF, and declares A and D to be equal angles as standing on equal

335

336

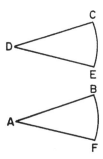

arcs. He must also assume that AB is equal to CD, in order that the circles may be equal. Now such a construction employing later theorems[164] we regard as alien to the nature of an elementary treatise and prefer that of our geometer because it is an orderly consequence of principles.

XXIV. *If two triangles have two sides equal to two sides respectively but have one of the angles contained by the equal straight lines greater than the other, they will also have one base greater than the other.*

He has gone over to theorems again and now presents reasonings concerning inequality in two triangles similar to those he has given about equality. Assuming two triangles with two sides equal respectively, he posits at one time that the vertical angles are equal and at another time that they are unequal [and the bases at one time equal and at another time unequal][165] and shows that the equality of the vertical angles implies the equality of the bases, that the equality of the bases implies the equality of the angles at the vertex, and that inequality implies inequality. The present theorem is therefore the opposite of the fourth; for that assumed equal angles at the vertex of the triangles and this unequal angles, and that proved the bases of the triangles to be equal, whereas this proves them to be unequal, like the angles. This theorem is also the leading theorem for the one following, for that proceeds from the bases to the angles which the bases subtend and infers their inequality, whereas this proceeds conversely

337

---

[164] 336.6 Apollonius' solution requires the use of III. 28 and 29.
[165] 336.19 From Barocius.

from the angles to the bases under them. So[166] the following proposition, being the converse of this in the sense just described, is the opposite of the eighth theorem; for the eighth proves from the equality of the bases that the vertical angles are equal, and this one shows from the inequality of the bases that the angles also are unequal. Of these four theorems—the fourth, eighth, twenty-fourth, and twenty-fifth—two, the fourth and the eighth, are concerned with equality; two, this and the following one, with inequality; two, the fourth and the one we are now considering, start from the angles; and two, the eighth and the one after this, start from the bases. But common to all four is the necessary assumption that the two triangles have two sides equal respectively; for if they are unequal, all inquiry is vain and subject to error.

338 So much in general about the propositions before us. Now let us examine the construction that the author of the *Elements* gives for this theorem and supply what it omits. He takes two triangles ABC and DEF having sides AB and AC equal respectively to DE and DF and the angle at A greater than the angle at D. To prove that BC is greater than EF, he constructs on DE and at point D upon it an angle EDH equal to the angle at A—the greater of the two angles A and D— and draws DH equal to AC.[167] Now when EF is produced,

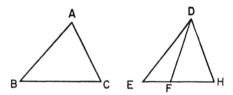

[166] 337.11 Reading with Schönberger ὥστε αὐτὸ τὸ instead of ὥσπερ αὖ τὸ in Friedlein.

[167] 338.11 Despite Proclus' assertions at 340.10 and 14 that he is following the diagram in Euclid's text, it is impossible to make either his diagram or his lettering accord with those in our text of Euclid. Proclus' diagram substitutes H (θ) everywhere for G (η); and to add to the complications the diagrams at 338.14-339.10 and at 340.7-343.10 are the reverse of those in our Euclid. Instead of trying to bring these two traditions into accord, I have left Proclus' text as it is, warning the reader that he will have to make some accommodations in com-

point H will lie either above the line, or on it, or below it. The author of the *Elements* takes it as lying above the line. But let us suppose it to be on the line and prove it again from this assumption. The two lines AB and AC are equal to lines DE and DH, and they contain equal angles. The base BC is then equal to EH. But EH is greater than EF, so that BC is greater than EF. Now let us suppose that it lies below EF. Then drawing line EH we shall say that, since AB and AC are equal to DE and DH and they contain equal angles,

339

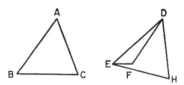

therefore BC is equal to EH. Now since within the triangle DEH lines DF and EF have been constructed on DE, they are less than the outer lines. But DH is equal to DF, for it is equal to AC; hence EH is greater than EF. But EH is equal to BC and hence BC is greater than EF. Thus the theorem has been demonstrated for every position.

Since in the fourth theorem our author proved also that

---

paring it with the Euclid text to be given in this footnote. Euclid's proof of XXIV is as follows: Given triangles ABC and DEF having sides AB and AC equal respectively to DE and DF and the angle at A greater than the angle at D, let there be constructed on DE at point D an angle EDG equal to angle BAC, and let DG be drawn equal to

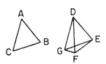

AC, and join EG and FG. Then since AB is equal to DE, AC to DG, and angle BAC to angle EDG, BC is equal to EG, by IV. Again since DF is equal to DG, angle DGF is also equal to angle DFG, by V. Therefore angle DFG is greater than angle EGF, and the angle EFG greater still than angle EGF. And since EFG is a triangle having angle EFG greater than angle EGF, side EG is also greater than side EF, by XIX. But EG is equal to BC, and therefore BC is also greater than EF, which is what was to be proved.

the areas of the triangles are equal, why did he not add to this theorem that the areas as well as the bases are unequal? To this difficulty let it be said that the same reasoning does not hold for unequal as for equal angles and bases. The equality of the angles and the bases implies the equality of the triangles, but when they are unequal, the inequality of the areas does not necessarily follow. The triangles can be either equal or unequal, and that which has the greater angle and base may be greater or it may be less. For this reason the author of the *Elements* omitted a comparison of the triangles, particularly since the investigation of these matters requires the doctrine of parallel lines. But if we must now make a comparison of the areas, let us do so, assuming in advance what is to be proved later. Using the diagram of this proposition in the *Elements*, we assert that, if the angles at A and D are equal to two right angles, the triangles are demonstrably equal; if they are greater than two right angles, that triangle which has the greater angle is less; and if they are less than two right angles,

340

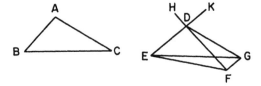

it is greater. Taking the constructions as given in the *Elements*, let ED and FD be produced [to K and H],[168] and let us suppose angles BAC and EDF equal to two right angles. Then since angle BAC is equal to angle EDG, angles EDG and EDF are equal to two right angles. But angles EDG and KDG are also equal to two right angles. Let the common angle EDG be subtracted, and the remainder, angle EDF, is equal to angle KDG. But angle EDF is equal to angle HDK (for they are at the vertex), and therefore angle KDG [is equal to HDK].[169] And since angle GDH is an exterior angle to triangle GDF, it is equal to the two opposite angles at G and F. But these are equal to one another, for DG is equal to DF.

341

[168] 340.15   From Barocius.
[169] 341.2   From Barocius.

Angle GDH is therefore double the angle at G. The angle at
G is then equal to angle KDG, and they are alternate angles;
DE is therefore parallel to FG. Consequently triangles GDE
and FDE are on the same base DE and between the parallels
DE and FG. Therefore they are equal. But triangle GDE is
equal to triangle ABC, and hence triangle DEF is equal to
triangle ABC.

You see that we needed three theorems from the doctrine
of parallels: one, that in every triangle the exterior angle is
equal to the two opposite interior angles; another, that if a
straight line falls on two straight lines making the alternate
angles equal, the straight lines are parallel; and the third,
that triangles on the same base and in the same parallels
are equal. The author of the *Elements* knew this and therefore omitted this comparison of the triangles.

Now let angles BAC and EDF be greater than two right
angles, and carry out the same constructions. Then since
angles BAC and EDF, that is, angles EDG and EDF, are
342 greater than two right angles and angles EDG and GDK are
equal to two right angles, if the common angle EDG is subtracted, angle EDF is greater than angle GDK; that is, angle
KDH is greater than angle GDK.[170] Angle GDH—that is, the

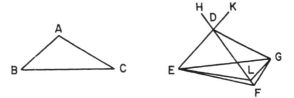

double of the angle at G—is therefore greater than the double
of angle GDK. Angle GDK is then less than the angle at G.
Let angle DGL be constructed equal to angle GDK, and let
EL be drawn. Then GL and DE are parallel. Triangles GDE
and LDE are therefore equal. But triangle LDE is less than
triangle FDE; therefore triangle GDE is less than triangle

[170] 342.4 The Greek text here is somewhat confused. I follow
Barocius, since his translation gives the sense required by the geometrical reasoning.

FDE. But triangle GDE is equal to triangle ABC. Therefore triangle ABC, which has the greater angle, is less than triangle FDE.

Now suppose, thirdly, that the unequal angles are less than two right angles, and complete the same constructions. Then since angles EDG and GDK are equal to two right angles, if we subtract the common angle EDG, the angle EDF (that is, KDH) is less than GDK; hence the whole angle GDH is less than the double of angle GDK. But GDH is also double the angle at G [that is, DGF]. Therefore the angle GDK is greater

343

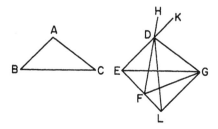

than the angle at G. Let angle DGL be constructed equal to angle GDK, let GL meet EF at L, and let DL be joined. GL is then parallel to DE; hence triangles GDE and DLE are equal to one another. But triangle DLE is greater than triangle FDE, and GDE is equal to ABC. Consequently triangle ABC is greater than triangle DEF.

Thus it has been proved that triangle ABC is either equal to, greater than, or less than triangle DEF when the angles at A and D are equal to, or greater than, or less than two right angles. And all these are possible assumptions. For suppose the angle at A is one and one-half times a right angle and that at D half a right angle. Are they not then equal to two right angles? Or if A is one and a half and D a right angle, are they not greater than two right angles? Or if A is one and a half and D a third of a right angle, are they not less than two right angles? And in each of these cases A is greater than D. All these comparisons[171] were possible for us only through

344

171 344.6 Friedlein's text has no substantive with πᾶσαι. I assume with Barocius that something like συγκρίσεις is implied.

the use of parallel lines, and this is why they are necessarily omitted by the author of the *Elements*.

XXV. *If two triangles have two sides equal to two sides respectively but have one base greater than the other, they will also have one of the angles contained by the equal straight lines greater than the other.*

This theorem is the opposite of the eighth and the converse of the one before it. The author of the *Elements* has presented these theorems in pairs, one pair concerning equality of angles and bases, another concerning inequality, in each pair taking a leading proposition and its converse and, for the leading proposition, using direct proof and for the converse reduction to impossibility. In this way he proceeds for each kind of triangle, now showing that the equality of the sides implies the equality of the subtended angles, and inequality inequality,[172] and again conversely showing that the equality of the angles implies the equality of the subtending sides, and inequality inequality.

345

Coming now to the present theorem, we leave it to the eager student to find out from the books how our geometer demonstrates it, since his procedure is quite clear.[173] But the proofs that others have produced for the same proposition we shall recount briefly, and first the proof discovered and set forth by Menelaus of Alexandria.[174] "Let ABC and DEF be two tri-

---

[172] 345.6 Reading τὴν ἀνισότητα after τῇ ἀνισότητι as in line 8.

[173] 345.12 Euclid proves XXV indirectly, and by division (cf. note at 318.16). Thus if the angle at A is not greater than the angle at B, it must be either equal to or less than D. But it cannot be equal to D, for then the bases of the two triangles would be equal, by IV; nor can it be less than D, for then the base of its triangle would be less than that of the other, by XXIV. Therefore the angle at A is greater.

[174] 345.14 Menelaus of Alexandria, who lived during the latter half of the first century, is important because of his contributions to spherical geometry and trigonometry. He wrote a *Spherica* in three books, preserved for us in an Arabic version, which defines the spherical triangle and demonstrates the basic propositions about it corresponding to Euclid's theorems about the plane triangle. He also wrote an *Elements of Geometry*. The proof cited by Proclus probably comes from this work. See Heath II, 260-273, and Van der Waerden, 274-276.

angles having the two sides AB and AC equal to the two sides
DE and DF and BC greater than EF. I say that the angle at A
is greater than the angle at D. Let a length BG be laid off on

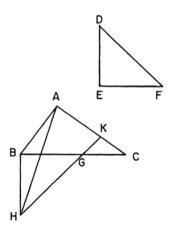

BC equal to EF, let an angle GBH be constructed at B equal
to DEF, and let BH be equal to DE. Join HG, let the line be
produced to K, and let AH be joined. Then since BG is equal
to EF and BH is equal to DE, the two sides are equal to two
sides, and they contain equal angles. Consequently GH is
equal to DF, and angle BHG to angle EDF. And since GH
is equal to DF and DF to AC, then HG is equal to AC. HK
is longer than AC, so that even more is it longer than AK.
And angle KAH is therefore greater than angle KHA. Again
since BH, being equal to DE, is equal to AB, angle BHA is
equal to angle BAH. The whole angle BAK is hence greater
than the whole angle BHK, and the whole angle BHK has
been demonstrated to be equal to the angle at D. Therefore
angle BAC is greater than the angle at D." Such is the proof
of Menelaus.

Heron the engineer proves the same theorem, without
using the reduction to impossibility, as follows. Let ABC and
DEF be triangles and the hypotheses the same as before.
Since BC is greater than EF, let EF be produced and EG be
made equal to BC. Similarly let ED be produced and DH be
made equal to DF. With D as center and distance DF let a

347

circle be described going through H, namely, FKH. And since AC and AB are greater than BC, and since they are equal to EH and BC is equal to EG, a circle described with center E and distance EG will cut EH. Let GK be the circle

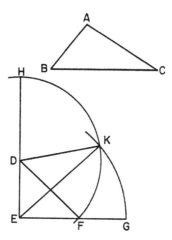

that cuts it, and let lines KD and KE be drawn to the centers of the circles from the point of their intersection. Now since D is the center of FKH, KD is equal to DH, that is, to DF and to AC. Again since E is the center of circle GK, KE is equal to EG, that is, to BC. Then since the two lines AB and AC are equal to the two lines DE and DK and BC is equal to KE, the angle BAC is equal to the angle EDK. Therefore angle BAC is greater than angle FDE.

XXVI. *If two triangles have two angles equal to two angles respectively and one side equal to one side, namely, either the side adjoining the equal angles or that subtending one of the equal angles, they will also have the remaining sides equal to the remaining sides and the remaining angle equal to the remaining angle.*

Anyone wishing to compare triangles with respect to their sides, angles, and areas must either assume equal sides only and inquire whether the angles are equal, or assume equal angles only and inquire about the equality of the sides, or take a combination of angles and sides. Now assuming equal

348

angles only, he was unable to prove that the sides of the triangles are also equal; for the smallest triangles may have equal angles with the greatest and yet have sides and included areas that fall short of the others, though they have each of their angles one by one equal to theirs. But positing the sides only as equal, he proved that all parts are equal by the eighth theorem, which assumes two triangles having two sides equal respectively and the base equal to the base and demonstrates that they have equal angles and enclose equal areas. The author of the *Elements* omits to add this last point, since it is a necessary consequence of the fourth theorem and needs no proof. Taking a combination of sides and angles, he had to assume either one side equal to a side and one angle equal to an angle, or one side and two angles of the triangles equal, or vice versa one angle and two sides, or one angle and three sides, or one side and three angles, or more than one side and more than one angle. But he did not try with one side and one angle to prove the proposed equality of the other parts. It is clearly possible for two triangles that have only one side and one angle equal to be unequal in all other respects. For exam-

349   ple, let AB be a straight line standing at right angles on CD, with BD greater than BC, and let AC and AD be joined. These triangles have a common side and an angle in one

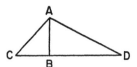

equal to an angle in the other, but they are in all other respects unequal. But it was possible to take one side and two angles and prove the other parts equal, and this he does in the present theorem. But to take one side and three angles equal was to posit too much, since the equality of the other parts can be proved from two equal angles only. Again, taking one angle and two sides, he proved in the fourth theorem that the other parts are equal.[175] But to take one angle and three sides

---

[175] 349.13   It is strange that neither here nor in his later comments at 350.14-24 on the fourth theorem does Proclus mention the case in

equal was superfluous, for the assumption of two equal sides was enough by itself to show the equality of the other parts Finally, to assume two equal sides and two equal angles, or two sides and three angles equal, or two angles and three sides, [or three angles and three sides][176]—all these were unnecessary, for what follows from fewer premises will certainly follow also from more, assuming that the hypotheses are set forth with the given qualifications.[177] Thus it is clear that three hypotheses only needed to be examined: the hypothesis of three sides only, the hypothesis of two sides and one angle, and its opposite, that which takes one side and two angles; and it is this last that our geometer now adds.[178] This is why we have only three theorems concerning equality in triangles with respect to their sides and their angles, all the other hypotheses being either insufficient to prove what is wanted or sufficient but redundant, since the same things can be proved on fewer assumptions.

350

Now just as our author, when he assumed two sides equal to two sides and an angle equal to an angle, did not take any chance angle but, as he there added, "the angle contained by the equal straight lines," so also when he assumes two angles equal to two angles and a side to a side, he does not take any chance side but "either the side adjoining the equal angles or a side that subtends one of the equal angles." For neither in the fourth theorem is it possible by taking any chance angle as equal, nor in this one by taking any side indifferently, to prove

---

which two sides and an angle opposite one of them are sufficient, under certain conditions, to make the two triangles equal. It cannot have been unknown to him, for Menelaus in his *Spherica* (I. 13) includes the corresponding theorem in spherical geometry (see previous note). For a discussion of this "ambiguous case" see Heath, *Euclid* I, 306.

[176] 349.18 From Barocius.

[177] 349.21 Reading with Grynaeus and Barocius δοθέντων instead of δεόντων in Friedlein.

[178] 349.26 I omit the details of Euclid's proof of XXVI, since Proclus makes no further reference to it. Euclid proceeds by division, showing by reduction to impossibility that each alternative other than that stated in the enunciation is false. This makes a lengthy and tedious proof and is a remarkable illustration of Euclid's desire to avoid the proof by superposition, which would have been very short and easy.

that the remaining parts are equal. For example, let BC in the equilateral triangle ABC be divided into unequal parts by the line AD. This produces two triangles having sides AB and AD equal to AC and AD and one angle, the angle at B, equal to

the angle at C. But the other parts, BD and DC, are not equal (for they were posited as unequal), nor are the other angles. The reason is that we have not taken as our equal angle the angle contained by the equal sides. For the same reason, then, the present theorem also will manifestly fail if we do not observe the restriction he lays down, to take as the equal side that which subtends one of the equal angles or the side adjoining the equal angles. Let ABC be a right-angled triangle having its angle at B right and side BC greater than AB. Let AB be produced, let there be constructed on BC and at point C on it an angle BCD equal to angle BAC, and let AB and CD when produced meet at D. ABC and BCD are then two triangles having a common side BC and two equal angles, angle ABC

351

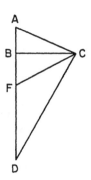

equal to CBD (for both are right angles), and angle BAC equal to BCD by construction. Therefore the triangles are equal, it seems; and yet it can be demonstrated that BCD is greater than ABC. The reason is that we have taken the common side BC, which in triangle ABC subtends one of the equal angles, the angle at A, but in triangle BCD adjoins the equal angles. It should have either subtended one of the equal angles or adjoined the equal angles, in both cases. Because we did not watch this, we have declared equal a triangle which

352 is necessarily greater. For how could BCD fail to be greater than ABC? At point C on straight line BC let an angle FCB be constructed equal to angle ACB, for angle BCD, like the angle at A, is greater than angle ACB. Then since there are two triangles ABC and BCF having two angles ABC and BCA equal respectively to the two angles CBF and FCB and a common side BC adjoining the equal angles, the triangles are equal. And since BCD is greater than FCB, it is greater than ABC. Our earlier proof that it is equal resulted from taking the side at random. This contribution to the precise understanding of the present matter comes from Porphyry; but Eudemus in his history of geometry attributes the theorem itself to Thales, saying that the method by which he is reported to have determined the distance of ships at sea shows that he must have used it.

From the analysis given above we can obtain a general view of the whole theory of equality between triangles and can explain the omission of certain hypotheses by proving that they are erroneous or superfluous. At this point we shall assume that the first section of the *Elements* comes to an end. The author has constructed triangles and compared them

353 with respect to their equality or inequality, establishing their existence by construction and their identity[179] and differences by comparison. For existence involves three factors: Being, Sameness, and Otherness, both quantitative and qualitative, according to the individual characters of the subjects concerned. Thus through these propositions as likenesses he has shown us that everything is both identical with itself and other than itself because of the plurality it contains; that is, all are the same with one another and other than each other, for equality and inequality have been discovered to exist in each single triangle and in two or more.

179 353.3   Reading with Barocius ταυτότητα for ἰσότητα in Friedlein.

PROPOSITIONS:
PART TWO

WE HAVE learned from the foregoing all that it is possible to say in an elementary treatise regarding the construction of triangles and their equality or inequality. Euclid next goes through the four-sided figures, primarily instructing us about parallelograms but including in the theory of them his teaching about trapezia. Earlier in our treatment of the Hypotheses[1] four-sided figures were divided into parallelograms and trapezia, and the parallelogram into its several species, and the trapezium likewise. Since the parallelogram is regular, through its participation in equality, while the trapezium has not the same nor a similar regularity, it is logical that he should first work out the doctrine of parallelograms and examine the trapezium in connection with them. For trapezia are revealed by the sectioning of parallelograms, a matter which will become clear as we proceed.

355

But again it is impossible to say anything about the construction of parallelograms, or about their equality, without the theory of parallel lines. For, as its name indicates, a parallelogram is the figure contained by parallel straight lines lying opposite one another. Hence of necessity he begins his instruction with parallel lines and, after proceeding a short way, turns from them to the theory of parallelograms, using as a connecting link between these two portions of the *Elements* a theorem that seems to be examining a property of parallel lines but in fact furnishes the primary genesis of the parallelogram. This theorem is "The straight lines joining equal and parallel straight lines in the same directions are themselves also equal and parallel."[2] Although this theorem considers a property of lines that are equal and parallel, yet by

[1] 354.8   I.e. Deff. XXX-XXXIV.
[2] 355.16   XXXIII. See note at 385.2.

the mention of "joining" it shows that a parallelogram is a figure that has its opposite sides equal and parallel.

From this it is clear that the doctrine of parallel lines must be taken up first. There are three inherent and essential properties of parallel lines to be considered, properties which are characteristic of them as such and convertible with them. We must examine them, not only all three together, but each of them separately from the others. One of them is that when a straight line cuts parallel lines their alternate angles are equal; another, that when a straight line cuts parallel lines the interior angles are equal to two right angles; and the third, that when a straight line cuts parallels the external angle is equal to the interior and opposite angle. Each of these properties when demonstrated is sufficient to show that the straight lines are parallel.

356

This is the way in which other mathematicians also are accustomed to distinguish lines, giving the property of each species. Apollonius, for instance, shows for each of his conic lines what its property is, and Nicomedes likewise for the conchoids, Hippias for the quadratrices, and Perseus for the spiric curves. After a species has been constructed, the apprehension of its inherent and intrinsic property differentiates the thing constructed from all others. In the same way, then, the author of the *Elements* first investigates the properties of parallel lines.

XXVII. *If a straight line falling on two straight lines makes the alternate angles equal to one another, the straight lines will be parallel to one another.*

It is taken for granted in this theorem,[3] or rather in all theorems in plane geometry, that straight lines are in one plane. I have added this remark because straight lines are not always parallel when alternate angles are equal, unless they lie in the same plane. If two straight lines are lying crosswise to one another, one in one plane and the other in

---

[3] 356.21 The text is intolerably elliptic here, if it has not been corrupted. I translate, following Barocius, as if it read ἐπὶ τοῦ προκειμένου ὡς ὁμολογούμενον προείληπται.

357

another, there is nothing to prevent a straight line falling upon them from making their alternate angles equal, while yet the lines lying thus are not parallel. Thus it is presupposed that everything that we write about in plane geometry we imagine as lying in one and the same plane. Hence this addition was not necessary.

But about the word "alternate" we should know that our geometer uses it in two senses, sometimes as referring to a certain position, sometimes as denoting a certain sequence of terms in proportion. It is in this latter sense that he uses "alternate" in Book V and in the arithmetical books;[4] but it has the former sense in this book and in all the others when the topic is parallel lines intersected by another line. Angles that are produced in different directions[5] and are not adjacent to one another, but separated by the intersecting line, both of them within the parallels but differing in that one lies above and the other below, he calls "alternate"[6] angles. Thus if AB and CD are straight lines and EF a line falling upon them, he

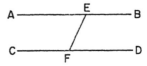

says that angles AEF and DFE, and again CFE and BEF, are alternate, because their positions are the reverse of one another.

358

We must also understand that, with the straight lines situated thus, analysis reveals six possible combinations of two angles; and of these our geometer assumes three only and ignores three. We can take the angles as lying in the same direction or not in the same direction; and if they lie in the same direction, we must take them as either both within the straight lines which the proof shows to be parallel, or both outside them, or one outside and the other inside. And likewise if they are not in the same direction, we must again take

4 357.13 Books VII-IX.
5 357.17 μὴ ἐπὶ τὰ αὐτά, i.e. on different sides of the transversal.
6 357.20 ἐναλλάξ. Cf. ἐνηλλαγμένως ἔχουσας in line 25 below.

them as either both outside the straight lines that are inter-sected, or both inside, or one inside and one outside. An identical diagram will make clear what I mean. Let AB and CD be straight lines, and let EF fall upon them and be produced to H and K. Now if you take the angles as lying in

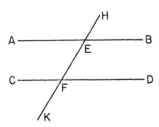

the same direction, you may take them as either both inside, like BEF and EFD, or AEF and EFC; or both outside, like HEB and DFK, or HEA and CFK; or one inside and one outside, like HEB and EFD, or KFD and FEB, or HEA and EFC, or KFC and AEF, for there are four pairs that meet this condition. And if you take the angles as not lying in the same direction, you may take them as either inside, like AEF and EFD, or CFE and FEB; or both outside, like AEH and DFK, or HEB and CFK; or one inside and one outside, and here again we have four pairs, either AEH and EFD, or HEB and EFC, or KFC and FEB, or KFD and FEA. Beyond these there is no other way of taking them.

Of these six ways in which the angles can be taken, our geometer selects three only, and their consequences reveal the characteristic properties of parallel lines.[7] One of the three assumes angles not lying in the same direction, and of these only angles that lie inside the parallel lines (the angles that he calls "alternate"), so that the combinations of two angles outside and of one outside and one inside are ignored. Of the hypotheses that take angles in the same direction he considers the supposition that both angles are inside (which he says are equal to two right angles) and that one angle is

359

_____

[7] 359.8ff. The text is puzzling here. I suspect that ταῦτα εἰς in line 8 is a corruption of ταύταις, and I have so translated it.

inside and the other outside (which he says are equal), omitting only the assumption that they are both outside.

Now the omitted hypotheses, we maintain, yield the same results. Let HEB and DFK be angles lying in the same direction and both outside. I say that they are equal to two right angles. For if EFD is equal to HEB and BEF to DFK, and if BEF and EFD are equal to two right angles, so also are DFK

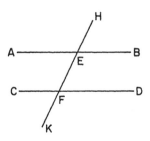

and HEB equal to two right angles. Again assume two angles, AEH and EFD, not[8] lying in the same direction, one inside and the other outside. I say that they are equal to two right angles. For if AEH is equal to BEF, and BEF and EFD are equal to two right angles, then AEH and EFD are equal to two right angles. Again let AEH and DFK be angles not lying in the same direction and both of them outside the straight lines. I say that they are equal to each other. For if AEH and BEF are equal to each other and DFK is equal to BEF, then AEH is equal to DFK. Consequently if one takes the results of the three hypotheses that our geometer has taken, all the same consequences follow as true which would follow from the other three hypotheses, except that in the cases our geometer has taken there are two which show the posited angles equal to each other and one showing them equal to two right angles; whereas, conversely, of the others two show them to be equal to two right angles and one shows them to be equal to each other. Thus of the six possible hypotheses it follows from three that the angles are equal to two right angles and from three that they are equal to each other. Hence it is natural that the omitted hypotheses should have a con-

360

---

[8] 359.27   Reading μή instead of μὲν in Friedlein.

verse relation to those that he has chosen to mention. It appears that our geometer chose from these hypotheses the ones that are more affirmative or simpler; and this is why he took from those not lying in the same direction only internal angles, the angles he calls "alternate," and from those lying in the same direction the case of two interior angles and the combination of one interior and one exterior, avoiding the other cases as requiring a more negative or a more complex expression. But whether this or some other cause should be assigned, it is clear from what we have said how many are the consequences they involve.[9]

361

XXVIII. *If a straight line falling on two straight lines makes the exterior angle equal to the interior and opposite angle in the same direction, or the interior angles in the same direction equal to two right angles, the straight lines will be parallel.*

The preceding theorem proved that straight lines are parallel if [alternate] angles lying in different directions and within the straight lines are equal to one another; this one proves the same conclusion by putting forward the two other hypotheses, one of them dividing the angles into one interior and one exterior, the other supposing them both to be interior. It would appear that the author of the *Elements* has divided his theorems in a strange fashion. Either he should have taken the

---

[9] 361.4 Proclus does not refer to Euclid's proof of XXVII, but because of its importance for the general theory of parallels it is well for us to have it before us. Given the straight line EF falling on the two straight lines AB and CD and making the alternate angles AEF and EFD equal to one another, to prove that AB is parallel to CD. If they are not parallel, AB and CD when produced will meet either

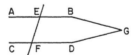

in the direction of B and D or towards A and C. Suppose them to meet in the direction of B and D at G. Then in triangle GEF the exterior angle AEF is equal to the interior and opposite angle EFG, which is impossible, by XVI. Similarly it can be shown that they cannot meet towards A and C. But straight lines which do not meet in either direction are parallel, by Def. XXXV (XXIII in Heiberg's text).

three hypotheses separately and made three theorems or included them all in one, as Aigeias of Hierapolis[10] does in his epitome of the *Elements*; or if he wished to divide them into two, he should have made his division orderly by dealing in one with the hypotheses that assume the angles to be equal to one another and in another with that which takes them to be equal to two right angles. But instead in one theorem he posits the alternate angles as equal, while in the other he takes an external angle as equal to an internal angle and the internal angles in the same direction as equal to two right angles. What is the reason for such a division? Clearly it was not the equality of the angles to each other, or alternatively their equality to two right angles, that concerned him, nor did he use this criterion for separating the theorems from one another; rather it was whether the angles are taken as lying in the same or in different directions. The preceding theorem takes them as lying in different directions, for that is what alternate angles are, whereas this one takes them as lying in the same direction, as the enunciation makes clear.[11]

How the author of the *Elements* proves that when the interior angles are equal to two right angles the straight lines are parallel is evident from his book.[12] But Ptolemy, in a book in

362

[10] 361.21 Nothing further is known of Aigeias.

[11] 362.11 As Heath points out (*Euclid* I, 311), the criterion of XXVII is that actually used to prove parallelism and is, moreover, the basis of the construction of parallels in XXXI, whereas XXVIII only reduces the other two hypotheses to that of XXVII. Thus precision of reference, as well as clearness of exposition, is better secured by the arrangement that Euclid adopts.

[12] 362.14 Euclid's proof of XXVIII is as follows: Let the straight line EF falling on the two lines AB and CD make the exterior angle EGB equal to the interior and opposite angle GHD, or the interior angles in the same direction, viz. BGH and GHD, equal to two right

angles. Then since angle EGB is equal to angle GHD and angle EGB to angle AGH, by XV, angle AGH is also equal to angle GHD, and they are alternate; therefore AB is parallel to CD, by XXVII. Again

which he proposes to prove that straight lines produced from
angles less than two right angles meet in the direction in which
lie the angles less than two right angles, begins by proving this
theorem, that when the interior angles are equal to two right
angles the straight lines are parallel, and does so as follows.

Let AB and CD be two straight lines cut by a straight line
EFGH in such a fashion as to make angle BFG and FGD
equal to two right angles. I say that the straight lines are
parallel, that is, nonsecant.[18] If possible, let FB and GD

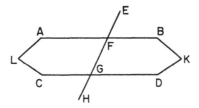

363

be produced to meet at K. Then since straight line GF
stands on line AB, it makes angles AFG and BFG equal to
two right angles. Likewise since GF stands on CD, it makes
angles CGF and DGF equal to two right angles. Consequently angles AFG, BFG, CGF, and DGF are equal to
four right angles, of which two, BFG and DGF, are posited
as equal to two right angles; hence the other two angles,
AFG and CGF, are also equal to two right angles.[14] If,
then, when the interior angles are equal to two right angles
the lines FB and GD when produced meet one another at
K, so also FA and GC when produced will meet, for

since angles BGH and GHD are equal to two right angles, then since
AGH and BGH are also equal to two right angles, by XIII, if we subtract angle BGH from each, the remaining angle AGH is equal to the
remaining angle GHD. And they are alternate; therefore AB is
parallel to CD, by XXVII.

[18] 362.27 ἀσύμπτωτοι.

[14] 363.12 "This seems to be a fallacy. Ptolemy wants to prove
that two lines satisfying certain conditions never meet. He cannot do
this by disproving the assumption that lines satisfying this condition
always meet. Ptolemy commits the same kind of fallacy again (366.1).
Proclus catches the second one (368.8)." (I.M.) See Heath, *Euclid* I,
204.

angles AFG and CGF are also equal to two right angles. The straight lines will meet either on both sides or on neither if these, like those, are equal to two right angles. Suppose, then, that FA and GC meet at L. Then straight lines LABK and LCDK enclose an area, which is impossible. It is therefore not possible that the lines should meet when the interior angles are equal to two right angles. Therefore they are parallel.

364

XXIX. *A straight line falling on parallel straight lines makes the alternate angles equal, the exterior angle equal to the interior and opposite angle in the same direction, and the interior angles in the same direction equal to two right angles.*

This theorem is the converse of both the preceding ones, for the conclusion of each of them is made the hypothesis here, and what is given in them is proposed for proof. We should note this additional difference among converses: a converse may be the converse either of a single theorem, as the sixth is of the fifth, or of more than one theorem, as this is of those preceding it. In this theorem the author of the *Elements* uses for the first time the postulate, "If a straight line falling on two straight lines makes the interior angles in the same direction less than two right angles, the straight lines if produced will meet in that direction in which are the angles less than two right angles."[15] As I said in the part of my

---

[15] 364.18   Post. V. It is used by Euclid in his proof of the first of the three elements in the conclusion of XXIX, i.e. the equality of the alternate angles, AGH and GHD in the diagram. If angle AGH is unequal to angle GHD, one of them is greater. Let it be AGH. Let angle BGH be added to each; then angles AGH and BGH are greater

than angles BGH and GHD. But angles AGH and BGH are equal to two right angles; therefore angles BGH and GHD are less than two right angles. But straight lines produced indefinitely from angles less than two right angles meet, by Post. V. Therefore AB and CD, if

exposition that precedes the theorems,[16] not everyone admits that this generally accepted proposition is indemonstrable. For how could it be so when its converse is recorded among the theorems as something demonstrable? For the theorem that in every triangle any two interior angles are less than two right angles is the converse of this postulate.[17] Since also[18] the fact that two straight lines when produced approach one another more and more nearly is not, as I said before,[19] a

365

sign that they will meet, because other lines have been discovered that converge towards one another more and more but never meet.

Hence others before us have classed it among the theorems and demanded a proof of this which was taken as a postulate by the author of the *Elements*. Ptolemy is thought to have proved it in his book entitled "That lines produced from angles less than two right angles meet one another." His proof employs many of the theorems established by the author of the *Elements* prior to this one. In order not to add to our labors, let us assume that these are all true and take it as a little lemma that they have been proved by the previous arguments. One of the propositions taken as previously proved is this, that lines produced from angles equal to two right angles never meet.[20]

---

produced indefinitely, will meet. But they do not meet, for they are by hypothesis parallel. Therefore angle AGH is not unequal to angle GHD, but equal to it. Next since angle AGH is equal to angle EGB, by XV, therefore angle EGB is also equal to angle GHD. Adding angle BGH to each, we see that the two angles EGB and BGH are equal to the two angles BGH and GHD. But angles EGB and BGH are equal to two right angles, by XIII. Therefore angles BGH and GHD are also equal to two right angles.

[16] 364.19   At 191.21ff.

[17] 364.25   XVII.

[18] 364.25   ἐπεὶ καὶ is puzzling here. I suspect a lacuna in the text, the loss of a sentence (such as "Nor is it self-evident") which this clause was intended to support.

[19] 365.4   192.1-193.2.

[20] 365.16   For Ptolemy's proof of this proposition see 362.20-363.18. Ptolemy now attempts to prove Euclid's XXIX without the use of Post. V.

I say, therefore, that the converse also is true, namely, that
when parallel straight lines are cut by a straight line the
interior angles in the same direction are equal to two right
angles. For it is necessary that the line cutting the parallel
lines make the interior angles in the same direction either
equal to two right angles or less or greater than two right

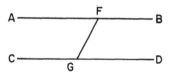

angles. Let AB and CD be parallel lines, and let GF fall
upon them. I say that it does not make the interior angles
in the same direction greater than two right angles. For if
angles AFG and CGF are greater than two right angles,
the remaining angles, BFG and DGF, are less than two
right angles. But these same angles are also greater than
two right angles; for AF and CG are no more parallel than
FB and GD, so that if the line falling on AF and CG makes
the interior angles greater than two right angles, so also
does the line falling on FB and GD make the interior angles
greater than two right angles. But these same angles are
less than two right angles (for the four angles AFG, CGF,
BFG, and DGF are equal to four right angles), which is
impossible. Similarly we can prove that the line falling on
the parallels does not make the interior angles in the same
direction less than two right angles. If, then, it makes them
neither greater nor less than two right angles, the only conclusion left is that the line falling on them makes the interior angles in the same direction equal to two right
angles. When this has been demonstrated, the proposition
before us[21] can indisputably be proved. I say that, if a
straight line falls upon two straight lines and makes the
interior angles in the same direction less than two right
angles, the straight lines if produced will meet in that direction in which are the angles less than two right angles. For

366

[21] 366.15 Euclid's Post. V, which was the subject of Ptolemy's
book.

let us suppose that they do not meet. But if they are non-secant in the direction in which are the angles less than two right angles, much more will they be nonsecant in the other direction in which are the angles greater than two right angles, so that the straight lines will be nonsecant in both directions; and if so, they are parallel. But it has been proved that the line which falls upon parallels will make the interior angles in the same direction equal to two right angles. The same angles are therefore both equal to two right angles and less than two right angles, which is impossible.

367

Having proved this, Ptolemy tries to add extra precision to it and reach the proposition before us by proving that, if a straight line falls upon two straight lines and makes the interior angles in the same direction less than two right angles, not only are the straight lines not nonsecant, as he has proved, but also they will meet in that direction in which are the angles less than two right angles, not in the direction in which they are greater.

Let AB and CD be two straight lines, and let the line EFGH fall upon them and make angles AFG and CGF less than two right angles. Hence the other angles are greater than two right angles. Now it has been demon-

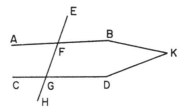

strated that the straight lines are not nonsecant. But if they meet one another, it will be either in the direction of A and C or in the direction of B and D. Let us assume that they meet in the direction of B and D at point K. Then since angles AFG and CGF are less than two right angles and angles AFG and BFG are equal to two right angles, if the common term, angle AFG, is subtracted, angle CGF will be less than angle BFG. It follows that the exterior angle of triangle KFG is less than the opposite interior one, which is impossible. Consequently they do not meet in this direction. But they do meet. Therefore they meet in the other direction, that in which are the angles less than two right angles.[22]

368

This is Ptolemy's proof. It is worth pausing to see whether there may not be a fallacy in the hypotheses that he has adopted. I mean in those which assert that, when a straight line cuts the nonsecant lines and makes four interior angles, the angles in the same direction on both sides[23] are either equal to two right angles or greater or less than two right angles.[24] His division is not exhaustive. There is no reason why one who calls nonsecant the lines produced from angles less than two right angles should not say that the angles lying in the same direction on one side are greater than two right angles and those in the same direction on the other side less than two right angles, that is, that no single principle can be

---

[22] 367.27 "This argument is very messy, and I am dubious that Proclus is reproducing Ptolemy accurately. The assumption (366.21) that, if the two lines don't meet on the one side, they can't meet on the other is unconvincing and unnecessary, since the next argument (367.3ff.) proves they can't meet on the other. The whole argument should be simply this: Using the diagram on 367, the two lines cannot meet at K for the reason given (367.10ff.). But they must meet, because if they don't they are parallel, and by the preceding argument (365.16) a straight line cutting parallel lines makes the interior angles in the same direction equal to two right angles. Therefore the two lines AB and CD meet on the side away from K." (I.M.) As to I.M.'s doubt, note the occurrence of λέγω at 362.26, 365.16, 366.16; i.e. Proclus professes to be quoting Ptolemy throughout.

[23] 368.5 αἱ ἐπὶ τὰ αὐτὰ κατ' ἀμφότερα, i.e. in the same direction on both sides of the transversal.

[24] 368.7 Ptolemy justifies this assumption by saying (366.2-6) that FA and GC are no more parallel in one direction than FB and GD are in the other; and this is equivalent, Heath says (*Euclid* I, 206), "to the assumption that through any point only one parallel can be drawn to a given straight line. That is, he assumes an equivalent of the very postulate he is endeavoring to prove." This equivalent is now known as Playfair's Axiom and is often substituted for Euclid's postulate in modern textbooks.

admitted to cover them.[25] Since his division is not exhaustive, the proposition under examination has not been demonstrated. Furthermore, this also must be said against the proof, that it does not show the impossibility to be one intrinsic to parallels. For it is not because a straight line cutting parallels makes the angles in the same direction on both sides greater or less than two right angles that the hypotheses are reduced to absurdity; it is because the four angles interior to the lines that are cut are equal to four right angles that each of the hypotheses becomes impossible, since even if one does not take the straight lines as parallel the same consequences follow from assuming these same hypotheses.

369   With these remarks we shall end our comments on Ptolemy, for the weakness of his proof is evident from what has been said. Now let us examine those who say it is impossible that lines produced from angles less than two right angles should meet. Taking two straight lines AB and CD and line AC falling upon them and making the interior angles less than two right angles, they think they can demonstrate that AB and CD do not meet. Let AC be bisected at E, and let a length AF equal to AE be laid off on AB, and on CD a length CG equal to EC. It is clear that AF and CG will not meet at any point on FG; for if they meet, two sides of a triangle will be

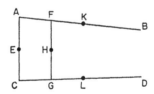

equal to a third, AC, which is impossible. Again let line FG be drawn and bisected at H, and let equal lengths be laid off. These likewise will not meet, for the same reasons as before. By doing this indefinitely, drawing lines between the non-coincident points, bisecting the connecting lines, and laying off

25 368.12   This sentence apparently refers to the reasoning of Ptolemy quoted at 366.16ff.; but I suspect that it has lost something that might have made its reference clearer.

on the straight lines lengths equal to their halves, they say they prove that lines AB and CD will not meet anywhere.

Such are their arguments. To them we must reply that what they say is true but that it does not prove as much as they think. It is true that it is not possible in this simple way to fix the point at which intersection occurs. It is not true, however, that the lines never meet at all. Let it be granted that AB and CD do not meet when angles BAC and DCA are defined by points F and G. But there is no reason why they should not come together at K and L, even if FK and GL are equal to FH and HG. For if AK and CL meet at K and L, the angles KFH and LGH are no longer the same; that is, some of FG has come to belong to AK and CL; and thus in turn the lines FK and GL are greater than the base by as much as they take away from within the line FG.²⁶ This also should be said. In affirming without qualification that lines produced from angles less than two right angles do not meet, they are overthrowing what they do not intend. Let the diagram be the same as

370

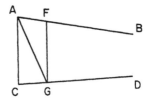

before. Now is it possible or not to draw a straight line from A to G? If they say it is not possible, they are denying not only the fifth postulate, but also the first, which claims the right to draw a straight line from any point to any point. If it is possible, let the line be drawn. Then since angles FAC and GCA are less than two right angles, it is clear even more that GAC and GCA are less than two right angles. Therefore

²⁶ 370.10 "Proclus certainly seems to have missed the point here, as his last remark on this topic shows (371.7). Points L and K cannot coincide, because if they did there would be a triangle, namely, FGL (or FGH), with two of its sides equal to its third. The real point is, of course, that the described process of extending AF and CG imposes a finite upper bound on their length. Their point of intersection is beyond this bound." (I.M.)

371

AG and CG meet at G, and they are produced from angles less than two right angles. It is consequently not possible to say without qualification that lines produced from angles less than two right angles do not meet. On the contrary, it is clear that some lines produced from angles less than two right angles do meet, though the argument proving this of all such lines is still to be found. Since "less than two right angles" is indeterminate, one could say that with such-and-such an amount of lessening the straight lines remain nonsecant, whereas with another amount less[27] than this they meet.

To anyone who wants to see this argument constructed,[28] let us say that he must accept in advance such an axiom as Aristotle used in establishing the finiteness of the cosmos:[29] If from a single point two straight lines making an angle are produced indefinitely, the interval between them when produced indefinitely will exceed any finite magnitude. At least he proved that, if the lines extending from the center to the circumference are infinite, the interval between them is infinite; for if it is finite, it is possible[30] to increase the interval between them, so that the straight lines are not infinite. Straight lines extended indefinitely, then, will diverge from each other a distance greater than any given finite magnitude. If this is

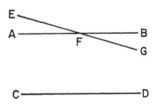

laid down, I say that, if a straight line cuts one of two parallel lines, it cuts the other also. Let AB and CD be parallel lines and EFG a line cutting AB. I say that it also

[27] 371.10 ἐλάσσονα, if it is not a slip on the part of the author, or the corruption of an original ἐλάττωσιν, must be taken proleptically, i.e. "another amount that still further decreases the angles."
[28] 371.11 This argument is the λόγος referred to at 371.6, viz. the proof of Euclid's Post. V.
[29] 371.14 De Caelo 271b28ff.
[30] 371.20 Reading with Barocius δυνατόν for ἀδύνατον in Friedlein.

372

cuts CD. For since there are two straight lines through point F, when FB and FG are extended indefinitely, they will have an interval between them greater than any magnitude and hence greater than the distance between the parallel lines. And so when they are separated from each other a greater distance than that between the parallel lines, FG will cut CD. Therefore if a straight line cuts one of two parallels, it cuts the other also.

Having proved this, we can demonstrate the proposition before us as a consequence of it. Let AB and CD be two straight lines and EF falling upon them and making angles BEF and DFE[31] less than two right angles. I say that the

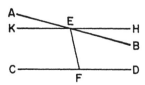

straight lines will meet in that direction in which are the angles less than two right angles. For since angles BEF and DFE are less than two right angles, let angle HEB be equal to the excess of two right angles over them, and let HE be produced to K. Then since EF falls upon KH and CD and makes the interior angles equal to two right angles namely, HEF and DFE, HK and CD are parallel straight lines. And AB cuts KH; it will therefore cut CD, by the proposition just demon-

373

strated. AB and CD therefore will meet in that direction in which are the angles less than two right angles, so that the proposition before us has been demonstrated.[32]

XXX. *Straight lines parallel to the same straight line are also parallel to one another.*

In propositions concerning relations it is our geometer's custom to show the identity pervading all things that have the

---

[31] 372.14   Reading ὅτε for δεῖ in Friedlein.

[32] 373.2   Heath (*Euclid* I, 208) points out that the axiom borrowed from Aristotle and the theorem proved by Proclus at 371.24ff. hold only for Euclidean space. The former is incorrect on the elliptic hypothesis, the latter incorrect on the hyperbolic hypothesis. Thus Proclus' proof begs the question by assuming Euclidean space, of which the postulate to be established is a criterion.

same relation to the same thing. Thus in the Axioms he asserted that things equal to the same thing are equal to each other, and in later propositions[33] he will affirm that things similar to the same thing are similar to each other and that ratios that are the same with the same ratio are the same with one another. So now also, in the same way, he proves that straight lines parallel to the same straight line are parallel to one another. But it happens that this principle is not valid for all relations. For things that are double the same thing are not also double one another, nor are things that are one and one-half times the same thing one and one-half times one another. It seems that this principle applies only to those relations that are unambiguously[34] convertible, namely, equality, similarity, identity, and the position of parallel lines. For a parallel is parallel to a parallel, as an equal is equal to an equal and a similar is similar to a similar; and parallelism is similarity of position, if we may so call it.

374

In his book, then, he asserts and proves that lines parallel to the same line are in every case so related as to be parallel to each other. Our author himself takes the lines that are parallel to the same line as extremes and the line to which they have a similar relation as lying between them, so that the assertion may also be evident to us from a common notion.[35] For if the outer lines meet one another, most certainly they will intersect the line lying between them and no longer be parallel to it. But it is possible also by interchanging the positions of the lines to demonstrate the proposition by the

[33] 373.9   I.e. V. 11 and VI. 21.

[34] 373.18   συνωνύμως. These relations are what modern logic calls symmetrical.

[35] 374.5   Euclid proves XXX as follows: Given AB and CD parallel to EF, as in the accompanying diagram, to prove that AB is also parallel to CD. Angle AGK is equal to angle GHF, by XXIX. Likewise

GHF is equal to angle GKD, by the same proposition. Hence angle AGK, being equal to angle GHF, is also equal to angle GKD; and they are alternate, so AB is parallel to CD.

same procedures as those used by our geometer here. Thus let us take AB, the line to which both CD and EF are parallel, as lying above the other two lines, and not between them.[36] For line HKL falling upon them will make each of the angles HKD and KLF equal to angle AHK, because they are alternate, so that it will also make angles HKD and KLF

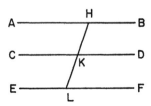

equal to one another. Therefore CD and EF are parallel. If someone should say, "Let AH and HB be parallel to CD, then they are parallel to each other," we shall reply that AH and HB are not two lines, but parts of one parallel line. For we must think of parallel lines as produced indefinitely, and AH when produced coincides with HB; it is therefore the same as it, and not another line. Therefore all the parts of a parallel line are themselves parallel to the straight line to which it is parallel, both to the whole of it and to its parts. Thus AH is parallel to KD and HB to CK; for when produced indefinitely they remain nonsecant.

We had to add these remarks[37] because of the difficulties raised by sophists and the immature attitudes of students. Most persons delight in turning up such fallacious inferences and giving unnecessary trouble to scientific expositors.

There is no need to convert this theorem and prove that lines parallel to one another are also parallel to the same straight line. For if we posit one of them as parallel to another

375

[36] 374.10ff. The text here is hopeless; neither Friedlein's nor Schönberger's suggested emendation commends itself to me, and Barocius gives little help. Fortunately the general sense is clear.

[37] 375.8 Reading ταῦτα for τούτοις in Friedlein. "It is interesting that what Proclus calls a fallacious inference involves a conception of parallel lines more like the modern one than Euclid's or Proclus'. For the modern every straight line is parallel to itself. See also 376.8." (I.M.)

line, that and the other parallel will both be parallel to the same line, and we come to the original theorem.

XXXI. *Through a given point to draw a straight line parallel to a given straight line.*

It was necessary for us not only to be taught in the *Elements* the essential properties of parallel lines, but also to investigate their construction by geometrical methods and ascertain how a straight line can be drawn parallel to another; for in many cases construction makes clearer to us the nature of the things investigated. This, then, the author of the *Elements* effects by means of the problem before us. He takes a point and a straight line and draws through the point a parallel to the straight line.[38] We must assume in advance that the point necessarily lies outside the straight line. For since he has said "through a given point," we cannot place the point on the straight line itself, because a parallel drawn through it will not be other than the straight line. So by mentioning separately the point and the straight line, he has shown that the point must be taken outside the straight line. This is just what he made clear in the case of the perpendicular by adding a qualification: "To a given infinite straight line, from a given point which is not on it, to draw a perpendicular."[39] This, then, is one feature common to these two problems. Another is that from the same point two perpendiculars cannot be drawn to the same straight line, nor through the same point can two parallels be drawn to the same straight line. This is why the author of the *Elements*,

376

[38] 376.4   Euclid solves XXXI as follows: Given a point A and a straight line BC, to draw through A a line parallel to BC. Take a point D at random on BC, draw AD, and on AD at A construct, by XXIII, an angle DAE equal to angle ADC. Let the straight line AF

be drawn in a straight line with EA. EF, then, is parallel to BC, as shown by the equality of the alternate angles EAD and ADC, by XXVII.

[39] 376.14   XII.

using the singular, says "to draw a straight line," in that case a perpendicular and here a parallel. There the uniqueness of the line was proved, but here it is evident from what has just been proved. For if two parallels to a straight line can be drawn through the same point, there will be parallels intersecting one another at the given point, which is impossible.[40]

But we must note the difference between the premises "from a given point" and "through a given point." In the one case the point is the origin of the line drawn, and the line is therefore drawn "from" it; in the other case it lies on the drawn line itself, and the line is therefore drawn "through" it. It is not as cutting the given point that the straight line is said to go through it, but as falling upon it and defining its own distance from the given straight line by the interval between the point and the straight line. For whatever distance separates the given point from the given straight line is the interval between the parallel and that straight line.

XXXII. *In any triangle, if one of the sides is produced, the exterior angle of the triangle is equal to the two interior and opposite angles, and the interior angles of the triangle are equal to two right angles.*

What was lacking in the sixteenth and seventeenth theorems our author adds in this. For we learn from it not only that the exterior angle of a triangle is greater than either of the opposite interior ones, but also how much greater; for, being equal to both angles, it is greater than either by the amount of the other. And not only do we know, as these theorems showed us, that any two of the angles of a triangle are less than two right angles, but also how much less, namely, by the amount of the third angle. Those earlier theorems were in a way less determinate, and this one gives scientific definiteness to them both. But we should not for this reason say that they are superfluous. They have helped us towards a number of proofs which we shall use in the present case. Furthermore, as our knowledge passes from imperfection to perfection, it

[40] 376.25 By XXX.

377

378

necessarily moves from imprecise results to determinate and irrefutable doctrines.

The author of the *Elements* proves each part of the conclusion by drawing a parallel outside the triangle.[41] But it is also possible to prove the same things by drawing a parallel

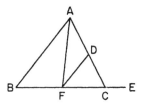

inside, changing only the order of the things demonstrated. He first proves that the exterior angle is equal to the opposite interior ones, and from this he establishes the rest; but let us proceed in the reverse order. Let ABC be a triangle, and let its side BC be produced to E. Let a point F be taken on BC and AF be joined, and through F let a line FD be drawn parallel to AB. Then since FD is parallel to AB and both AF and BC intersect them, the alternate angles are equal, and the exterior angle is equal to the interior one.[42] The whole angle AFC is therefore equal to angles FAB and ABF. Similarly we can prove, by drawing a parallel, that angle AFB is equal to angles FAC and ACF. Thus the two angles AFB and AFC are equal to the three angles of triangle ABC. The three angles are consequently equal to the two right

[41] 378.4 Euclid's proof of XXXII: Given triangle ABC with side BC produced to D, Euclid draws CE parallel to AB. Then angles

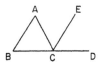

BAC and ACE are equal, being alternate angles between parallels AB and CE; and ECD is equal to the interior and opposite angle ABC; therefore the whole angle ACD is equal to the two interior and opposite angles BAC and ABC. Let ACB be added to each; then BAC, ABC, and ACB are equal to ACD and ACB, which are equal to two right angles.

[42] 378.17 I.e. FAB and AFD are equal as alternate angles within parallels, and DFC is equal to the corresponding interior angle ABF.

angles, namely, AFB and AFC. But angles ACF and ACE are also equal to two right angles. Let the common angle ACF be subtracted; then the remainder, the exterior angle, is equal to the interior and opposite angles. This, then, is the way in which the theorem is demonstrated.

379

Eudemus the Peripatetic attributes to the Pythagoreans the discovery of this theorem, that every triangle has internal angles equal to two right angles, and says they demonstrated it as follows. Let ABC be a triangle, and through A draw a line DE parallel to BC. Then since BC and DE are parallel,

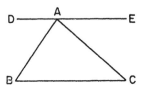

the alternate angles are equal, and angle DAB is therefore equal to ABC and EAC to ACB. Add the common angle BAC. Then angles DAB, BAC, CAE—that is, angles DAB and BAE, which are two right angles—are equal to the three angles of the triangle ABC. Therefore the three angles of a triangle are equal to two right angles. Such is the proof of the Pythagoreans.

But we must also examine the converses of our author's theorem. The theorem is one, but its converses are two, since the theorem is compound, with regard both to the conclusion and to what is given. The hypothesis is twofold: a triangle, and one of its sides produced. The conclusion is also a double one: one part is that the exterior angle is equal to the interior and opposite angles, the other that the three interior angles are equal to two right angles. If, then, we posit that the exterior angle is equal to the opposite interior angles, we prove that the produced side is in a straight line with one of the sides of the triangle; and if we posit that the three interior angles are equal to two right angles, we prove that the figure is a triangle. Thus the conclusion as a whole is the converse of

380

the whole of the given.[43] Then let ABC be a triangle, with its exterior angle ACD equal to the interior and opposite angles.

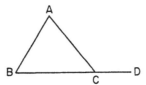

I say that BC has been produced to D and BCD is one straight line. For since angle ACD is equal to the interior and opposite angles, let angle ACB be added to both. The angles ACD and ACB are then equal to the three angles of the triangle ABC. But the three angles of the triangle ABC are equal to two right angles; [and therefore angles ACD and ACB are equal to two right angles].[44] And if upon a straight line and at a point on it two successive straight lines not on the same side make adjacent angles equal to two right angles, the straight lines are on a straight line with one another.[45] BC therefore is on a straight line with CD. Again let ABC be a [rectilinear][46] figure having only three angles, A, B, C, equal to two right angles. I say that it is a triangle and AC is one straight line. For let BD be joined. Then since the angles in

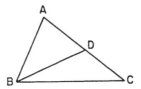

each of the triangles ABD and DBC are equal to two right angles, and since of these the angles in triangle ABC are equal to two right angles, the remaining angles ADB and CDB are equal to two right angles; and they are on the line

381

[43] 380.6   This is the first of the three kinds of conversion distinguished at 409.1-6. Cf. 252-253.
[44] 380.16   From Barocius.
[45] 380.20   By XIV.
[46] 380.21   From Barocius.

BD, so that DC and DA are on a straight line with each other. [Similarly we can prove that AB and BC are straight lines.][47] Therefore if the figure be a rectilinear figure which has its interior angles equal to two right angles, it is necessarily a triangle.

Not that a figure is necessarily a triangle if it has its interior angles equal to two right angles. For you can find a figure bounded by circular sides with its interior angles equal to two right angles. Let ABCD be a square; on one side AB let a semicircle AEB be described inside it, and on the other

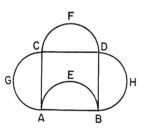

sides, and outside, the semicircles G, F, H. The figure bounded by the semicircles will have two angles, GAE and EBH, equal to angles CAB and DBA (this was proved in the Postulates),[48] and these are the only angles in this figure. It is possible, then, for a figure not a triangle to have its interior angles equal to two right angles.

So much about the converses. We can now say that in every triangle the three angles are equal to two right angles. But we must find a method of discovering for all the other rectilinear polygonal figures—for four-angled, five-angled, and all the succeeding many-sided figures—how many right angles their angles are equal to. First of all, we should know that every rectilinear figure may be divided into triangles, for the triangle is the source from which all things are constructed, as Plato teaches us when he says, "Every rectilinear plane face is composed of triangles."[49] Each rectilinear figure is

382

---

[47] 381.4   From Barocius.   [48] 381.17   At 189.17ff.

[49] 382.5   *Tim.* 53c: ἡ ὀρθὴ τῆς ἐπιπέδου βάσεως ἐκ τριγώνων συνέστηκεν. For an explanation of this cryptic text see A. E. Taylor, *A Commentary on Plato's Timaeus*, Oxford, 1928, 362.

divisible into triangles two less in number than the number of its sides: if it is a four-sided figure, it is divisible into two triangles; if five-sided, into three; and if six-sided, into four. For two triangles put together make at once a four-sided figure, and this difference between the number of the constituent triangles and the sides of the first figure composed of triangles is characteristic of all succeeding figures. Every many-sided figure, therefore, will have two more sides than the triangles into which it can be resolved. Now every triangle has been proved to have its angles equal to two right angles. Therefore the number which is double the number of the constituent triangles[50] will give the number of right angles to which the angles of a many-sided figure are equal. Hence every four-sided figure has angles equal to four right angles, for it is composed of two triangles; and every five-sided figure, six right angles; and similarly for the rest.

This, then, is one inference that we can draw from this theorem with regard to all figures that are polygonal and rectilinear. Let us briefly state another that follows from it: When all the sides of a rectilinear figure are produced at one time, the exterior angles constructed are equal to four right angles.[51] For the angles in both directions must be equal to right angles double the number of the sides, since on each of the extended sides angles are constructed equal to two right angles; and if we subtract the right angles to which the interior angles are equal, the remaining angles, the exterior ones, are equal to four right angles. For example, if the figure is a triangle and all its sides are produced at once, the interior and exterior angles produced are equal to six right angles, and of these the interior angles are equal to two, so that the remaining angles, the exterior ones, are equal to four. If it is a four-sided figure, the sum of them all will be eight right angles, double the number of sides; and of these the interior angles are equal to four, and therefore the exterior ones are equal to the other four. If it is five-sided, all the angles will

---

[50] 382.15   Following Barocius and ignoring τῶν γωνιῶν in Friedlein.
[51] 383.3   This diagram taken from Heath, *Euclid* I, 322, may be helpful.

equal ten right angles, the interior ones being equal to six and the exterior to the other four. And so on indefinitely in the same way.

Besides these, let us list the following consequences of this theorem: that every equilateral triangle has each of its angles equal to two-thirds of a right angle; that an isosceles triangle whose vertical angle is a right angle has each of the other angles half a right angle, as in the half-square; and that the scalene half-triangle produced by dropping a perpendicular from any angle of an equilateral triangle to the side which subtends it has one of its angles right, another two-thirds of a right angle (the angle already in the equilateral triangle), and the third angle therefore one-third of a right angle, for the three must together be equal to two right angles. I do not mention these matters without a purpose, but because they prepare us for the teaching of the *Timaeus*.[52]

Finally, we should say that the property of having its interior angles equal to two right angles is an essential property of the triangle as such. This is why Aristotle, in his treatise on apodictic reasoning, when discussing intrinsic attributes uses this as a ready example.[53] Just as a primary and intrinsic property of every figure is to be bounded, so also is it an intrinsic property of every rectilinear triangle, though not of every figure, to have its interior angles equal to two right angles. The truth of this theorem seems to coincide with our common notions. For if we think of a straight line with perpendiculars standing at its extremities and then think of these perpendiculars as coming together to produce a triangle, we see that in proportion to their convergence they reduce the size of the right angles which they made with the straight line, so that the amount which they took away from the original right angles they gain at the vertical angle as they converge and so of necessity make the three angles equal to two right angles.

384

[52] 384.4   Cf. *Tim.* 53d-54b.
[53] 384.9   *Post. Anal.* 73b31ff.

385   XXXIII. *The straight lines joining equal and parallel straight lines in the same directions*[54] *are themselves also equal and parallel.*

This theorem we called[55] a kind of boundary between the study of parallels and the study of parallelograms. It appears to state a property of equal and parallel lines but it also gives us, without openly doing so, the method of constructing a parallelogram. For a parallelogram is formed by the equal and parallel lines assumed and the lines which connect them and which are demonstrated likewise to be equal and parallel.[56] Hence the next theorem proceeds at once, as if the parallelogram had been constructed, to investigate the essential properties of such areas.

This is evident; but we must observe also the precision of the enunciation. First, it was not sufficient to say that the lines connected are equal, for lines that connect equal lines are not always equal unless the assumed two lines are parallel as well. In an isosceles triangle, if a point is taken on one of its equal sides and a line drawn through it parallel to the base, this parallel to the base and the base itself connect equal sides but are not themselves equal, for the lines that meet at the vertex of the triangle were not parallel. Secondly, neither does our author suppose that, when the posited straight lines are

386   parallel though unequal, he can make the lines connecting

[54] 385.2 ἐπὶ τὰ αὐτὰ μέρη. This phrase is not immediately clear to an English reader. But Proclus' commentary (386.12ff.) shows that for a Greek geometer it meant "at the corresponding extremities."

[55] 385.5   At 355.9ff.

[56] 385.13   Euclid demonstrates XXXIII as follows: Given AB and CD as equal and parallel lines connected at their extremities in the same directions by straight lines AC and DB; let BC be joined. Then since AB and CD are parallel and BC has fallen upon them, the alternate angles ABC and BCD are equal, by XXIX. And since the two triangles have sides AB and BC equal respectively to sides DC and BC and angle ABC equal to angle BCD, the base AC is equal to the base BD, the triangle ABC is equal to the triangle DCB, and angle ACB equal to angle CBD, by IV. And since the line BC falling upon the two lines AC and BD has made the alternate angles equal to one another, AC is parallel to BD, by XXVII. And it was also proved equal to it.

them parallel. This also is evident in the construction just described for the isosceles triangle. The line drawn and the base are parallel, but the lines connecting them are not parallel, for they are parts of the sides of the isosceles triangle. Obviously it is necessary for the equality of the connecting lines that the position of the lines they connect be parallel, and for the position of the parallel lines we need the equality of the connecting lines. For this reason the author of the *Elements* includes both properties in his statement of the lines that are connected—that is, that they are equal to one another and are parallel—in order that he may demonstrate both of them also with respect to the connecting lines. Thirdly, we should add that, when the straight lines are given as equal and parallel, the lines connecting them are not always equal and parallel. For if we do not make these lines con-

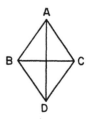

nect extremities "in the same directions," it is impossible that the connecting lines be parallel; instead they will intersect one another, and they can sometimes be equal and sometimes not. If you take a square or an oblong area, such as ABCD, and join AD and BC, the diagonals are indeed equal but not parallel; and yet they connect equal and parallel lines, namely, the opposite sides of the area described. And if you take a rhombus or a rhomboid figure, the diagonals, besides not being parallel, are not even equal. For since AB is equal to CD, and AC is a common side, and the angle BAC is unequal to angle ACD, so also the bases will be unequal. Quite properly, then, the author of the *Elements* requires that the lines connecting equal and parallel lines make the connections "in the same directions," in order that, supposing the equal and parallel lines to be AC and BD, we do not take AD and BC

387

as the connecting lines, but rather AB and CD. For these we can demonstrate to be equal and parallel, but those we could never prove to be parallel, and although we could show them to be equal in the case of squares and oblong figures, we could never demonstrate this for rhombi and rhomboids. Rather the opposite can be demonstrated, that is, that they are unequal because of the inequality of the interior angles on the same side.

XXXIV. *In parallelogrammic areas the opposite sides and angles are equal to one another, and the diameter bisects them.*

388

Taking the parallelogram as already constructed by the preceding theorem, our author now examines the properties that belong to it primarily, that is, the characteristics of its special structure. They are these: the opposite sides are equal, the opposite angles are equal, and the areas are cut in half by the diameter. For it is to areas that the words "and the diameter bisects them" refer, meaning that the area is the whole that is bisected, not the angles through which the diameter passes. These three traits, then, belong to parallelograms as such: the equality of their opposite angles, the equality of their opposite sides, and the bisection of their areas by their diameters. You see that he has obtained these specific properties of the parallelogram from all its parts—from its sides, from its angles, and from its area.[57]

[57] 388.13   Euclid's proof of XXXIV takes as given the parallelogrammic area ABCD, with diameter BC. Since AB is parallel to CD, the alternate angles ABC and BCD are equal, by XXIX. Similarly since AC is parallel to BD, the alternate angles ACB and CBD are equal. Therefore in triangles ABC and DBC the two angles ABC and BCA are equal respectively to angles DCB and CBD, and the side BC adjoining the two angles in each is common; therefore, by XXVI, side AB is equal to side CD, and AC to BD, and angle BAC to angle CDB. Thus the whole angle ABD is equal to the whole angle ACD. And the angle BAC was also proved equal to angle CDB; so the opposite sides and angles are equal. And since in triangles ABC and DCB the two sides AB and BC are equal to the two sides CD and BC respectively and angle ABC is equal to angle BCD, triangle ABC is equal to triangle DCB. Therefore the diameter bisects the parallelogram ABCD.

There are four kinds of parallelograms, which we distinguished in the Hypotheses,[58] namely, the square, the oblong, the rhombus, and the rhomboid. It is worth noting further that, if we divide these four species into rectangular and nonrectangular, we shall find not only that the diameters of rectangular parallelograms bisect the areas, but also that their diameters themselves are equal when the angles are right angles and unequal when they are not, as was said in the preceding theorem; and if we distinguish equilateral from nonequilateral, again we shall find that in the equilateral parallelograms not only are the areas bisected by the diameters, but also the angles through which they are drawn; for in the square and the rhombus the diameters bisect the angles as well as the areas, whereas in the oblong and the rhomboid

389   they bisect the areas only. Let ABCD be a square or a rhombus and AD its diameter. Then since AB and BD are equal to AC and CD (for the sides are equal), and angles ABD and ACD are equal (for they are opposite), and they

have a common base, therefore all corresponding parts are equal, so that angles BAC and CDB are bisected. Now let the figure be oblong or rhomboid. Then if angle CAB is bisected, while angle CAD is equal to angle ADB,[59] the result will be that angle BAD[60] is equal to angle ADB and AB equal to BD. But they are unequal. [Therefore angle CAB is not bisected by the diameter. Likewise angle CDB, which is equal to it, is not bisected.][61] Let us sum it all up as follows. In the square the diameters are equal because of the rightness of the

58 388.14   At 169.10ff.
59 389.11   As an alternate angle between parallels.
60 389.12   As equal to CAD on the assumption that CAB is bisected.
61 389.13   These two sentences in brackets come from Barocius.

angles, the angles are bisected by the diameters because of the equality of the sides, and the area is divided into equal parts by the diagonal because of the property common to all parallelograms. In the oblong the diameters are equal [by virtue of its being a rectangle], the angles are not bisected by the diameters [because it is not equilateral], but the division of the area into equal parts is present, and this by virtue of its being a parallelogram. In the rhombus the diameters are
390   unequal [since it is not rectangular],[62] but they bisect not only the areas, because it is a parallelogram, but also the angles, because it is equilateral. And in the remaining figure, the rhomboid, the diameters are unequal because the figure is not rectangular, they divide the angles unequally because it is not equilateral, and only the areas on both sides of the diagonal are equal because it is a parallelogram.

We have said this in order to bring out the differences that exist between the four species of parallelograms. The following technical point also comes to light in this theorem and should not be passed over, namely, that some theorems are universal and others not. What we mean by each of these statements will come to mind if we divide[63] the conclusion of this theorem into a part that is universal and a part that is not. Yet it would seem that every theorem is universal, that is, that every attribute demonstrated by the author of the *Elements* is a universal one. For example, this theorem seems to say not only that to have opposite sides and angles equal is a universal character of all parallelograms, but also that each of them is bisected by the diameter. Nevertheless we affirm that the former properties have been demonstrated universally, the last not. In one sense the term "universal" is used to denote a statement true of all instances of its subject, in another sense
391   to mean a statement about everything to which the same attribute belongs.[64] Every isosceles triangle has its three angles equal to two right angles" is universal, because it is

[62] 390.1   All material in brackets comes from Barocius.
[63] 390.14   μετέρχεσθαι appears to be a misplaced dittograph of μεμερίσθαι. I follow Barocius in ignoring it.
[64] 391.1   For this distinction see Arist. *Post. Anal.* 73a25-74b4.

true of all isosceles triangles. But universal also is "Every triangle has its three angles equal to two right angles," because it embraces everything to which as such this attribute belongs. Hence we say we have proved that it belongs primarily to the triangle to have its angles equal to two right angles. It is in this sense of the term that we say some theorems are universal and others not universal and say of this theorem that one part of its conclusion has universality and the other not. That parallelograms have their opposite sides and angles equal is universal, for this character belongs only to parallelograms; but that the diameter of a parallelogram bisects the area is not a universal statement, because it does not embrace all the things in which this character is observed, for it belongs also to circles and to ellipses.[65] Our earliest conceptions of things appear to be of this partial sort, and only as inquiry proceeds do we take in the whole. Thus the ancients, having perceived that the diameter bisects the ellipse, that it bisects the circle, and that it bisects the parallelogram, proceeded to investigate what was common in these cases.

A man may mistakenly suppose, Aristotle says,[66] that he is proving something universally when he is not, because the common subject to which the character primarily belongs has no name. For instance, it is not possible to say what the common element is in numbers, magnitudes, motions, and sounds, to all of which the rule of alternate proportion applies. And it is difficult to set forth what is common to the ellipse, the circle, and the parallelogram; for one is rectilinear, another is circular, and the other is bounded by a mixed line. This is why, when a man proves that the diameter bisects the parallelogram, we think he is proving universally, because we have not grasped the common subject of which this statement is true. Such a statement, then, about parallelograms is not universal, for the reason stated; but the statement that every parallelogram has its opposite sides and angles equal is universal. For if we posit a figure that has its

392

---

[65] 391.17 *Sc.* to be bisected by their diameters.
[66] 391.23 *Post. Anal.* 74a5-74b5.

opposite sides and angles equal, it can be proved that it is a parallelogram. Let ABCD be such a figure with a diameter AD. Then since AB and BD are equal to AC and CD, and the angles contained by them are equal, and they have a

common base, all corresponding parts are equal. If, then, angle BAD is equal to angle ADC and angle ADB to CAD, it follows that AB is parallel to CD and AC to BD, so that ABCD is a parallelogram.

So much for this. It seems also that this very term "parallelogram" was coined by the author of the *Elements* and that it was suggested by the preceding theorem. For when he had shown that the straight lines connecting equal and parallel lines in the same directions are themselves equal and parallel, he had clearly shown that both pairs of opposite sides, the connecting and the connected lines, are parallel; and he rightly called the figure enclosed by parallel lines a "parallelogram," just as he had designated as "rectilinear" the figure enclosed by straight lines.

It is clear also that the author of the *Elements* puts the parallelogram among four-sided figures. But it is worth reflecting whether every rectilinear figure with an even number of sides, when it is both equilateral and equiangular, ought not to be called a parallelogram. For such a figure has its opposite sides equal and parallel and its opposite angles equal—for example, the hexagon, the octagon, the decagon. For if you think of a hexagon ABCDEF and join A and C, you can prove line AF to be parallel to CD. For since the angle at B (and every other angle of the hexagon, if it is equiangular) is one and one-third of a right angle, and since AB is equal to BC (for the figure is posited as equilateral),

393

then angles BAC and BCA will each be one-third of a right angle, and consequently angles FAC and ACD are right angles, so that AF is parallel to CD. Similarly we can prove

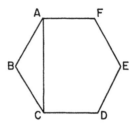

that the other opposite sides are parallel; and likewise for the octagon and the others. If, then, a parallelogram is a figure included within opposite sides that are parallel, there will be parallelograms that are not four-sided. But evidently for the author of the *Elements* every parallelogram is four-sided. This is clear above all in the theorem[67] in which he asserts that a parallelogram on the same base as a triangle and in the same parallels is double the triangle, for this is true only of four-sided figures.

394

XXXV. *Parallelograms which are on the same base and in the same parallels are equal to one another.*

As we said that some theorems are universal and others particular, and added the meaning of this distinction, and that some theorems are simple and others compound, and explained what each of these types is, so now, following another distinction, we say that some theorems are locus-theorems and others not. I call "locus-theorems" those in which the same property occurs throughout the whole of a certain locus, and I call "locus" a position of a line or a surface producing one and the same property. Some locus-theorems refer to lines, others to surfaces; and since some lines are plane and others solid—plane lines being those which, like the straight line, lie in a plane and whose genera-

[67] 394.3   XLI below.

395

tion[68] is simple, and solid lines those which are produced by some sectioning of a solid figure, like the cylindrical helix[69] and the conic lines—I should say further that of locus-theorems referring to lines some have a plane and others a solid locus. The theorem before us is therefore a locus-theorem, it is one of the locus-theorems referring to lines, and it has a plane locus. For the whole space between the parallel lines is the locus of the parallelograms constructed on the same base which the author of the *Elements* shows to be equal to one another. An example of the so-called solid locus-theorems is the following: "The parallelograms inscribed in the asymptotes and the hyperbola are equal";[70] for the hyperbola is clearly a solid line, since it is a section of the cone.

Chrysippus,[71] so Geminus tells us, likened theorems of this sort to the Ideas. For just as the Ideas embrace the generation of an indefinite number of particulars within determinate limits, so also in these theorems an indefinite number of cases are comprehended within determinate loci. Their equality is shown to result from this limitation; for the height of the parallels,[72] which remains the same while an indefinite number of parallelograms can be thought of on the same base, shows all these parallelograms to be equal to one another.

The present theorem is the first locus-theorem that the author of the *Elements* has presented. In his evident intent to give us the utmost variety of theorems compatible with an elementary work, he rightly did not omit this particular spe-

[68] 394.22 Adopting ver Eecke's suggestion that νόησις in Friedlein is a copyist's error for γένεσις. Cf. γένεσις in the parallel statement in line 23.

[69] 394.25 "The inclusion by Proclus of the cylindrical helix among solid loci, on the ground that it arises from a section of a solid figure, would seem to be . . . due to some misapprehension" (Heath, *Euclid* I, 330).

[70] 395.11 This is, as ver Eecke notes, Prop. XII of the second book of Apollonius' *Conics*.

[71] 395.14 Chrysippus of Tarsus, of the third century B.C., the successor to Cleanthes as head of the Stoa. He does not appear to have made any contributions to mathematics, but this comment on the Platonic Ideas is well worth Proclus' mention.

[72] 395.19 I.e. the distance between them.

396

cies. But here, since this book is about rectilinear figures, he gives us plane locus-theorems that refer to straight lines, whereas in the third book, when he is concerned with circles and their properties, he will teach us the circular lines involved in plane locus-theorems. Such is the theorem in that book, "The angles inscribed in the same segment of a circle are all equal to one another," and the theorem, "The angles inscribed in a semicircle are right angles."[73] For of the indefinitely numerous angles that may be constructed within a segment of a circle on the same base all are proved to be equal, [and of the angles contained by the base and the circumference of a semi-circle all are proved to be right angles].[74] These figures are analogous to triangles and parallelograms constructed on the same base [and between the same parallels].[75] This, then, is the species of theorems that are next to be investigated, named "locus-theorems" by the ancient mathematicians.

It may seem a great puzzle to those inexperienced in this science that the parallelograms constructed on the same base [and between the same parallels][76] should be equal to one another. For when the sides[77] of the areas constructed on the same base can be increased indefinitely—and we can increase the length of these sides of the parallelograms as far as we can extend the parallel lines—we may well ask how the areas can remain equal when this happens. For if the breadth is the same (since the base is identical) while the side becomes greater, how could the area fail to become greater? This theorem, then, and the following one about triangles belong among what are called the "paradoxical" theorems in mathe-

397

matics. The mathematicians have worked out what they call the "locus of paradoxes," as the Stoics have done in their dogmas,"[78] and this theorem is included among them. Most

[73] 396.5   The theorems referred to here are III. 21 and 31.
[74] 396.7   From Barocius.
[75] 396.9   From Barocius.
[76] 396.14   From Barocius.
[77] 396.15   μῆκος denotes the length of the sides other than the base, πλάτος the length of the base. See 397.11.
[78] 397.2   δείγματα. The reference here is, so von Arnim (III, 547ff.) thinks, to the Stoic paradoxes regarding the wise man—that

people at least are immediately startled to learn that multiplying the length of the side does not destroy the equality of the areas when the base remains the same. The truth is, nevertheless,[79] that the equality or inequality of the angles is the factor of greatest weight in determining the increase or decrease of the area. For the more unequal we make the angles, the more we decrease the area, if the side and base remain the same; hence if we are to preserve equality, we must increase the side. Take any parallelogram,[80] say ABCD, and let AC be produced indefinitely. Suppose it to be a rectangular figure, and on the base BD let another parallelogram BEFD be constructed. Clearly the side has been increased, for

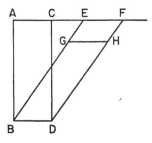

BE is longer than AB, since the angle at A is a right angle. And this increase was necessary, for the angles of the parallelogram BEFD have become unequal, some acute, the others obtuse; and this has happened because side BE is, as it were, folded back on BD and contracts the area. Let a line BG be taken equal to AB and GH be drawn through G parallel to BD. Then the side of parallelogram BDGH is equal to the side of ABCD, and the breadth is the same, but its area is less, namely, less than that of BEFD. The inequality of the angles has clearly made the area less, and the increase in the side, by adding as much as the inequality of the angles has

398

he alone is free, beautiful, rich, happy, etc. But I find no hint of δείγματα in these numerous fragments, whereas these paradoxes are often called δόγματα. Our text should probably be amended, despite the absence of any evidence of a variant reading.

[79] 397.6   Reading with Barocius ὅμως instead of ὁμοίως in Friedlein.
[80] 397.13   For τὰ in Friedlein read τὸ.

taken away, preserves the equality of the areas; and the limit
of increase for the side is the locus of the parallels. The
square is demonstrably greater than the oblong, when both
are rectangular [and have equal perimeters];[81] and when both
are equilateral [and have equal perimeters], the rectangular
figure is demonstrably greater than the nonrectangular. For
the rightness of the angles and the equality of the sides are the
all-important factors affecting the increase of the areas; and
this is why the square is manifestly greater than all others with
an equal length of boundaries, and the rhomboid is the least
of all.

But these matters we shall prove elsewhere, for they are
more appropriate to the hypotheses of the second book. With
regard to the theorem before us we must realize that, when it
says the parallelograms are equal, it means that the areas, not
the sides, are equal, for the statement is about the included
spaces, the areas; and also that in the demonstration of this
theorem our author for the first time mentions trapezia. This
shows that he was right in the Hypotheses[82] when, in explaining what the trapezium is, he said that it is a species of four-
sided figure, but not a parallelogram. For a figure that does
not have both its opposite sides and its opposite angles equal
falls outside the class of parallelograms.

Now the author of the *Elements* demonstrates this theorem
by selecting the most difficult of the cases.[83] But if someone

[81] 398.10   The bracketed words here and in the next clause come
from Barocius.
[82] 398.25   I.e. in Def. XXXIV, 169.8ff.
[83] 399.5   Euclid proves XXXV as follows: Let ABCD and EBCF
be parallelograms on the same base BC and within the same parallels
AF and BC. As opposite sides of parallelograms, AD is equal to BC

and EF to BC, so that AD is equal to EF; and since DE is common,
the whole AE is equal to the whole DF. But AB is also equal to DC.
Therefore triangles EAB and FDC have two sides AE and AB equal

399

should say, "Let ABCD and BCDE be parallelograms on the same base DB, so constructed that DC is the diameter of parallelogram AB," then we can prove at once that they are equal. For triangle BCD is half of each of the two parallelograms, since DC is the diameter of AB and BC the diameter of DE. The diameters of parallelograms bisect them, and hence AB is equal to DE. Again if we suppose that DC cuts the side of parallelogram AB and that the parallelograms

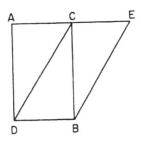 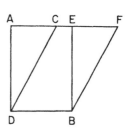

have the position of ABDE and BCDF in the diagram, we can demonstrate that these also are equal. For since AE equals CF (for each is equal to DB, the opposite side), let us subtract the common portion CE. AC is then equal to EF. But AD is equal to EB, and angle CAD is equal to angle FEB, for AD is parallel to EB. Therefore the base CD is equal to the base FB, and the whole triangle ADC is equal to triangle EBF. Let the trapezium CB be added to them both. AB as a whole is then equal to DF. You see that these three are the only cases.[84] For CD either cuts EB, as in the case assumed by the author of the *Elements*, or falls upon E, as in the preceding diagram,[85] or cuts AE, as we assumed just now;

400

---

respectively to two sides DF and DC and angle FDC equal to angle EAB, by XXIX; therefore the base EB is equal to the base FC, and the two triangles are equal. Let DGE be subtracted from each; then the trapezium ABGD which remains is equal to the trapezium EGCF which remains. Let triangle GBC be added to each; then the whole parallelogram ABCD is equal to the whole parallelogram EBCF.

[84] 400.7   Reading πτώσεις instead of πως in Friedlein.

[85] 400.9   This is somewhat confusing, since the point designated E in the last diagram corresponds to what was designated as C in the preceding one; but the meaning is clear.

and the theorem has been proved true for all cases. But we must note that of the two kinds of trapezia, those that have no side parallel to another and those that have one side parallel to another, it is only the second species of trapezia that is used by our geometer and appears in the diagram here: for CE is parallel to DB.

XXXVI. *Parallelograms which are on equal bases and in the same parallels are equal to one another.*

The preceding theorem took the bases as identical, whereas this one takes them as equal though distinct from one another; but common to both theorems is the assumption that the parallelograms are in the same parallels. They must, then, lie neither inside nor outside the given parallel straight lines. Parallelograms are said to lie in the same parallels when their bases and the sides lying opposite them coincide with the same parallels. The author of the *Elements* proves the theorem by assuming that the bases are completely separate from one another.[86] But there is nothing to prevent our

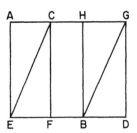

[86] 401.6 Euclid proves XXXVI as follows: Let ABCD and EFGH be parallelograms on equal bases BC and FG and within the same parallels AH and BG. Join BE and CH. Then since BC is equal to FG and FG is equal to EH, BC is also equal to EH. But they are also parallel; hence, by XXXIII, the lines that join them, BE and CH, are

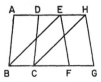

equal and parallel, and EBCH is a parallelogram, by XXXIV. And it is equal to ABCD, by XXXV, and also equal to EFGH, by the same proposition. So ABCD is also equal to EFGH.

401

assuming that they so lie as to have a common segment. Thus let AB and CD be parallelograms on equal bases EB and FD [having a common segment and in the same parallels].[87] I say that they are equal. Let EC and BG be drawn. Then since EF is equal to BD (for EB is equal to FD), CF equal to DG, and angle EFC equal to angle BDG (for CF is parallel to DG), EC is also equal to BG and also parallel to it; and CB is a parallelogram. And it has the same base as each of the parallelograms AB and CD and is in the same paral-

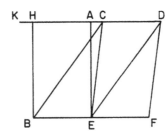

lels. AB is therefore equal to CD. And if we suppose the bases of the parallelograms to have no common segment, nor to be separate from one another but—the only remaining alternative—touching one another at a point, as do AB and ED, we shall say that BE is equal to EF and to CD, so that CB is also equal to DE [and parallel to it].[88] For the lines joining equal and parallel lines are themselves equal and parallel. BD, then, is a parallelogram on the same base and in the same parallels as AB and DE. Therefore parallelograms AB and DE are equal.

402

Thus on our first approach we have distinguished alternative constructions for the theorem, saying that the bases either have a common segment, or only touch one another, or are separate from one another. But it is also possible, if they touch, like BE and EF, to assume that the whole of DE lies outside AE, or that the side CE coincides with AE, or that CE cuts AH, or that CE falls as a diameter on HE (and then DF will also be the same as AF), or, with

[87] 401.10 From Barocius.
[88] 401.23 From Barocius.

AH produced to K, that CE cuts it beyond H, with DF either cutting AH or coinciding . . . [89]

403

[XXXVII. *Triangles which are on the same base and in the same parallels are equal to one another.*][90]

. . . they show. For it has been proved that areas can be unequal when they[91] are equal and equal when they are unequal. Such a misconception is held by geographers who infer the size of a city from the length of its walls. And the participants in a division of land have sometimes misled their partners in the distribution by misusing the longer boundary line; having acquired a lot with a longer periphery, they later exchanged it for lands with a shorter boundary and so, while getting more than their fellow colonists, have gained a reputation for superior honesty. Let us suppose two isosceles triangles, one of them having each of its equal sides five

---

[89] 402.19   There is a lacuna in our text extending from this point until after the beginning of Proclus' comment on XXXVII. None of the MSS notes this gap, ἐφαρμόζουσαν at this point being followed without a break by ἀποφαίνονται in 403.4. This suggests the loss of several pages from the archetype from which our MSS are derived. Barocius noted this gap and supplied the missing part of the commentary on the present proposition with figures and explanations of his own. Since the omitted matter, so far as we know, was only a further elaboration of the various cases of this theorem, there is little reason for translating Barocius' conjectural supplement. But for convenience of reference the enunciation of the next proposition in Euclid's text is included in my translation, and Euclid's proof appended in a footnote.

[90] 403.3   Euclid's proof of XXXVII is as follows: Let ABC and DBC be triangles on the same base BC and within the same parallels AD and BC. Let AD be produced in both directions to E and F. Through B let BE be drawn parallel to CA, and through C let CF be drawn parallel to BD. Then each of the figures EBCA and DBCF is a parallelogram, and they are equal, by XXXV. Moreover, the triangle

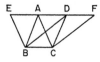

ABC is half of the parallelogram EBCA, by XXXIV, for the diameter AB bisects it. Likewise triangle DBC is half of the parallelogram DBCF. Therefore triangle ABC is equal to triangle DBC.

[91] 403.4   The context shows that ἐκείνων refers to the perimeters.

units—cubits or fingerbreadths—in length and a base measuring six of the same units, and the other with each of its equal sides of five units and a base of eight of the same units. In choosing between them the inexperienced person is likely to be completely deceived. The one has a perimeter of eighteen, the other of sixteen of the same units. But the geometer will realize that their areas are equal, even though their perimeters are unequal; for the area of each is twelve. That is, if you drop a perpendicular from the vertex, you will bisect the bases and in one have a half-base of three, in the other of four, and conversely the perpendicular itself in the first case will be four, in the other three; for the square on the side with a length of five units must be equal to the square on the perpendicular and the square on half the base. Now when the half-base is three, the perpendicular is four, and when the half-base is four, the perpendicular is three. Then if you multiply half the base by the perpendicular, you get the area of the triangle; and this is the same in either case, whether you make it three times four or four times three.

This has been said to show that we cannot at all infer equality of areas from the equality of the perimeters; and so we should not be amazed to learn that triangles on the same base may have their other sides lengthened indefinitely in the same parallels, while yet the equality of their areas remains unchanged. But we can only regard triangles as in the same parallels when, with their bases on one of the parallels, they extend their vertices to the other, that is, when their vertices lie on a single straight line parallel to the bases lying on a single straight line.

XXXVIII. *Triangles which are on equal bases and in the same parallels are equal to one another.*

This also is a locus-theorem like those about parallelograms, positing the triangles as having equal bases. It appears to me that of these four theorems—of which two establish properties about parallelograms and two about triangles, and some assume that the base is the same and some assume the bases to be equal—our author has given a single proof in the

404

405

first theorem of the sixth book and that most people have failed to notice that he does this. For when in that theorem he proves that triangles and parallelograms with the same height have to one another the ratio of their bases, he does nothing other than demonstrate all these theorems more generally from the principle of proportion. For to have the same height is the same thing as to be in the same parallels, since all figures in the same parallels have the same height, and conversely. For the height is the perpendicular from one of the parallels to the other. In that proposition, then, it is demonstrated by means of proportion that triangles and parallelograms with the same height, that is, lying in the same parallels, are related to one another as their bases are. When their bases are equal, their areas are equal; when one is double the other, so is its area; and whatever other ratio the bases may have to one another, the areas will stand in that same ratio. But here, since he could not use proportion, because he has not yet taught us its principles, he is satisfied with equality only, inferring it from the equality or identity of the bases. So these four theorems are surpassed by that one, not only because it uses a single proof for all that these four theorems contain, but also because it adds something more, the identity of ratios even when the bases are unequal.[92]

406

So much for that. This theorem also has many cases. It is possible to assume, as in the case of the parallelograms, that the bases of the triangles have a common segment, or have no segment in common but meet one another at a single point, or are completely separate and have a line between them.[93]

[92] 406.9   Heath (*Euclid* I, 334) qualifies Proclus' comments here by noting that Euclid's VI. 1 does not prove the propositions of the first book, as Proclus seems to imply; they are in fact assumed in order to prove VI. 1.

[93] 406.15   I.e. the segment CE in Euclid's diagram below. The third of the three alternatives mentioned by Proclus is the case chosen by Euclid for demonstrating XXXVIII. His proof is as follows: Let ABC and DEF be triangles on equal bases BC and EF and within the same parallels BF and AD. Let AD be produced in both directions to G

It takes but little understanding to see this. It is also evident that in all the cases, however the bases or the vertices may be placed, we should follow the same procedure, that is, draw parallels to the sides and make each of the triangles a parallelogram and through them establish the equality of the triangles.

407 XXXIX. *Equal triangles which are on the same base and on the same side are also in the same parallels.*

When our purpose was to demonstrate equality, we constructed theorems four in number, two for parallelograms and two for triangles,[94] assuming them as lying either on the same base or on equal bases. But now, in converting these theorems, we have passed over the converses regarding parallelograms and considered only the two about triangles as needing attention. The reason is that, since the method of proof is exactly the same for parallelograms, using reduction to impossibility and a similar construction, it is enough for us to handle the simpler cases, that is, the triangles, exhibiting the method and leaving it for the abler minds to carry through the same reasoning in the other cases, since it is easy to see that the same method is applicable to them; that is, we assume equal parallelograms on the same or equal bases and affirm that they are in the same parallels. For if they are not, one figure will fall either inside or outside the produced parallels bounding the other. In either case we shall assume it and the parallels that bound it and prove,[95] as in the case of the triangles, that the whole is equal to a part of itself, which is impossible.

---

and H. Through B draw BG parallel to CA, and through F draw FH parallel to DE. Then GBCA and DEFH are parallelograms and are equal to each other, by XXXVI. Triangle ABC is half of GBCA, for the diameter AB bisects it; and likewise triangle FED is half of DEFH, for the diameter DF bisects it. Therefore triangle ABC is equal to triangle DEF.

[94] 407.7 Punctuating with a comma after τριγώνων, not after λαμβάνοντες, as in Friedlein.

[95] 407.23 Reading with Barocius δείξομεν instead of ἐδείξαμεν in Friedlein.

408

It is clear that the author of the *Elements* is right in adding "and on the same side." For on a single base it is possible to take equal triangles, one on one side and one on the other. But such triangles never lie in the same parallels; nor need they have the same height. This is the reason he added this phrase.

In the hypothesis that leads to absurdity there are two possible ways of drawing the parallel, [either inside or outside].[96] Our author draws it inside,[97] but we shall draw it outside and prove the same result. Let ABC and DBC be [equal][98] triangles on the same base and on the same side. I say that they are in the same parallels and that the line joining their vertices is parallel to the base. Let AD be drawn. If it is not parallel to the base, let AE, outside it, be the parallel. Let CD be produced to E, and draw EB. Triangle ABC is then

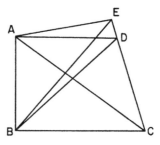

equal to triangle EBC; but triangle ABC is equal to triangle DBC. Therefore EBC is equal to DBC, the whole to the part; but this is impossible. Therefore the parallel does not fall outside AD. And it has been demonstrated by the author of the *Elements* that neither does it fall inside. Hence AD itself is parallel to BC. Therefore equal triangles on the same side

[96] 408.7 From Barocius.

[97] 408.7 Euclid proves XXXIX as follows: Let ABC and DBC be equal triangles which are on the same base BC and on the same side. Join AD. If AD is not parallel to BC, draw AE parallel to BC, and join EC. Then triangle ABC is equal to triangle EBC, by XXXVII. But by hypothesis ABC is equal to DBC; therefore DBC is also equal to EBC, the greater to the less, which is impossible. Therefore AE is not parallel to BC; and in the same way we can prove that neither is any other straight line, except AD.

[98] 408.9 From Barocius.

are within the same parallels. Thus the other half of the proof by reduction to impossibility has also demonstrated that they lie in the same parallels.

409   It is worth remarking that there are three kinds of converse theorems: the whole theorem may be the converse of a whole theorem, such as the eighteenth of the nineteenth, as we said; or the whole may be the converse of a part, as the sixth is of the fifth; or a part may be the converse of a part, like the eighth and the fourth, where it is not the whole of what is given[99] in the one that is the conclusion of the other, nor the whole of the conclusion in one the given in the other, but a part only. These theorems about triangles[100] appear to be of this character. In the theorems preceding them the conclusion was that the triangles are equal. But these not only take equality as given, but add a part of what was hypothesis in the former; for "on the same, or equal, bases" was given in those as well as in these. But in these hypotheses he adds something which does not appear in those, either as conclusion or as hypothesis: "on the same side" is an extra assumption here.

XL. *Equal triangles which are on equal bases and on the same side are also in the same parallels.*

The type of conversion is the same in this, the proof is similar, and what the author of the *Elements* leaves out in his reduction to impossibility can be proved similarly, so that there is no need to go over it again.[101] Since there are three

410   parts in the preceding enunciations—that the figures are on

[99] 409.6   Reading with Barocius δεδομένον for δεδειγμένον in Friedlein.

[100] 409.9   I.e. XXXIX and XL, which are the partial converses of XXXVII and XXXVIII.

[101] 409.24   In his proof of XL Euclid draws his alternative line inside the triangle, as in the previous proof, ignoring the case where the line falls outside. Let ABC and CDE be equal triangles on equal bases BC and CE and on the same side, and let AD be joined. If AD is not parallel to BE, let AF be drawn parallel to BE, and join FE. Then triangle ABC is equal to triangle FCE; hence DCE is also equal to FCE, the greater to the less, which is impossible. Therefore AF is not parallel to BE. Similarly we can prove that neither is any other straight line, except AD.

equal bases or on the same base, that they are in the same parallels, and that they are equal, whether triangles or parallelograms—clearly we can convert in various ways by taking two together and leaving out the third. We can suppose the bases to be the same or equal and the triangles and parallelograms to be in the same parallels and thus construct four theorems; or take the figures as equal and the bases either equal or the same and construct four others, of which the author of the *Elements* omits the two dealing with parallelograms and proves the two for triangles; or take the figures as equal and in the same parallels and prove the third condition, that they are either on the same base or on equal bases, and thus make four others. These the author of the *Elements* entirely omits, for the proof is the same for them, except that two of the four are not true by themselves, for equal parallelograms or triangles in the same parallels are not necessarily on the same base. But the combined conclusion from these hypotheses is true, namely, that they are on the same base or on equal bases, though neither alternative necessarily follows from the hypotheses adopted.

So of the ten theorems[102] in all, our geometer has included six and omitted four, and this to avoid repeating himself, the proof being the same. For example, let us prove about triangles that, if they are equal and in the same parallels, they will be on the same base or equal bases. For suppose it is not so, and if possible, let triangles ABC and DEF with these characteristics have unequal bases BC and EF, and let BC be the greater. Subtract a length BH equal to EF, and draw AH.

411

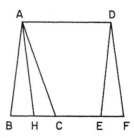

---

[102] 410.23   There were three groups of four enumerated, but the last group reduces to two, as explained above.

Then since ABH and DEF are on equal bases, BH and EF, and in the same parallels, they are equal. But ABC and DEF are posited as equal. ABC and ABH are therefore equal, which is impossible. Consequently the bases of the triangles ABC and DEF are not unequal. The same method of proof can be used for parallelograms. Then since the method of proof is the same and the resulting impossibility the same, namely, that the whole is equal to the part, they are rightly omitted by the author of the *Elements*. Thus we have said that there are necessarily ten theorems, and we have shown which are omitted and have given the reason for his silence about them. Now let us go on to the next theorems.

412   XLI. *If a parallelogram has the same base with a triangle and is in the same parallels, the parallelogram is double the triangle.*

This also is a locus-theorem. It combines the structures of triangles and parallelograms that have the same [base and][103] height; so since we have considered parallelograms separately, and again triangles, let us now take them both together and consider in what ratio they stand to one another when they have the same properties[104] as in the preceding theorems. In them it is the ratio of equality that is shown, for all triangles on the same bases and in the same parallels are equal, and likewise all parallelograms. But in these theorems what is demonstrated is the first of the unequal ratios, the double; for the parallelogram is proved to be double the triangle on the same base and of the same height.

In proving the present theorem the author of the *Elements* assumes the vertex of the triangle to lie outside the parallelogram,[105] but we shall prove the same thing by taking it on the other side of the parallelogram, that is, the side parallel

---

103 412.7   The bracketed terms add a condition tacitly presupposed, as is shown in line 13.

104 412.9   Reading ταὐτά for ταὐτόν in Friedlein.

105 412.20   Euclid proves XLI as follows: Let the parallelogram ABCD have the same base BC with triangle EBC, and let it be within the same parallels BC and AE. Join AC. Then triangle ABC is equal to triangle EBC, by XXXVII. But ABCD is double the triangle ABC, by XXXIV. Therefore ABCD is double the triangle EBC.

to their common base. Since the base is the same for both, these are the two cases of this theorem, for the triangle must have its vertex either inside or outside the parallelogram.[106] Therefore let ABCD be a parallelogram and EDC a triangle, and let E lie between A and B, and let AD be joined. Then since the parallelogram is the double of ACD and ACD is equal to triangle EDC, the parallelogram is double the triangle EDC. Thus it is clearly shown that the paral-

413

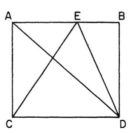

lelogram is double the triangle when they have the same base. And if the bases are equal, we can demonstrate the same thing by drawing diameters of the parallelograms. For when two triangles are equal, the double of one will be double the other, and the triangles are equal because they have equal bases and the same height. Our geometer has rightly left out these cases, for the proof is the same; they will either have an identical segment, or touch only at a point, or be separate from one another; and regardless of these variations there is one proof for all cases.

Finally, we can demonstrate the converses of this theorem in the same way. One of the converses is "If a parallelogram is double a triangle and they have the same base with each other or equal bases, and if they are on the same side, they are in the same parallels." For if this is not true, the whole is equal to the part, that is, the same argument as above will

414

[106] 412.26 "It is curious that Proclus does not mention (as he usually would) the case where the vertex of the triangle and a vertex of the parallelogram coincide. Probably the reason is that, as Proclus points out (413.20), the position of the vertex of the triangle does not matter in the proof." (I.M.)

hold. For necessarily the vertex of the triangle will fall in the parallels or outside them. Whichever it is, when a line is drawn through the vertex parallel to the base, the same impossibility follows. Another converse is "If a parallelogram is double a triangle and in the same parallels, they will have the same base or equal bases." For if they are on unequal bases, we assume figures with equal bases and show that the whole is equal to the part. Hence all the proofs of these theorems end in this common impossibility. For this reason the author of the *Elements* has left it to us to track down the variety of cases here and has given his attention to the simpler and more fundamental ones.

Now that we have made these comments, let us for the sake of practice take not a parallelogram, but a trapezium which has only two parallel sides, having the same base with a triangle and lying in the same parallels, and let us see what ratio it has to the triangle. Clearly it will not have the double ratio, for then it would be a parallelogram, since it is a quadrilateral. I say that it is either more or less than double. Of the two parallel sides one must be greater, the other less, since if they are equal the lines that join their extremities will be parallel. Now if the triangle has the greater side as base, the quadrilateral will be less than double the triangle, and if it has the shorter side as base, it will be more than double. Let ABCD be a quadrilateral, with AB less than CD, and let AB be produced indefinitely, and let triangle ECD have the same base CD as the quadrilateral, and let DF be drawn through D parallel to AC. The parallelogram ACDF is then double

415

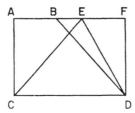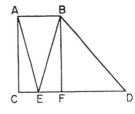

the triangle ECD, so that the quadrilateral ABCD is less than double. Again let the triangle have AB as base, and let

BF be drawn parallel to AC. Then ABCF is double the triangle, so that quadrilateral ABCD is more than its double.

With these propositions demonstrated we assert that if, in a quadrilateral which has only two of its opposite sides parallel, straight lines are drawn from the midpoint of one of the parallel lines to the other, the quadrilateral will be either greater than the double of the resulting triangle or less than its double; but if straight lines are drawn from the midpoint of one of the lines joining the parallels to the other, the quadrilateral will always be double the resulting triangle. Let us demonstrate the latter theorem. Let ABCD be the quadrilateral, with AD parallel to CB, and let DC be bisected at E. Let lines EA and BE be joined, and let BE be produced and fall upon AD at F. Now since the angles at E are equal (for

416

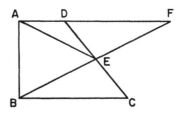

they are at the vertex) and angle FDE is equal to angle BCE,[107] FE will be equal to BE, and triangle DEF equal to triangle BCE.[108] Let triangle ADE be added to each. The whole of triangle AFE is therefore equal to the two triangles ADE and BCE. But triangle AFE is equal to triangle AEB, for they are on equal bases, BE and EF, and are in the same parallels.[109] Therefore triangle AEB is equal to triangles ADE. . . .[110]

[107] 416.9 As alternate angles between parallels AF and BC.
[108] 416.10 As having side ED equal to side EC and the two adjacent angles respectively equal.
[109] 416.13 "If the parallel to BF through A is drawn." This addition comes from the MS in which Barocius found the completion of the proof lacking in our MSS. See next note.
[110] 416.14 The remainder of Proclus' commentary on this proposition, together with the whole of his commentary on XLII and the beginning of his commentary on XLIII, are missing from all the MSS

[XLII. *To construct, in an angle equal to a given rectilinear angle, a parallelogram equal to a given triangle.*][111]

which Barocius consulted, except one which contained the following completion of the commentary on XLI:

> . . . and BCE, and the quadrilateral ABCD is double the triangle AEB, which is what was to be proved. In the same way we can prove that for the case in which connecting lines are drawn to CD from the midpoint of AB, the quadrilateral is double the resulting triangle. Therefore when from the midpoint of either of the lines that join the parallels, lines are drawn to the extremities of the other, the quadrilateral is double the resulting triangle. This has been demonstrated for the sake of practice. Now let us turn to the propositions that follow.

Barocius infers from its contents that this supplement is not the work of Proclus; Proclus' commentary, he thinks, would have been much more extensive and would have dealt with cases not mentioned here. Barocius' opinion is confirmed on more mechanical grounds when we compare this lacuna with the earlier one at the end of XXXVI. Neither lacuna is noted in the MSS; in each case the text that precedes is followed without a break by the words following the lacuna (see note at 402.19). This indicates the loss in each case of several pages from the codex from which our MSS are derived; and in the present case this loss must have included the concluding portion of the commentary on XLI. Since these identical defects occur in all our MSS, the loss of these two groups of pages must have been sustained by their archetype at an early date. Another lacuna suggesting a similar loss in the archetype of some of our MSS occurs at 82.23 (see note at that point); but in that case the missing portion of the text is fortunately supplied by other extant MSS.

[111] 416.17 The following is Euclid's proof of XLII, which is included here for convenience of reference: Given the triangle ABC and the rectilinear angle D, to construct in angle D a parallelogram equal to triangle ABC. Let BC be bisected at E and AE be joined. On EC and at point E on it let angle CEF be constructed, by XXIII, equal to angle D. Let AG be drawn parallel to EC and CG parallel to EF,

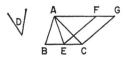

by XXXI. Then FECG is a parallelogram. And triangle ABE is equal to triangle AEC, by XXXVIII. Therefore triangle ABC is double the triangle AEC. But parallelogram FECG is also double the triangle AEC, by XLI. Therefore parallelogram FECG is equal to triangle ABC, and it has the angle CEF equal to the given angle D.

[XLIII. *In any parallelogram the complements of the parallelograms about the diameter are equal to one another.*][112]

417

. . . that the parallelograms do not touch one another at a point.[113] Because the complements are not quadrilaterals, we must also expound this case in order to see that the same consequence follows. In the parallelogram AB let parallelograms CK and DL be inscribed about the same diameter, with the straight line KL, a segment of the diameter, between

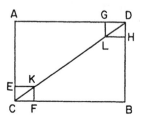

them. Now you can say the same things as before. Triangle ACD is equal to BCD, triangle ECK to KCF, and triangle DGL to DHL; therefore the remainder, the five-sided figure AGLKE, is equal to the five-sided figure BFKLH. And these are the complements. Again if the parallelograms neither meet at a point nor are separate from one another but overlap, the same proof can be used, as follows. Let AB be a parallelogram with diameter CD and parallelograms inscribed about it, one of them ECFL and the other DGKH cutting the former. I

418

say that the complements FG and EH are equal. For since

[112] 416.21 The following is Euclid's proof of XLIII: Let ABCD be a parallelogram and AC its diameter, and about AC let EH and FG be parallelograms and BK and KD the so-called complements, which are to be proved equal to each other. Triangle ABC is equal to triangle ACD, by XXXIV. Again triangle AEK is equal to triangle AHK; and for the same reason triangle KFC is also equal to KGC. Now since AEK is equal to AHK, and KFC to KGC, AEK together with KGC is equal to AHK together with KFC. And the whole triangle ABC is equal to the whole triangle ADC; therefore the complement BK which remains is equal to the complement KD which remains.

[113] 417.1 As the proof in the preceding note shows, Euclid has dealt with the case in which the inner parallelograms touch one another. Evidently Proclus has announced that he will consider the other possible cases of the problem. "So let us assume . . . ."

triangle DGK as a whole is equal to triangle DHK and a part of it, triangle KLM, is equal to triangle KLN (for LK is a parallelogram), the trapezium DLNH which remains is equal

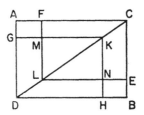

to the trapezium DLMG. But triangle ADC is equal to BDC, and triangle FCL in the parallelogram EF is equal to triangle ECL, and trapezium DLMG is equal to DLNH; therefore the remaining quadrilateral GF is equal to the quadrilateral EH. The theorem has therefore been demonstrated for all cases. There are three cases only, no more nor less; for the parallelograms about the same diameter will either cut one another, or touch one another at a point, or be separated from one another by a segment of the diameter.

The term "complements" was derived by the author of the *Elements* from the thing itself, since complements fill the whole of the area outside the two parallelograms. This is why he does not regard it as deserving of special mention in the Definitions. It would have required a complicated explanation to make us understand what a parallelogram is and what are the parallelograms that are constructed about the same diameter as the whole; for only after these had been explained would the meaning of "complement" have become clear. Those parallelograms are about the same diameter which have a segment of the entire diameter as their diameter; otherwise they are not about the same diameter. For when the diameter of the whole figure cuts a side of the interior parallelogram,

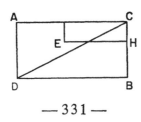

then this parallelogram is not about the same diameter as the whole. For example, in the parallelogram AB the diameter CD cuts the side EH of the parallelogram CE. Hence CE is not about the same diameter as CD.

XLIV. *To a given straight line to apply, in an angle equal to a given rectilinear angle, a parallelogram equal to a given triangle.*

Eudemus and his school tell us that these things—the application (παραβολή) of areas, their exceeding (ὑπερβολή), and their falling short (ἔλλειψις)—are ancient discoveries of the Pythagorean muse. It is from these procedures that later geometers took these terms and applied them to the so-called conic lines, calling one of them "parabola," another "hyperbola," and the third "ellipse," although those godlike men of old saw the significance of these terms in the describing of plane areas along a finite straight line. For when, given a straight line, you make the given area extend along the whole of the line, they say you "apply" the area; when you make the length of the area greater than the straight line itself, then it "exceeds"; and when less, so that there is a part of the line extending beyond the area described, then it "falls short." Euclid too in his sixth book[114] speaks in this sense of "exceeding" and "falling short"; but here he needed "application," since he wished to apply to a given straight line an area equal to a given triangle, in order that we might be able not only to construct a parallelogram equal[115] to a given triangle, but also to apply it to a given finite straight line. For example, when a triangle is given having an area of twelve feet and we posit a straight line whose length is four feet, we apply to the straight line an area equal to the triangle when we take its length as the whole four feet and find how many feet in breadth it must be in order that the parallelogram may be equal to the triangle. Then when we have found, let us say, a breadth of three feet and multiplied the length by the

420

---

[114] 420.7 VI. 27-29.
[115] 420.12 Reading with Barocius and Schönberger ἴσου instead of ἴσον in Friedlein.

breadth, we shall have the area, that is, if the angle assumed is a right angle. Something like this is the method of "application" which has come down to us from the Pythagoreans.[116]

421

There are three things given in this problem: a straight line along which the area is to be applied so that the line as a whole becomes a side of the area itself, a triangle to which the area applied must be equal, and an angle to which the angle of the area must be equal. Again it is clear that, when the angle is a right angle, the applied area will be either a square or an oblong; and when it is acute or obtuse, the area will be either a rhombus or a rhomboid. The straight line obviously must be finite, for it is not possible to apply an area to an infinite line; so in saying that we are to apply an area to "a given straight line," he makes it clear that the line is necessarily finite. He uses for the construction in this problem the construction of a parallelogram equal to the given triangle.[117] Application and construction are not the same

[116] 420.23 The reader not familiar with Greek geometry should supplement Proclus' brief account of this Pythagorean discovery with the more extensive exposition in Heath I, 150-153, 394-396. It is an essential part of what has appropriately been called "geometrical algebra." On the reasons for the appropriation of the terms "parabola," "hyperbola," and "ellipse" to designate the conic sections see Heath II, 134-139, esp. 138f. Cf. note at 111.8.

[117] 421.12 Euclid's impressive solution of XLIV is as follows: Given a straight line AB, a triangle C, and a rectilinear angle D, to apply to AB in an angle equal to D a parallelogram equal to C. Let parallelogram BEFG be constructed, by XLII, equal to triangle C in the angle EBG which is equal to D, let it be placed so that BE is in a straight line with AB, let FG be drawn through to H, let AH be drawn parallel to BG and EF, and let HB be joined. Then since line

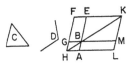

HF falls on parallels AH and EF, angles AHF and HFE are equal to two right angles, by XXIX. Therefore angles BHG and GFE are less than two right angles, and straight lines produced indefinitely from angles less than two right angles meet; hence HB and FE when produced will meet. Let them be produced and meet at K. Let KL be drawn parallel to EA and FH, and let HA and GB be produced to L and M. Then HLKF is a parallelogram; HK is its diameter, and

thing, as we have said. Construction brings the whole figure into being, both its area and all its sides, whereas application starts with one side given and constructs the area along it, neither falling short of the length of the line nor exceeding it, but using it as one of the sides enclosing the area.

But why, you may ask, did he use theorems when demonstrating the equality of triangles to triangles, but problems when making triangles equal to parallelograms? Because, we reply, the equality of things of the same species is natural and can be determined by simple inspection, whereas equality between [dissimilar][118] things, because of the difference in species, needs construction and artifice, since by itself it is difficult to discover.

422

XLV. *To construct, in a given rectilinear angle,*
a parallelogram equal to a given rectilinear figure.

This problem is more general than the two in which he investigates the construction and the application of parallelograms equal to a given triangle. For whether it be a triangle that is given, or a square, or a quadrilateral in general, or any other sort of multilateral figure, this problem will enable us to construct a parallelogram equal to it. For any rectilinear figure, as we said earlier, is as such divisible into triangles, and we have given the method by which the number of its triangles can be found.[119] Therefore by dividing the given rectilinear figure into triangles and constructing a parallelogram equal to one of them, then applying parallelograms equal to the others along the given straight line—that line to which we made the first application—we shall have the parallelogram composed of them equal to the rectilinear figure

---

AG and ME are parallelograms about HK; and LB and BF are the so-called complements. Therefore LB is equal to BF. But BF is equal to triangle C; therefore LB is also equal to C. And since angle GBE is equal to angle ABM, angle ABM is also equal to angle D. Therefore the parallelogram LB equal to the given triangle C has been applied to the given straight line AB in the angle ABM which is equal to D.

[118] 421.22   After τῶν δὲ some word or words have dropped out, such as ἄλλων or μὴ ὁμοειδῶν.

[119] 422.13   At 381.23ff.

composed of the triangles, and the assigned task will have been accomplished. That is, if the rectilinear figure has ten sides, we shall divide it into eight triangles, construct a parallelogram equal to one of them, and then by applying in seven steps parallelograms equal to each of the others, we shall have what we wanted.[120]

It is my opinion that this problem is what led the ancients to attempt the squaring of the circle. For if a parallelogram can be found equal to any rectilinear figure, it is worth inquiring whether it is not possible to prove that a rectilinear figure is equal to a circular area. Indeed Archimedes proved that a circle is equal to a right-angled triangle when its radius is equal to one of the sides about the right angle and its perimeter is equal to the base.[121] But of this elsewhere; let us proceed to the next propositions.

423

XLVI. *On a given straight line to describe a square.*

Our author particularly needs this problem for the establishment of the following theorem, but it seems that he also wishes to give us the construction of the two best of the rectilinear figures,[122] the equilateral triangle and the square. He

[120] 422.23 Euclid's solution of XLV, abbreviated, is as follows: Given the rectilinear figure ABCD and the rectilinear angle E, to construct in angle E a parallelogram equal to ABCD. Dividing ABCD into two triangles, he constructs parallelogram FH equal to triangle

ABD in angle HKF which is equal to E, by the method shown in XLII; he then applies parallelogram GM equal to triangle DBC to the line GH in the angle GHM which is equal to E, by the method shown in XLIV. KFLM is a parallelogram (by XIV, XXIX, XXXIII, XXXIV) and is the parallelogram whose construction is required.

[121] 423.5 In the *Measurement of a Circle*, Prop. I.

[122] 423.12 Reading with Barocius εὐθυγράμμοις instead of εὐθυγράμμῳ in Friedlein. Euclid's construction in XLVI is obvious. Given a straight line AB on which it is required to describe a square, he draws AC at right angles to AB, takes a point D on it such that AD is equal to AB, through D draws DE parallel to AB, and through B draws BE parallel to AD. ADEB is an equilateral parallelogram by construction, and it is shown to be right-angled by XXIX and XXXIV.

obviously needs these rectilinear figures for constructing the cosmic figures, and especially the four that are subject to generation and destruction; for the icosahedron, the octahedron, and the pyramid are composed of equilateral triangles, and the cube of squares.[123] This is why, I think, he prefers to speak of "constructing" the triangle and "describing" the square.[124] These terms he obviously finds appropriate to these figures, for the triangle, being put together of many parts, requires to be constructed, while the square, since it is generated from one of its sides, requires to be described. We get the square by multiplying the number of the given straight line by itself, but it is not so with the triangle; we draw lines from elsewhere to the extremities of the straight line and put them together into one equilateral triangle, and the drawing of a circle is needed to find the point from which the straight lines must be drawn to the extremities of the given straight line.[125]

424

This, then, is clear. But we must show that, when the straight lines are equal on which squares are described, the squares themselves are equal. Let lines AB and CD be equal, and on AB let a square ABEG be described, and on CD the

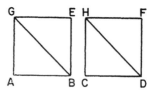

square CDFH, and let GB and HD be joined. Then since AB and CD are equal, and also AG and CH, and they include equal angles, then GB is equal to HD, triangle ABG is

---

[123] 423.18 These are the four figures used in the *Timaeus* to explain the generation and transformations of water, air, fire, and earth respectively.

[124] 423.20 See note at 82.22. In the following pages of the Greek text squares are always described as drawn *from* (not *on*) a line. My translation ignores this peculiarity of Greek diction and conforms to the modern mathematical idiom.

[125] 424.6 "These remarks make no sense to me. Obviously Post. III is presupposed in the construction of the square as well as in that of the triangle." (I.M.)

equal to triangle CDH, and the doubles of them are equal. Therefore AE is equal to CF. And the converse is also true, for if the squares are equal, the lines on which they are described will be equal. Let AF and CG be equal squares, and let them so lie that AB is on a straight line with BC. Since the angles are right angles, FB is also on a straight line with

425   BG. Let FC and AG be joined. Then since the square AF is equal to the square CG, triangle AFB is equal to triangle CBG. Let triangle BCF be added to each. Then the whole of triangle ACF is equal to triangle CFG, and consequently AG is parallel to FC. Again since angle AFG and angle CGF are

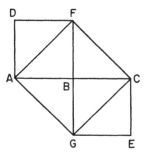

each half a right angle, AF is parallel to CG. Therefore line AF is equal to line CG, for they are opposite sides of a parallelogram. Then since ABF and BCG are two triangles having alternate angles equal, lines AF and CG being parallel, and one side AF equal to side CG, side AB will be equal to side BC and side BF to side BG. Consequently it has been demonstrated that the sides on which the squares AF and CG have been described are themselves equal when the squares are equal.

426   XLVII. *In right-angled triangles the square on the side subtending the right angle is equal to the squares on the sides containing the right angle.*

If we listen to those who like to record antiquities, we shall find them attributing this theorem to Pythagoras and saying that he sacrificed an ox[126] on its discovery. For my part,

---

[126] 426.8   Reading with Barocius βουθυτεῖν for βουθύτην in Friedlein.

though I marvel at those who first noted the truth of this theorem, I admire more the author of the *Elements*, not only for the very lucid proof by which he made it fast, but also because in the sixth book he laid hold of a theorem even more general than this and secured it by irrefutable scientific arguments. For in that book he proves generally that in right-angled triangles the figure on the side that subtends the right angle is equal to the similar and similarly drawn figures on the sides that contain the right angle.[127] Every square is of course similar to every other square, but not all similar rectilinear figures are squares, for there is similarity in triangles and in other polygonal figures. Hence the argument establishing that the figure on the side subtending the right angle, whether it be a square or any other kind of figure, is equal to the similar and similarly drawn figures on the sides about the right angle, proves something more general and scientific than that which shows only that the square is equal to the squares. For there the cause of the more general proposition that is proved becomes clear: it is the rightness of the angle that makes the figure on the subtending side equal to the similar and similarly drawn figures on the containing sides, just as the obtuseness of the angle is the cause of its being greater and the acuteness of the angle the cause of its being less.

How he proves the theorem[128] in the sixth book will be evident there. But now let us consider how he shows the theorem before us to be true, remarking only that he does not prove the universal proposition here, since he has not yet explained similarity in rectilinear figures, nor proved anything in general about proportion. Hence many of the things here proved in a partial fashion[129] are proved in that book more generally through the use of the above method. In the present

427

---

[127] 426.18 In VI. 31.

[128] 427.10 Reading with Barocius θεώρημα for θεωρήματι in Friedlein.

[129] 427.14f. Something has evidently been lost here. I agree with Schönberger, following Barocius' translation, that the text must have originally read πολλὰ τῶν ἐνταῦθα μερικώτερον δεδειγμένων ἐν ἐκείνῳ δέδεικται καθολικώτερον.

proposition the author of the *Elements* proves his conclusion by means of the ordinary theory of parallelograms.[130]

There are two sorts of right-angled triangles, isosceles and scalene. In isosceles triangles you cannot find numbers that fit the sides; for there is no square number that is the double of a square number, if you ignore approximations, such as the square of seven which lacks one of being double the square of five. But in scalene triangles it is possible to find such numbers,[131] and it has been clearly shown that the square on the side subtending the right angle may be equal to the squares on the sides containing it. Such is the triangle in

[130] 427.18   This "very lucid proof" of XLVII is as follows: Given a right-angled triangle ABC, with angle BAC right, and squares inscribed on each of its sides according to the method shown in XLVI, to prove that the square on BC is equal to the squares on BA and AC. Through A let AL be drawn parallel to BD and CE, and let AD and FC be drawn. Since BAC and BAG are right angles, it follows, by XIV, that CA is on a straight line with AG; and for the same reason BA is on a straight line with AH. Since angle DBC is equal to angle FBA (for each is a right angle), let angle ABC be added to each.

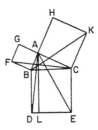

Then the whole angle DBA is equal to the whole angle FBC. Then in triangles ABD and FBC sides AB and BD are equal respectively to sides FB and BC, and angle ABD is equal to angle FBC; therefore the base AD is equal to the base FC, and the two triangles are equal, by IV. Now the parallelogram BL is double the triangle ABD, by XLI, and the square GB is double the triangle FBC. Therefore the parallelogram BL is equal to the square GB. Similarly if AE and BK be joined, the parallelogram CL can also be proved equal to the square HC. Therefore the whole square BDEC is equal to the two squares GB and HC; i.e. the square on BC is equal to the squares on BA and AC.

[131] 427.25   There is an unindicated lacuna in Friedlein's text after λαβεῖν, long enough at least to contain a connective with the following δείκνυται. Barocius either had such a particle or saw the need and supplied it.

428

the *Republic*,[132] in which sides of three and four contain the right angle and five subtends it, so that the square on five is equal to the squares on those sides. For this is twenty-five, and of those the square of three is nine and that of four sixteen. The statement, then, is clear for numbers.

Certain methods have been handed down for finding such triangles, one of them attributed to Plato, the other to Pythagoras. The method of Pythagoras begins with odd numbers, positing a given odd number as being the lesser of the two sides containing the angle, taking its square, subtracting one from it, and positing half of the remainder as the greater of the sides about the right angle; then adding one to this, it gets the remaining side, the one subtending the angle. For example, it takes three, squares it, subtracts one from nine, takes the half of eight, namely, four, then adds one to this and gets five; and thus is found the right-angled triangle with sides of three, four, and five. The Platonic method proceeds from even numbers. It takes a given even number as one of the sides about the right angle, divides it into two and squares the half, then by adding one to the square gets the

429

subtending side, and by subtracting one from the square gets the other side about the right angle. For example, it takes four, halves it and squares the half, namely, two, getting four; then subtracting one it gets three and adding one gets five, and thus it has constructed the same triangle that was reached by the other method. For the square of this number is equal to the square of three and the square of four taken together.[133]

These remarks are somewhat outside our subject. But since the proof given by the author of the *Elements* is clear, I do not think I should add anything superfluous but should be content with what he has written, especially since those

[132] 428.1   Probably a reference to *Rep.* 546c.

[133] 429.8   By the Pythagorean method we get the three numbers a (assumed to be odd), $\frac{a^2-1}{2}$, and $\frac{a^2+1}{2}$, which satisfy the equation $a^2+\left(\frac{a^2-1}{2}\right)^2=\left(\frac{a^2+1}{2}\right)^2$. The Platonic method yields 2a, a^2-1, a^2+1, which satisfy the equation $(2a)^2+(a^2-1)^2=(a^2+1)^2$. How "Pythagoras" and Plato respectively discovered these methods is discussed by Heath, *Euclid* I, 356-360.

who have made additions, such as the disciples of Heron and Pappus, have been obliged to assume something proved in the sixth book, and for no material purpose.[134] Let us then proceed to what follows.

XLVIII. *If in a triangle the square on one of the sides is equal to the squares on the remaining two sides of the triangle, the angle contained by the remaining two sides of the triangle is right.*

430

This theorem is the converse of the one before it and is a whole-to-whole converse. For if the triangle is right-angled, the square on the subtending side is equal to the squares on the other sides, and if the square on the subtending side is equal to those on the other sides, the triangle is right-angled, and its right angle is that which is contained by the other sides. The proof given by the author of the *Elements* is clear. He assumes a triangle ABC having the square on AC equal to the squares on AB and BC and draws a line on this triangle from B at right angles to BC.[135] If someone says that the line at right angles should not be drawn in the direc-

---

[134] 429.15 Heath (*Euclid* I, 366-368) gives an account of Pappus' extension of XLVII and of Heron's proof that lines AL, BK, and CF in Euclid's diagram meet in a point. The former is an "elegant" theorem, and the latter, as proved by Heron, involves no use of anything beyond Book I. It is likely, then, that Proclus' criticism is directed against other members of their schools.

[135] 430.9 As Proclus says, Euclid assumes in the proof of XLVIII a triangle ABC having the square on AC equal to the squares on AB and BC; he draws BD at right angles to BC, makes BD equal to AB, and draws CD. (The lettering on the diagram in Heiberg is slightly different from that in Proclus' description, which I follow here.) Then since BD is equal to AB, the square on BD is equal to the square on AB. Let the square on BC be added to each. Then the squares on DB and BC are equal to the squares on AB and BC. But by XLVII the square on DC is equal to the squares on BD and BC; and the square on AC is by hypothesis equal to the squares on AB and BC. Therefore the square on DC is equal to the square on AC, so that side DC is equal to AC. Then in the two triangles sides DB and BC are equal respectively to AB and BC, and the base DC is equal to the base AC; therefore angle DBC is equal to angle ABC, by VIII. But DBC is a right angle by construction; therefore ABC is also a right angle, which is what was to be proved.

tion in which the author of the *Elements* draws it, but in the opposite direction, we shall reply that the idea is impossible, since the line[136] cannot fall either within the triangle or outside it but is identical with AB. For, if possible, let it fall as does BE. Then since EBC is a right angle, CFB is acute, so that the other angle, AFB, is obtuse. AB then is greater than BF.

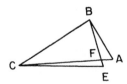

Therefore let BE be supposed equal to AB, and let EC be joined. Then since EBC is a right angle, the square on EC is equal to the squares on BE and BC. But BE is equal to AB, and consequently the square on EC is equal to the square on AB and BC. But the square on AC was equal to the same squares. Consequently the square on EC is equal to the square on AC, and EC is therefore equal to AC. But BE was equal to AB. Therefore the two lines BE and EC have been constructed on BC equal respectively to the lines AB and AC, which is impossible.[137] Consequently the line drawn at right angles does not fall within the triangle. But neither can it fall outside, that is, on the other side of the line AB. If possible, let it fall as does BG, and let BG be equal to AB, and let CG be joined. Then since angle GBC is a right angle, the

431

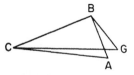

square on CG is equal to the squares on BG and BC. But BG is equal to AB, and hence the square on CG is equal to those on AB and BC. But the square on AC was also equal to the squares on AB and BC. Therefore CG is equal to AC. But BG was also equal to AB on the same straight line BC,

[136] 430.14   Reading with Barocius αὐτὴν for αὐτόν.
[137] 431.2   By VII.

which is impossible.[138] Therefore the line drawn from B at right angles to BC will fall neither inside nor outside. It will therefore coincide with AB itself, and angle ABC is consequently a right angle. In this way the objection is answered.

With this theorem the author of the *Elements* completes his first book. He has presented many species of conversion, with numerous examples of whole-to-whole, whole-to-part, and part-to-part converses; he has devised a variety of problems, showing us how to bisect lines and angles, how to place lines, construct figures, and apply areas; he has touched on the so-called locus of paradoxes in mathematics and acquainted us generously with locus-theorems themselves; he has introduced us to both general and partial theorems and taught the difference between determinate and indeterminate problems—all of which we, following his lead, have systematically expounded. He has directed the entire book to one end, an introduction to the study of the simplest rectilinear figures, finding their constructions and examining their essential properties. As for us, if we are able to go through the remaining books in the same fashion, we shall have much to thank the gods for; but if other concerns draw us aside, we ask those who are admirers of this science to expound the remaining books by the same method, aiming always at what is important and can be clearly divided, since the commentaries now in circulation contain great and manifold confusion and contribute nothing to the exposition of causes, to dialectical judgment, or to philosophical understanding.

432

138 431.10 Again by VII.

Supplementary Note
on the Prologue: Part One

It was supposed by Tannery (II, 302-310) that Proclus wrote a commentary on Nicomachus' *Introduction to Arithmetic* which has since been lost. If this supposition is true, then we could plausibly assume that this first part of the prologue was originally intended to serve as the introduction to a more comprehensive treatment of mathematics, of which the supposed commentary on Nicomachus and the extant commentary on Euclid were to be parts. But this supposition has been effectively challenged recently by Leonardo Tarán; see his *Asclepius' Commentary on Nicomachus' Introduction to Arithmetic* (Philadelphia, 1969), pp. 7-8. Tarán's edition shows, however, that Asclepius' commentary was based on a commentary (or a course of lectures) by Ammonius, who was a pupil of Proclus; and this makes it natural to suppose that Proclus himself at one time proposed to write a commentary on Nicomachus, even if he did not actually do so. The concluding words of this first prologue may well reflect such an intention.

There are other features of this first prologue that should be noted. In the topics that it considers, in the arrangement of these topics, and in the concepts it uses in dealing with them, it follows closely the first half of a work of Iamblichus, περὶ τῆς κοινῆς μαθηματικῆς ἐπιστήμης, written more than a century before Proclus' time. Both works deal first with the status of mathematical being as intermediate between νοητά and αἰσθητά; both assert that the Limit and the Unlimited are the primary principles of mathematical as of all other being, and then discuss the principles and theorems common to all the mathematical sciences and the peculiar principles that distinguish each of them, particularly arithmetic and geometry. Both ask about the κριτήριον, the ἔργον, and the δυνάμεις of general mathematics and give similar answers. Both emphasize the contributions of mathematics to the other sciences and arts, and particularly its value for philosophy and education. Each deals with the detractors of the science, citing the same objections and meeting them in similar ways; each deals at length with the "unifying bond" (42.11) of the mathematical sciences and includes an explanation of the meaning of the name "mathematics." Yet there are few verbal similarities, other than the concepts employed, and no borrowing that I have been able to detect by Proclus of phrases or sentences from the earlier work. In general Proclus' treatment is the more concise, lucid, and philosophical; and it is graced by stylistic qualities to which Iamblichus can make no claim. Unquestionably we have here an improved and tightened version of the material in Iamblichus' treatise. Yet Proclus nowhere mentions Iamblichus, either in this first prologue or elsewhere in the *Commentary*, whereas his *Commentary on the Timaeus* is liberally sprinkled with references to the words and thoughts of Iamblichus, who is called ὁ θεῖος, ὁ θειότατος, and ὁ μέγας Ἰάμβλιχος.

Professor Brumbaugh has suggested to me that Proclus' failure to refer to Iamblichus in this prologue may have been due to his recognition that Iamblichus in this work had himself drawn heavily upon others, i.e. that his was a "scissors-and-paste" compilation. And this to me suggests that the several topics common to both Iamblichus and Proclus, and possibly the substance of the comments attached to each, were so familiar in the instruction of the school that they could be regarded as, in a sense, common school doctrine. It is possible that other Neoplatonists had written similar essays on general mathematics, Iamblichus' being the only one of these that happens to have survived. The words in the opening paragraph of the second prologue referring to "other sources" from which Proclus had "collected" his material for the first part suggest some such situation.

Index

(Numbers refer to pages in this volume)

Milton Keynes UK
Ingram Content Group UK Ltd.
UKHW030031180324
439604UK00002B/379